Textile advances in the automotive industry

The Textile Institute and Woodhead Publishing

The Textile Institute is a unique organisation in textiles, clothing and footwear. Incorporated in England by a Royal Charter granted in 1925, the Institute has individual and corporate members in over 90 countries. The aim of the Institute is to facilitate learning, recognise achievement, reward excellence and disseminate information within the global textiles, clothing and footwear industries.

Historically, The Textile Institute has published books of interest to its members and the textile industry. To maintain this policy, the Institute has entered into partnership with Woodhead Publishing Limited to ensure that Institute members and the textile industry continue to have access to high calibre titles on textile science and technology.

Most Woodhead titles on textiles are now published in collaboration with The Textile Institute. Through this arrangement, the Institute provides an Editorial Board which advises Woodhead on appropriate titles for future publication and suggests possible editors and authors for these books. Each book published under this arrangement carries the Institute's logo.

Woodhead books published in collaboration with The Textile Institute are offered to Textile Institute members at a substantial discount. These books, together with those published by The Textile Institute that are still in print, are offered on the Woodhead web site at: www.woodheadpublishing.com. Textile Institute books still in print are also available directly from the Institute's website at: www.textileinstitutebooks.com.

A list of Woodhead books on textile science and technology, most of which have been published in collaboration with The Textile Institute, can be found pages on xi-xiii.

Woodhead Publishing in Textiles: Number 79

Textile advances in the automotive industry

Edited by

R. Shishoo

The Textile Institute

CRC Press
Boca Raton Boston New York Washington, DC

WOODHEAD PUBLISHING LIMITED
Cambridge, England

Published by Woodhead Publishing Limited in association with The Textile Institute
Woodhead Publishing Limited, Abington Hall, Granta Park, Great Abington
Cambridge CB21 6AH, England
www.woodheadpublishing.com

Published in North America by CRC Press LLC, 6000 Broken Sound Parkway, NW, Suite 300, Boca Raton, FL 33487, USA

First published 2008, Woodhead Publishing Limited and CRC Press LLC

British Library Cataloguing in Publication Data
A catalogue record for this book is available from the British Library.

Library of Congress Cataloging in Publication Data
A catalog record for this book is available from the Library of Congress.

Woodhead Publishing ISBN 978-1-84569-331-2 (book)
Woodhead Publishing ISBN 978-1-84569-504-0 (e-book)
CRC Press ISBN 978-1-4200-9000-0
CRC Press order number: WP9000

The publishers' policy is to use permanent paper from mills that operate a sustainable forestry policy, and which has been manufactured from pulp which is processed using acid-free and elementary chlorine-free practices. Furthermore, the publishers ensure that the text paper and cover board used have met acceptable environmental accreditation standards.

Project managed by Macfarlane Book Production Services, Dunstable, Bedfordshire, England (email: macfarl@aol.com)
Typeset by Replika Press Pvt Ltd, India
Printed by T J International Limited, Padstow, Cornwall, England

Contents

Part III: Safety applications of automotive textiles

Part IV: Use of textiles in automotive bodywork

Contributor contact details

(* = main contact)

Editor and Introduction

Professor R. Shishoo
Shishoo Consulting AB
Svartlavsvägen 18
43639 Askim
Sweden

E-mail: roshan.shishoo@shishooconsulting.se

Chapter 1

E. Söderbaum
Volvo Car Corporation (VCC)
Gothenburg
Sweden

E-mail: erik@soederbaum.se

Chapter 2

Professor N. B. Powell* and R. Barnhardt
CB 8301
College of Textiles
North Carolina State University
Raleigh, NC 27695
USA

E-mail: nancy_powell@ncsu.edu

Professor R. Handfield
CB 7229
College of Management
North Carolina State University
Raleigh, NC 27695
USA

E-mail: robert_handfield@ncsu.edu

Chapter 3

Dr T. Stegmaier,* J. Mavely, M. Schweins, Dr V. von Arnim, G. Schmeer-Lioe, P. Schneider, H. Finckh and Professor H. Planck
ITV – Institute of Textile Research and Process Engineering
Koerschtalstr 26
D-73770 Denkendorf
Germany

E-mail: thomas.stegmaier@itv-denkendorf.de

Chapter 4

Professor S. J. Russell* and M. J. Tipper
Nonwovens Innovation & Research Institute (NIRI)
Centre for Technical Textiles
Woodhouse Lane
University of Leeds
Leeds LS2 9JT
UK

E-mail: s.j.russell@leeds.ac.uk

Chapter 5

X. Normand
Institut Français du Textile et de l'Habillement
France

E-mail: Xnormand@ifth.org

Chapter 6

Professor N. B. Powell
CB 8301
College of Textiles
North Carolina State University
Raleigh, NC 27695
USA

E-mail: nancy_powell@ncsu.edu

Chapter 7

Dr H. Erth* and B. Gulich
STFI
Saxon Textile Research Institute
Annaberger Strasse 240
D09125 Chemnitz
Germany

E-mail: holger.erth@stfi.de
bernd.gulich@stfi.de

Chapter 8

Dr V. T. Bartels
Hohenstein Institutes
Department of Clothing Physiology
D-74357 Boennigheim
Germany

E-mail: v.bartels@hohenstein.de

Chapter 9

F. Boussu,* C. Cochrane, M. Lewandowski and Professor V. Koncar
ENSAIT – l'Ecole Nationale Supérieure des Arts et Industries Textiles
Laboratory GEMTEX
Roubaix
France

E-mail: vladan.koncar@ensait.fr

Chapter 10

Professor Jonathan Y. Chen
School of Human Ecology
Louisiana State University
Agricultural Center
Baton Rouge, LA 70803
USA

E-mail: chenyan@lsu.edu

Chapter 11

Dr M. Meyer,* Dr H. Schulz and Professor M. Stoll
Research Institute of Leather and Plastic Sheeting (FILK)
Meißner Ring 1-5
D-09599 Freiberg/Sa
Germany

E-mail: michael.meyer@filkfreiberg.de

Chapter 12

Dr Samir K. Mukhopadhyay
Sans Fibres (Pty) Ltd
Peter Barlow Drive & Sacks Circle
Bellville 7530
Cape Town
South Africa

E-mail: mukhopadhyays@sans.co.za

Chapter 13

W. K. Westgate* and Dr J. G. Gillick
Corporate Research Division
The Goodyear Tire & Rubber Company
142 Goodyear Boulevard
Akron, OH 44305
USA

E-mail: kwestgate@goodyear.com
jim_gillick@goodyear.com

Chapter 14

Professor T. Gries,* J. Stueve, Dr T. Grundmann and D. Veit
Institut fuer Textiltechnik
RWTH Aachen University
52056 Aachen
Germany

E-mail: thomas.gries@ita.rwth-aachen.de

Chapter 15

Dr F. K. Ko
Advanced Materials and Process Engineering Laboratory
Department of Materials Engineering
University of British Columbia
Vancouver, British Columbia
Canada

E-mail: frank.ko@ubc.ca

Woodhead Publishing in Textiles

1 **Watson's textile design and colour Seventh edition**
Edited by Z. Grosicki

2 **Watson's advanced textile design**
Edited by Z. Grosicki

3 **Weaving Second edition**
P. R. Lord and M. H. Mohamed

4 **Handbook of textile fibres Vol 1: Natural fibres**
J. Gordon Cook

5 **Handbook of textile fibres Vol 2: Man-made fibres**
J. Gordon Cook

6 **Recycling textile and plastic waste**
Edited by A. R. Horrocks

7 **New fibers Second edition**
T. Hongu and G. O. Phillips

8 **Atlas of fibre fracture and damage to textiles Second edition**
J. W. S. Hearle, B. Lomas and W. D. Cooke

9 **Ecotextile '98**
Edited by A. R. Horrocks

10 **Physical testing of textiles**
B. P. Saville

11 **Geometric symmetry in patterns and tilings**
C. E. Horne

12 **Handbook of technical textiles**
Edited by A. R. Horrocks and S. C. Anand

13 **Textiles in automotive engineering**
W. Fung and J. M. Hardcastle

14 Handbook of textile design
J. Wilson

15 High-performance fibres
Edited by J. W. S. Hearle

16 Knitting technology Third edition
D. J. Spencer

17 Medical textiles
Edited by S. C. Anand

18 Regenerated cellulose fibres
Edited by C. Woodings

19 Silk, mohair, cashmere and other luxury fibres
Edited by R. R. Franck

20 Smart fibres, fabrics and clothing
Edited by X. M. Tao

21 Yarn texturing technology
J. W. S. Hearle, L. Hollick and D. K. Wilson

22 Encyclopedia of textile finishing
H-K. Rouette

23 Coated and laminated textiles
W. Fung

24 Fancy yarns
R. H. Gong and R. M. Wright

25 Wool: Science and technology
Edited by W. S. Simpson and G. Crawshaw

26 Dictionary of textile finishing
H-K. Rouette

27 Environmental impact of textiles
K. Slater

28 Handbook of yarn production
P. R. Lord

29 Textile processing with enzymes
Edited by A. Cavaco-Paulo and G. Gübitz

30 The China and Hong Kong denim industry
Y. Li, L. Yao and K. W. Yeung

31 **The World Trade Organization and international denim trading**
Y. Li, Y. Shen, L. Yao and E. Newton

32 **Chemical finishing of textiles**
W. D. Schindler and P. J. Hauser

33 **Clothing appearance and fit**
J. Fan, W. Yu and L. Hunter

34 **Handbook of fibre rope technology**
H. A. McKenna, J. W. S. Hearle and N. O'Hear

35 **Structure and mechanics of woven fabrics**
J. Hu

36 **Synthetic fibres: nylon, polyester, acrylic, polyolefin**
Edited by J. E. McIntyre

37 **Woollen and worsted woven fabric design**
E. G. Gilligan

38 **Analytical electrochemistry in textiles**
P. Westbroek, G. Priniotakis and P. Kiekens

39 **Bast and other plant fibres**
R. R. Franck

40 **Chemical testing of textiles**
Edited by Q. Fan

41 **Design and manufacture of textile composites**
Edited by A. C. Long

42 **Effect of mechanical and physical properties on fabric hand**
Edited by Hassan M. Behery

43 **New millennium fibers**
T. Hongu, M. Takigami and G. O. Phillips

44 **Textiles for protection**
Edited by R. A. Scott

45 **Textiles in sport**
Edited by R. Shishoo

46 **Wearable electronics and photonics**
Edited by X. M. Tao

47 **Biodegradable and sustainable fibres**
Edited by R. S. Blackburn

48 Medical textiles and biomaterials for healthcare
Edited by S. C. Anand, M. Miraftab, S. Rajendran and J. F. Kennedy

49 Total colour management in textiles
Edited by J. Xin

50 Recycling in textiles
Edited by Y. Wang

51 Clothing biosensory engineering
Y. Li and A. S. W. Wong

52 Biomechanical engineering of textiles and clothing
Edited by Y. Li and D. X-Q. Dai

53 Digital printing of textiles
Edited by H. Ujiie

54 Intelligent textiles and clothing
Edited by H. Mattila

55 Innovation and technology of women's intimate apparel
W. Yu, J. Fan, S. C. Harlock and S. P. Ng

56 Thermal and moisture transport in fibrous materials
Edited by N. Pan and P. Gibson

57 Geosynthetics in civil engineering
Edited by R. W. Sarsby

58 Handbook of nonwovens
Edited by S. Russell

59 Cotton: Science and technology
Edited by S. Gordon and Y-L. Hsieh

60 Ecotextiles
Edited by M. Miraftab and A. Horrocks

61 Composite forming technologies
Edited by A. C. Long

62 Plasma technology for textiles
Edited by R. Shishoo

63 Smart textiles for medicine and healthcare
Edited by L. Van Langenhove

64 Sizing in clothing
Edited by S. Ashdown

65 Shape memory polymers and textiles
J. Hu

66 Environmental aspects of textile dyeing
Edited by R. Christie

67 Nanofibers and nanotechnology in textiles
Edited by P. Brown and K. Stevens

68 Physical properties of textile fibres Fourth edition
W. E. Morton and J. W. S. Hearle

69 Advances in apparel production
Edited by C. Fairhurst

70 Advances in fire retardant materials
Edited by A. R. Horrocks and D. Price

71 Polyesters and polyamides
Edited by B. L. Deopora, R. Alagirusamy, M. Joshi and B. S. Gupta

72 Advances in wool
Edited by N. A. G. Johnson and I. Russell (forthcoming)

73 Military textiles
Edited by E. Wilusz

74 3D fibrous assemblies: Properties, applications and modelling of three-dimensional textile structures
J. Hu

75 Medical textiles 2007
Edited by J. Kennedy, A. Anand, M. Miraftab and S. Rajendran (forthcoming)

76 Fabric testing
Edited by J. Hu

77 Biomimetics and textiles
Edited by A. Abbott and M. Ellison (forthcoming)

78 Friction in textiles
Edited by B. S. Gupta

79 Textile advances in the automotive industry
Edited by R. Shishoo

Introduction

R SHISHOO, Shishoo Consulting AB, Sweden

Automotive textiles represent the most valuable world market for technical textiles and within this segment there is a broad spectrum of products comprising novel textile structures with performance properties and attractive design. Total global production of cars and light trucks in 2006 was 67,723,891 units (*Automotive News*, 2007). The EU is the world's largest automotive manufacturer with about 16 million new passenger cars and 1.8 million light commercial vehicles in 2006. The global market for automotive interiors was worth an estimated US$165 billion in 2005 and is expected to grow to an estimated US$210 billion by 2015. The demand and production of vehicles will occur at increasing pace in developing countries – mainly China and India. Quite a large number of different textiles are used in the production of a car.

Textile advances in the automotive industry have been spearheaded by advances in science and technology of fibres and fabric/web forming technologies. These advances have led to the development of textiles and textile-based components for a broad variety of automotive applications which are capable of meeting the industry's tough specifications regarding high performance during use. Automotive textiles represent the most valuable world market for technical textiles and within this segment there is a broad spectrum of products comprising novel textile structures and high quality design. Automotive technical textiles cover a broad range of applications that includes: upholstery and seating, floor covering, trunk liners, headliners, door and side-panel coverings, pillar coverings, sun-visors as well as safety belts, airbags, thermal and sound insulators, filter fabrics, battery separators, hose/belt products, tyres and a variety of textile-reinforced flexible and hard composites. The textiles for interior furnishing are primarily made of woven, weft knitted, warp knitted, tufted and laminated fabrics and nonwovens. The design, aesthetics, feel and comfort are important considerations for automotive textiles. They are designed and developed according to strict technical specifications set up by the industry and must be able to meet many end-use requirements such as: high resistance to different types of mechanical actions under severe external environmental conditions of temperature, humidity, sunlight and soiling. Since seat covers are not washed, these must show strong soil masking behaviour. The design and performance properties of technical textiles present in automotive interiors are among the most important criteria for consumer satisfaction.

There have been interesting developments taking place in the production of 3D fabrics using all forms of manufacturing technologies *viz.* weaving, knitting, braiding and nonwoven processes. The 3D textile structures have great potential as automotive components in both load-bearing and non-load bearing applications. For example, spacer fabrics can be used as substitutes for polyurethane (PUR) foam in car seats and thereby solve the recycling problems with PUR foam. 3D fabrics are used as reinforcement fibre network in textile composites in both load bearing and non-load bearing automotive components. They are also used as preforms for moulding of components and component parts.

The supply chain for textiles in the production process of a vehicle has a distinctive feature and is unlike textile supply chains in other application sectors (see Fig. I.1). In the automotive supply chain, original equipment manufacturers (OEMs) are the vehicle producers, and the textile component manufacturers do not sell their products directly to OEMs. They have to position their entry in the supply chain depending on the nature of their products and the general specifications passed down and acceptance granted by the OEMs. In the automobile industry, suppliers to OEMs do not have much power and the cost pressures of OEMs have intensified, resulting in long-term contracts being opened for new more competitive bidding from global suppliers.

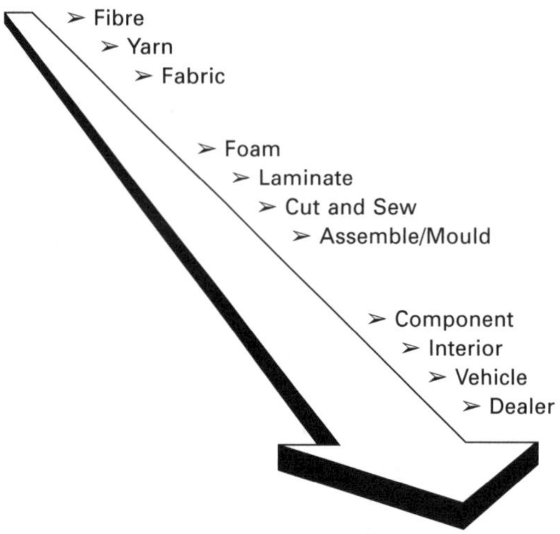

I.1 Supply chain.

The diversity in textiles for automotive end uses offers plenty of opportunities for growth in the automotive textiles industry because of the ever increasing demands on aesthetics, comfort and safety of passengers as well as the environmental issues such as low weight, lower energy consumption and recycling after a vehicle's life cycle. Many industrial sub-sectors in the automotive textile supply chain are continuously looking for specialised fibre and fabric/web-forming technologies, specialised finishing and coating processes, specialised testing specifications and specialised component-forming technology.

The environmental requirements are putting pressure on the automotive industry to reduce a vehicle's fuel consumption and thereby reduce the CO_2 emission. One way to achieve this goal is by reducing the weight of the vehicle. Textiles can play a big part in this if structural textile composites can be developed to replace the heavy metal-based load-bearing components. Besides, strong requirements do exist in the European automotive sector due to the European Directive on end-of-life vehicles. This necessitates some design changes upstream in order to develop new textile products which are 100% recyclable. Nowadays, many automotive textile products are composed of two or more types of raw materials, which are difficult to separate, and are only partially recyclable.

Airbags and seat belts constitute a major tool for consumer safety. The airbag market has had strong growth over the last decade as a result of increasing market penetration not only by the frontal airbag system but also by the side impact and side curtain airbags. Seat belt development has reached maturity and these are now an integral part of all cars and trucks.

Leather interiors have been perceived in the market place as a luxury level option. Leather and its substitutes are used in the automotive industry as upholstery material for seats, headrests, as covering material for dash boards, back shelves and door linings, for centre consoles, gear shifts, steering wheels and ceilings as well as for accessories. Coated textile substrates – summarised under the name artificial leather – compete with 'real leather' in the many applications in car interiors. Whereas leather consists of a meshwork of collagen fibres, artificial leathers are compound materials made of textile substrates and polymeric layer/s.

Tyres constitute the largest application for textiles in automotive applications. Tyre markets differ from those for other automotive components in that end users consume as much as two-thirds of total supply. Only one-third of tyres are supplied to OEMs.

About the book

The objective of this book is to give the reader a comprehensive description of the advances made in textile materials and technologies from the point of view of their application in the automotive industry. The contents of the book have been planned to cover many topics of interest ranging from: technical requirements and specification of automotive textiles; mapping of the automotive supply chain; design management for performance and style in automotive textiles; advances in woven, knitted and nonwoven materials for varied automotive applications, and requirements of safety, comfort, noise reduction and recycling. This book comprises four parts: Part I: General, Chapters 1–5; Part II: Automotive interiors, Chapters 6–11; Part III: Safety applications of automotive textiles, Chapters 12–13; and Part IV: Use of textiles in automotive bodywork, Chapters 14–15.

The contributors to this book represent a team of international experts and specialists from both academic and industrial worlds.

In Chapter 1, Erik Söderbaum writes about the requirements for automotive textiles from a car producer's perspective. He describes Volvo Car Corporation's criteria for selection of technical textiles which include the technical requirements, the design requirements, the purchase requirements and the purchase engineering requirements. Volvo purchases upholstery material which includes textiles for seats. Other textiles in the interior are purchased by system suppliers but developed by Volvo. Volvo has some specific technical demands – design prerequisites, master samples, drawings, technical regulations, etc., where they also specify testing methods. There are some unique and extra tough demands for textiles including light-fastness, ageing, fogging/emissions, dirt resistance/cleanability. Volvo's development work needs early involvement by textile suppliers. The suppliers must have a creative design/styling department of their own and a flexible production capacity. The normal lead time is 12–24 months to develop a new textile. As Volvo has warranty periods up to four years in some markets, the textile supplier will also be involved in regress issues – continuous improvement is a must.

In Chapter 2, Nancy Powell, Robert Barnhardt and Robert Handfield map the automotive textile supply chain and write about the market drivers, trends in automotive textiles, quick response and just in time concepts, effect of globalisation and future trends. The sourcing of textile products for vehicles is typical of the complex supplier network for the large-scale global automotive industry. The demand for rapid prototyping, cost effectiveness and streamlining of supply systems is critical in all component parts, but particularly in automotive textile materials. In the tiered structure of the automotive supply chain the original equipment manufacturer (OEM) is responsible for initial overall design, assembly and marketing of the complete vehicle and brand through dealerships to the consumer. A complete system component such as an interior compartment would be supplied directly to the OEM by a Tier One supplier. A Tier Two company would supply subassemblies to the interior system. The individual parts of these subassemblies would be sourced to a

Tier Three supplier, and the materials would be provided by a Tier Four company. Some companies function in roles across tier levels. A textile company may be considered a supplier at the third or fourth tier. In order to improve performance in the automotive supply chain and create a truly integrated value system, a number of tactical and strategic changes will be required throughout the automotive textile supply chain.

In Chapter 3, Thomas Stegmaier, and his co-authors write about the advances made in woven, weft-knitted and warp-knitted fabrics used as textile components in automotive interiors. Owing to their function in the application, automotive textiles have to fulfil comprehensive lists of requirements regarding properties for aesthetics, comfort, security and fastnesses. So they have to pass corresponding test procedures to determine the relevant mechanical, physical and chemical properties. The finishing of textile plays an important role for the mechanical and surface-related properties, and for the final product a textile material often has to pass processes such as lamination, coating, deep drawing and make up. The authors describe various types of textile components in car interiors followed by description of advances in woven and knitted fabrics and their finishing, lamination and other processing resulting in composite structures. They then describe different test procedures and standards which are specific to automotive textiles. In the future, new fibre developments will extend the resources and application areas, and new numerical simulation tools will be used in construction, processing and application for vehicles.

In Chapter 4, Stephen Russell and Matthew Tipper write about the advances made in nonwoven materials for a number of applications in automobiles.This chapter considers the utilisation of nonwoven fabrics in OEM automotive interiors and in automotive filtration. The importance of nonwovens in the automotive industry continues to be fuelled by the ability of such fabrics to be customised and engineered to meet the requirements of modern vehicle assembly operations and technical performance in use. According to the authors, nonwovens are particularly attractive in the automotive industry because of their ability to:

(i) be integrated into multi-layer, modular components with other materials including foam,
(ii) be incorporated into light-weight and low-density modules,
(iii) meet stringent cost-performance targets in what is a highly price-sensitive industry,
(iv) be deep draw moulded into complex shapes at relatively low temperature,
(v) contain recycled raw materials and still meet performance requirements, and
(vi) be compatible with emerging recycling processes.

In Chapter 5, X Normand writes about recycling of automotive textiles.

He describes the legislative considerations with regard to the disposability of used vehicles; shortcomings in textile materials used today; examples of recyclable automotive textiles and their performance properties. The author then presents the results of an experimental study where the principal aims were: (i) to replace the classical engine air filter manufactured in several raw materials, which are difficult to separate, and partially recyclable, by a filter composed of a thermoplastic (recyclable or not) mono material. The recycling of this mono material was studied for automotive applications or for other fields; and (ii) to replace classical acoustic insulation products manufactured in several raw materials, which are difficult to separate, and partially recyclable, by textile structures (nonwovens) composed of a thermoplastic (recyclable or not) mono material. The recycling of this mono material was studied for automotive applications or for other fields.

In Chapter 6, Nancy Powell writes about design management for performance and style in automotive interior textiles. This chapter considers the challenges for designers across the automotive supply chain for textile products, where art meets engineering through the design process. The range of basic textile characteristics developed within high performance and low cost parameters to be considered are fibre characteristics, fabric constructions, pattern, colour, hand, lustre, and recyclability. The author describes contributing factors such as cost pressures, speed to markets, quality, sustainable products, design strategy, trim selection process, impact of seat design requirements, communicating value and design, service and innovative materials. Furthermore, the influences of consumer lifestyles and fashion trends continue to grow in the differentiation of brands in the wide choice of vehicle offerings.

In Chapter 7, Holger Erth and Bernd Gulich write about the progress made in developing 3D textiles and nonwoven structures for PUR foam substitution in car seats. They describe some key technologies for replacing PUR foams in car seats and explain the resulting environmental advantages. Some selected and summarised requirements for seat cover composite and cushioning layers are described, followed by a discussion of the substitution of PUR foam in upholstery components by polyester nonwovens. The properties and performance requirements are set and evaluated for both the complete composite of seat cover fabric and the cushioning layer. These requirements, including passenger comfort, manufacturer's guidelines, environmental regulations and total cost, must be met by a complete seat containing a textile substituted foam cushion.

In Chapter 8, Volkmar Bartels writes about the developments in physiologically optimised car seats. He describes in detail the topics of thermophysiological comfort of car seats and the methodology of seat comfort evaluation. In this chapter it is demonstrated how the described physiological test methods can be used to improve the comfort of car seats. Some of the recent experiments and results in this area are described. The advantages of modern technical textiles in comparison to traditional materials are discussed. Using a number of examples, it is demonstrated how the physiological seat comfort of vehicle seats can be improved by using modern technical textiles. Only if the material and design of a seat are optimally coordinated will the driver enjoy optimum seat comfort. Otherwise, heat and moisture can accumulate, which feels unpleasant and which can have a negative physical effect on the driver, thereby limiting driving capability and concentration.

In Chapter 9, Francois Boussu and his co-authors write about the advances made in the development of smart textiles in automotive interiors. They describe, in general, the textile structures in car interiors and, in particular, the technical elements enabling the production of smart textiles. Different building blocks, i.e. textile sensors and textile actuators, have been presented in order to show development and integration possibilities of smart textile structures designed for car interiors. Furthermore, they describe the design of textile heating elements and shape memory alloys for use in car seats. Sensors based on conductive composites may be used to detect fabric lengthening, passenger presence or to evaluate safety belt ageing

In Chapter 10, Yan Chen writes about advances made in reduction of interior noise in vehicles. This chapter introduces noise reduction issues involved in the design and production of vehicle interior parts. The focus is on the principles of noise control and engineering approaches for vehicle noise reduction using acoustic textile materials. The issues of vehicle noise sources and noise propagation are discussed and two major noise paths, airborne noise path and structure-borne noise path, are addressed in particular. The concept of a noise-path-receiver model in vehicle noise systems is also described. The author then lays down some theoretical fundamentals that help understand why textile materials can be used as acoustic materials for noise absorption and noise insulation in vehicle interiors. The main effects of textile structure, fibre type, and physical parameter on the material acoustic performance are explained. Major standard methods (ISO, ASTM and SAE) for testing noise absorption and noise transmission loss of acoustic materials are described. Some typical cases of material approaches for controlling interior noise absorption and insulation are exhibited. Finally future trends in the new surface vehicle design and production are discussed and new material requirements for vehicle interiors are reviewed. New developments in acoustic interior manufacture are highlighted, including applications of new textile materials, new design of interior parts in both substrate and surface structure, and new computing tools for interior noise mapping and simulation.

In Chapter 11, Michael Meyer, Haiko Schulz and Michael Stoll write about advances made in leather and coated textiles in car interiors. They describe the market value of leather upholstery and the latest developments in producing leather upholstery and coated textiles as well as the design of car seats with leather and leather/textile face fabrics. They describe the technologies of making leather and leather substitutes, their specialities, and requirements to be performed when used as automotive interior materials. The technological trends for the production of automotive leather are described. There is a tendency to eliminate toxic, mutagenic and other health hazardous chemicals from the tanning processes. Furthermore, automotive leather will be optimised as regards its utilisation properties such as abrasion resistance and non-soiling behaviour as well as water vapour permeability. The authors then describe various technologies for producing coated textiles and artificial leather for automotive interiors. Testing of physical and chemical requirements of leather and coated textiles for automotives is now carried out according to recently developed EN/ISO methods.

In Chapter 12, Samir Mukhopadhyay writes about technical developments and market trends of automotive airbags. The fabric used in making airbags needs to be strong, tough, hard wearing with low and controlled in air/gas permeability behaviour. He describes the characteristics of various polymers for airbag yarns, airbag deployment and performance criteria as well as airbag key technical parameters and contributing factors. In the event of an accident, the critical performance factors of an airbag in saving an occupant's life and injuries are described. The author gives an example of what an airbag supply chain looks like and mentions some of the leading players in this chain. He then describes some future market trends and technological challenges and lists some of the technical areas on which the industry is expected to put more emphasis.

In Chapter 13, Kevin Westgate and James Gillick write about the key technology developments in textiles for use in automotive tyres. They describe in detail the properties and performance requirements of automotive tyres, tyre designs in use today and how the various reinforcement materials and components contribute to tyre performance. They then review recent developments in fibre/textile reinforcements used in tyre manufacturing including the new, novel, advanced fibre reinforcement technology. They also describe the important issue of fibre-rubber adhesion in tyres and recent advances in tyre designs. The authors also take a look at trends in the automotive world and write about how they may impact future tyre development and hence fibre reinforcement opportunities.

In Chapter 14, Thomas Gries, Jan Stueve, Tim Grundmann and Dieter Veit write about development of textile structures for load-bearing applications in vehicles. They describe the role of high performance textiles in automotive engineering and the latest developments in technologies for producing 2D and 3D textile load-bearing structures for automotive applications. Complex textile reinforcement structures in near-net-shape form (textile preforms) can be produced in one-step or multi-step processes, using one-step preforming technologies such as 3D braiding, 3D weaving, contour warp knitting or net-shape weft knitting. Usually, at least sub-preforms are manufactured that are processed in further preforming steps. In many cases 2D textile structures are used to create structures of higher complexity by forming and assembling several plies in multi-step preforming processes. An alternative technology is the use of 3D textile structures and preforming technologies. The relevance of textile and preforming technologies for use in automotive applications is discussed.

In Chapter 15, Frank Ko writes about advances in textile composites as automotive components. He describes the benefits of using textile composites in the automotive industry and the advances in products and technologies. In this chapter, after a review of textile structures for composites, the role of fibre architecture in the manufacturing of automotive composite parts is discussed. Various textile preforms are classified according to fibre architecture and manufacturing technology. The author then reports the results of a special programme – The Sunrise^{TM} electric vehicle programme – which is used as an example to illustrate the design procedure and feasibility for large-scale manufacturing of automotive structural composite components at targeted weight and affordable cost. Specifically, composite pillars and wheel wells are used as examples to demonstrate the procedure in the integrated design for manufacturing (IDM) of automotive structural composites based on the Fabric Geometry Model (FGM). The chapter concludes with a summary and discussion of future trends in the use of textile composites for automotive structural components.

This book is intended for a broad spectrum of readers including students, scientists, researchers, designers, technologists, product development staff and company strategists. It gives in-depth information about the design and development of automotive textiles along the whole supply chain and provides the state of the art knowledge about the recent advances made in technical textiles for a variety of automotive applications. The book should also be very useful for companies thinking of entering into the automotive textiles market.

Part I

General

1

Requirements for automotive textiles – a car producer's view

E SÖDERBAUM, Volvo Car Corporation, Sweden

Abstract: The development of automotive textiles is a very complex and complicated process. Some parts are developed and purchased by Volvo, and some parts by the first tier supplier. The number of technical demands are huge. The number of applications also. The same textile can be applied on different surfaces in the car, resulting in different technical solutions for the applications, but still with the same visual appearance demands. The cooperation with the design department starts very early and remains very close during the whole development process.

Key words: automotive textiles, master sample, photo book, flammability, light fastness, colour changes, fogging, metamerism.

1.1 Introduction

This chapter will describe the textile materials used in the interiors of Volvo cars. Volvo purchases some upholstery material, which includes the textiles used for seats, whereas other textiles for the interior are purchased by the system supplier but developed by Volvo. We use technical requirements specifications – design prerequisites, master samples, drawings, technical regulations, etc. – in which we also specify testing methods. There are some unique and extra tough requirements for the properties of textiles such as: light-fastness, ageing, fogging/emissions, and dirt resistance/cleanability, therefore the textile supplier needs to be involved at an early stage in our development work. The supplier must have a creative design/styling department of their own and a flexible production capacity, since the normal lead time for developing a new textile is 12–24 months. As Volvo has warranty periods of up to four years in some markets, the textile supplier will also be involved in regress issues, and therefore continuous improvement is a necessity.

1.2 Automotive textiles

Automotive textiles are applied on several surfaces of the car interior (see Fig. 1.1). An automotive textile is a complex and complicated product with a great many demands placed upon it. The fabrics used are:

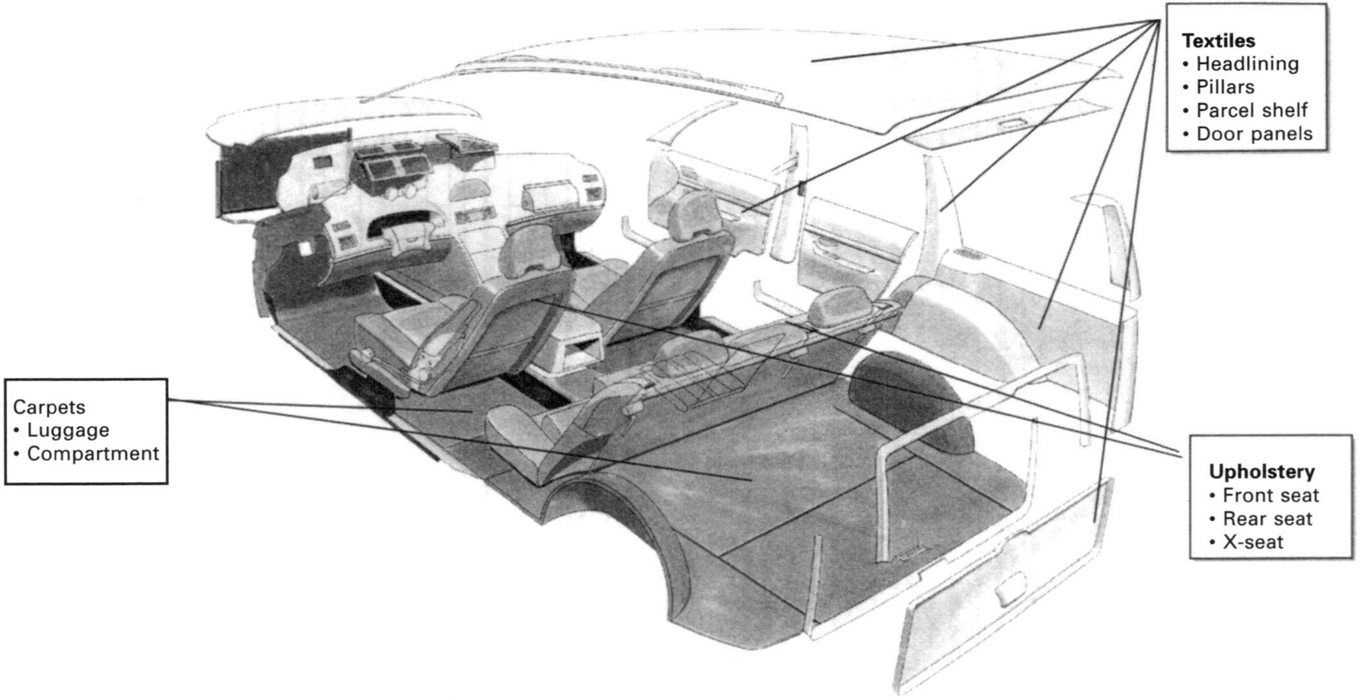

1.1 Diagram of a car interior.

- Knitted, warp-knitted or woven fabric for upholstery and panel applications.
- Tufted or non-woven fabric for carpet applications.

An upholstery textile consists of the textile plus laminate and scrim (backing), which are normally bonded together by a flame-lamination process, but a process of glue-lamination has also been developed. Unlaminated textiles are specified for panel and insert applications, but for carpet applications, different laminations and backings are specified.

1.3 Volvo Car Corporation development process

It is a challenge for suppliers to work with the car industry, with its highly complex development process. Suppliers are involved from a very early stage and need to remain so over a long period since the normal lead time is 12–24 months. The first challenge is to be nominated, followed normally by taking part in Volvo's competitive selection process to become a development supplier. After that, the textile supplier starts to work with Volvo's engineering section either directly or via a system supplier. When the technology and initial appearance are approved, the development of the colour can begin. The normal procedure is that the supplier sends samples for design approval using a light cabinet at Volvo. When the design is approved, which can involve 1–3 loops, testing starts at Volvo and its suppliers. Once the relevant testing is acceptable, Volvo releases a master technical document.

Requirements are often contradictory, which means that not all of them can be fulfilled at the same time, in the same project and on the same product. Decisions on balancing the various demands are made during the development process. The work is normally performed by an assignment team, led by a Volvo-person, together with representatives from Purchasing, Purchase Engineering, Testing, Preparation Manufacturing, the System Supplier and the Material Supplier.

The following three major documents show the input from all the other parties involved:

- Job-split list
- Time schedule
- Interior colour matching process.

1.4 Technical demands

The automotive textile is specified using the following technical demands/ requirements documentation.

Primary documents

- Part drawings
- Technical regulations (TR)
- Master samples
- Standards
- Final Requirements Picture (photo book) and Natural Characteristics Standard (NC Std)

Complementary documents

- Design prerequisite (DPR)
- Colour matching basis
- A-master
- Property description (PD) (or EB in Swedish)
- Function requirement description (FRD) (or FKB in Swedish)

1.4.1 Master sample

What is a master sample?

- It is a physical sample distributed by Volvo.
- It is also a primary document specifying the design of the material.

There are three different masters that are used at different times:

- *A-master* is a master sample that shows only the initial idea of the material's appearance.
- *C/D-master* (Component Development) shows the approved appearance, i.e. colour, pattern, grain/structure, gloss and technology.
- *P-master* (Procurement) is the final master, which shows the approved appearance and the approved technical demands.

1.4.2 Material drawing

The material drawing specifies width, weight, thickness, etc. The drawing also specifies references to TR, Part no. Colour and Master no. Normally the textile is delivered on a roll.

For all other interior surface materials there are normally tolerances on colour and gloss, although an ongoing AE-job will introduce tolerances on textiles in a couple of years.

1.4.3 Design prerequisite

Purpose

The purpose of this document is to show what the design prerequisites are so that suppliers can fulfil Volvo Car Corporation (VCC)'s demands and VCC product design can select suitable surface materials.

The requirements in a design prerequisite (DPR) are valid for all processed materials but individual material requirements can be found in the specific TR. The DPR is a Volvo document and also contains details of what is required when testing methods are inadequate.

General demands

- The interior and the materials will exude quality and harmony.
- The interior should look the same as that of a new car even after three years' normal usage in the harshest market in which the car is sold.
- The interior must be easy to clean.
- The materials will age in a homogeneous way.
- The materials must not have an unpleasant smell.
- The surface materials will last for the four years of the guarantee without any problems.
- The materials will be as rattle- and squeak-free as possible.
- The upholstery materials should be positioned in a way that minimises squeak and rattle.

Material demands

- The aim should be to fulfil all requirements without the use of any chemical substances. The material may only be treated if no other solution can be found to solve a problem.
- The EPS (Environmental Priority Strategy) is to be used systematically in the engineering design and choice of concepts/materials in order to apply the environmental holistic approach that we stand for. EPS calculations will be part of the respective decision parameters in the project.
- All textile and leather materials will be certified according to OEKO TEX 100 in product class 4. Features to be used by children should be certified according to product class 2. A copy of each certificate will be submitted to the environmental co-ordinator at the Interior and Climate Engineering department.
- Materials and chemicals containing substances listed in the 'Restricted Substance Management Standard' STD WSS-M99P9999-A1 must not be used.

- The use of rare metals must be avoided. If they have to be used, they must be documented so that it will be possible to take the appropriate action for the metals at the end of the lifecycle of the car. In addition, EPS calculations must be done on the system and alternative solutions must be analysed.
- Surface coatings containing chromium may not be used.
- Cadmium may only occur as a trace element or contaminant and its content must fall well below the limits of legal regulations. In surface treatments, the content must not exceed 50 mg/kg.
- PVC should be avoided in interior applications. If there are no cost-effective alternatives, the PVC applications must not contain:
  - phthalate plasticisers other than linear C>10
  - lead and organotin stabilisers
  - antimony oxide (SbO_3) flame retardant.
- PUR (polyurethane) must be used with caution because isocyanates present a hazard during production and their uncontrolled incineration during the end-of-life phase can release unwanted gases.

1.5 Technical regulation

The technical regulation (TR) document contains a combination of requirements and well-defined testing methods. Testing is done both at Volvo and the supplier via a job-split list (defined in the DPR).

Most of the requirements are familiar to all textile suppliers, but there are some demands specifically developed by the car industry. The TR reflects in a sense 'the soul of the company' – the requirement, its importance and the testing method to be used is an empiric, long-term, experience-based result connected to the product – the car.

For the most typical automotive requirements, a description is given of both the background of the test and the testing method as follows.

Flammability

Background of test

The flammability test is an American legal requirement which must be performed successfully on all materials of the car interior. This test is carried out on every production batch.

Test method

To fulfil the legal requirements, the test must be carried out according to standard MVSS 302. This document specifies what parts and materials are to be tested, what equipment to use, testing procedure, etc.

The test piece is placed horizontally in a small cabinet and exposed to a flame. The burning rate of the material is calculated, measured in mm/min. The legal requirement is 102 mm/min, but some car manufacturers specify a lower value.

Light-fastness

Background of test

During the life of a car, the interior materials will be exposed to a large amount of sunlight, especially in areas like Australia or the southern USA. Fading and deterioration of the materials leads to a poor impression of the quality of the interior.

Test method

Testing is normally performed according to international test methods like ISO 105-B02 or ISO 105-B06, where the test pieces are exposed to a filtered Xenon arc lamp for a certain length of time. A maximum value for a change of colour is specified and usually the requirement is blue-scale rating class 7 or higher.

Ageing

Background of test

As cars are sold and used all over the world, they will be exposed to different temperatures and climates with temperatures ranging from –30°C in winter to +90°C in warmer climates, in dry or humid conditions, with a year-to-year variation. Cars are shipped by boat from Gothenburg to Singapore via the Suez Canal, therefore the cars are exposed to big differences in temperature and humidity during this three-week period; and during the summertime in places like Arizona and Saudi Arabia, areas such as the dashboard, head rests and parcel shelf can reach temperatures of 110°C or more.

Test method

Deterioration and changes of colour through ageing are evaluated by accelerating the ageing process in a climate chamber for a number of weeks. The climate chamber might be set at a fixed temperature and humidity level, or it might be set to change over a cycle of, normally, 24 h.

Fogging

Background of test

Materials release substances, especially when heated, and these substances, including additives, process enhancers and flame retardants, condense on the cold windscreen, creating a film on the inside called fogging. Not only is fogging difficult to remove, requiring solvents and therefore causing annoyance to the customer, but it also reduces vision and is a safety issue.

Test method

Fogging is tested for by putting a piece of the material in a glass beaker, which is placed into hot oil for several hours. The beaker opening is covered with a glass plate or aluminium foil, which is cooled by a cooling device placed above it. The purpose of the cooling device is to cause condensation of the fogging onto the glass plate or aluminium foil. The fogging is then determined by measuring the gloss of the glass plate before and after the test (reflectrometric fogging) or weighing the aluminium foil before and after the test (gravimetric fogging). A maximum level of fogging is specified as a percentage (reflectrometric) or in mg (gravimetric). Testing is usually done according to ISO 6452.

Soiling and cleanability

Background of test

People spend more and more time in their cars and it is not unusual for them to eat burgers and drink coke, therefore items such as cup-holders are featured in most vehicles. The surface materials are also exposed to dirty shoes and dust from the streets when people enter the car. In addition, people often have their coats and jackets on in a car, unlike when they are at home. As a result of customer demand and fashion trends, many car manufacturers, including Volvo, tend to offer lighter-coloured interiors, which makes them more sensitive to soiling and staining.

Test method

During testing, the materials are exposed to artificial standard soils and also to stains from coke, coffee, ketchup, etc. The appearance of the soiling or staining is evaluated, as is the possibility of cleaning the material using conventional cleaning methods.

Abrasion

Background of test

The materials in a car should last as long as the rest of the car, meaning at least 10–15 years, which requires them to be abrasion-resistant.

Test method

Abrasion is usually tested using well-known methods like Martindale (ISO 12947) or Taber (ISO 5470). The material can be subjected to up to 50 000 cycles in a Martindale tester.

Metamerism

Background of test

The light in a showroom differs from natural sunlight. When the light source is changed, the difference in wavelengths must not lead to a change in colour appearance.

Test method

Metamerism is not allowed. The procedure for obtaining design approval involves the use of a light cabinet.

Generally speaking, testing is done by both Volvo and the supplier. What is done and when is defined in the job-split list.

Volvo performs complete car tests at their test facilities. For example, cars are left for one year at the Arizona Proving Ground in Phoenix for long-term, real-time ageing. Another example is the TAXI-test for seats, which normally lasts for nine months, covering 100 000 km and involving 15 000 ingress/egress cycles.

1.5.1 Additional functional demands

The application of a textile can add some additional automotive demands that are hard to fulfil or cause design changes to be made to the textile because it is easier to change the textile than the form of the part or the function.

A panel textile is specified to be a *non-laminated* material. This means that the panel supplier is responsible for selecting and carrying out the lamination, and also means that the textile supplier must be competent enough to know how a laminate affects the textile and vice versa.

1.6 Design demands

Often a supplier's first contact with Volvo is with a designer during company visits, exhibitions and conferences, where the designer is looking for new trends, new ideas, new appearances, etc. However, there is a long development period before the start of production (SOP).

The design department selects a textile supplier on the basis of a textile's appearance, which must portray the following features of the Volvo brand:

- core values
- image
- long-term fashion and trends
- identity
- heritage
- the spirit of time
- type of car
- type of level.

This means that it is essential that the supplier understands design trends, fashion and the Volvo image.

The first critical 'key-hole' is to attract the attention of the design department. In the concept and pre-study phases, the prerequisites from the design department are normally shown as *life-style boards*. To be able to work with design, each textile supplier must have a design department of its own that is extremely creative. VCC often chooses textiles from the supplier's own collection, which means that the collection must already be developed up to an early production status and its feasibility must be satisfactory. The development of VCC colours and adaptation to VCC's demands remain important criteria.

1.7 Purchase demands

Volvo sources the following upholstery materials for application on both seats and doorpanels:

- textiles
- leather
- tricot-plastic.

All other surface materials including:

- carpets
- cloth on pillars, headlining, parcel shelf

are sourced by the system supplier but developed by VCC.

1.7.1 Volvo's general purchasing conditions

- The supplier must be third party (an outside-accredited institute) certified according to Automotive Industry Quality Standards, e.g. QS9000/ TS16949, EAQF or VDA 6.1.
- For environmental requirements, the supplier must be certified as conforming to environmental management systems, e.g. EMAS or ISO 14001.
- The supplier must fulfil requirements as detailed in the PQP – 'Purchasing Quality Procedures' manual.
- Demonstrable capability in the production process.
- 4-year normal warranty.
- 15-year post-sales service.
- Flexibility in development.
- Flexibility in technologies.
- Flexibility in production.
- Testing facilities.

1.7.2 Supplier evaluation model

The supplier evaluation model (SEM) enables Volvo to make a formal decision to accept the supplier (see Fig. 1.2).

1.7.3 Cost

A typical price for a material is 5–15 Euro per square metre, depending on its application and quality level. Normally, annual cost reduction figures are included in the business agreement and during the last decade the square metre price has dropped by approximately 20%. Table 1.1 shows some normal coverage figures.

When it comes to warranty issues there are two main sources.

Customer surveys

'JDP' and 'GQRS' represent customer surveys which gather customer opinions with reference to different systems in the car. One result is often a request for product changes.

Warranty cost

With regard to regress issues with the supplier, in most markets there is a four-year warranty, but in the US market there are two additional circumstances:

VOLVO

Evaluation matrix

Supplier Evaluation

		Evaluation						
		0 pts	1 pt	2 pts	3 pts	4 pts	Res.	Max
1	**Company profile**						0	8
	1.1 ownership							
	1.2 global ability							
2	**Management**						0	12
	2.1 management							
	2.2 customer satisfaction							
	2.3 TQM work procedure							
3	**Environment**						0	8
	3.1 environmental management system							
	3.2 company-level assessment							
4	**Quality**						0	12
	4.1 quality system							
	4.2 quality performance, deliveries							
	4.3 reliability							
5	**Logistics**						0	8
	5.1 logistic system							
	5.2 delivery precision							
6	**After Market**						0	4
	6.1 co-operation and support							
7	**Competence**						0	12
	7.1 product and ind technology							
	7.2 industrial engineering							
	7.3 customer support, comm.							
8	**Product development**						0	12
	8.1 prod developement process & proj. support							
	8.2 engineering experience & technology							
	8.3 support (prototypes, design changes)							
9	**Finance**						0	4
	9.1 finance evaluation D&B							
10	**Productivity**						0	12
	10.1 internal cost reduction							
	10.2 investments							
	10.3 cost targets							
11	**Sourcing**						0	8
	11.1 sourcing process							
	11.2 sub-contractor performance							

stopping parameter

Res	Max
0	100

Suppl no | Supplier name

Date | Examinator

Grading:

A	World class	>=90
B	Excellent	>=75
C	Good	>=60

1.2 Volvo supplier evaluation model (SEM).

2
Mapping the automotive textile supply chain

N POWELL, R HANDFIELD and
R BARNHARDT, North Carolina State University, USA

Abstract: Information visibility is an essential element to improved supply chain performance. Without adequate sharing of information, firms are required to supplement the available information with other costly alternatives. This chapter follows the development of a convertible fabric material in the supply chain for a large automotive assembler, beginning with a large textile manufacturer in fabric formation to final assembly of the convertible top. The current level of information sharing in the supply chain is assessed and the effects of inefficient information flows between the various tiers of suppliers are determined.

Keywords: supply chain, automotive textiles, information visibility, sourcing in the value system, case study.

2.1 Introduction

The sourcing of textile products for vehicles is typical of the complex supplier network for the large-scale global automotive industry. Nieuwenhuis and Wells describe the complex automotive market as characterized by the conflict between the production capabilities and the market demand (2003). The auto industry is considered the epitome of the mass production system, affecting the global balance in trade and commerce, environmental issues, safety, employment, manufacturing and assembly methods, technology, and style (Cagan and Vogel, 2002). As Cagan and Vogel describe the automotive industry, 'No other market develops products in an atmosphere where competition is as fierce, the product is as complex, and the challenge of constant refinement and innovation is as great' (2002, p. 254).

The Society of Automotive Engineers lists more than 10,000 companies in their 2007/2008 supplier directory (2007). Materials and components are estimated to contribute up to 80% of the value of a finished vehicle. Global Insight (2007) estimated the occupant restraint equipment business alone to be worth US$17 billion in 2005 and it is expected to grow (p. 13).

The demand for rapid prototyping, cost effectiveness and streamlining of supply systems is critical in all component parts, but particularly in automotive textile materials. In the tiered structure of the automotive supply chain the original equipment manufacturer (OEM), such as General Motors, is responsible for initial overall design, assembly and marketing of the complete vehicle and brand through dealerships to the consumer. A complete system component such as an interior compartment would be supplied directly to the OEM by a Tier One supplier, such as Johnson Controls. A Tier Two company would supply subassemblies to the interior system. The individual parts of these subassemblies would be sourced to a Tier Three supplier, and the materials would be provided by a Tier Four company. Some companies function in roles across tier levels. A textile company may be considered a supplier at the third or fourth tier.

In order to improve performance in the automotive supply chain and create a truly integrated value system, a number of tactical and strategic changes will be required throughout the automotive textile supply chain. Without the buy-in of supply chain members, asking them to react faster in response to real-time ordering systems is an exercise in futility. A supplier who receives an order via the Internet, prints it out and puts it on a stack on their desk is not contributing to supply chain value. Instead, when information is shared via the supply chain and made available to all parties, such that demand requirements, capacity limitations, inventory positioning, and collaboration between partners is established through *a priori* agreements, then the power of the Internet can begin to be fully exploited. This means providing an Internet-accessible, real time forum where buyers and their suppliers can communicate and share inventory and forecasting information, and allow for the effective dissemination of engineering change orders (ECOs) throughout the supply chain. Nowhere is this more important than in the automotive supply chain, where textile manufacturers are often several tiers removed from the final OEM assembler, and often subject to the vagaries of poor forecasts and lack of information. Visibility of information within the supply chain is a critical starting point in considering supply chain management in the automotive industry. An automotive textile case study tracks the development of a convertible top material (fabric) in the supply chain for an automotive components manufacturer for a large Big Three automotive beginning with a large textile manufacturer in fabric formation to final assembly assembler. The current level of information sharing in the supply chain is assessed and the effects of inefficient information flows between the various tiers of suppliers are determined. A discussion of the approach used to map the supply chain follows, and then summary observations conclude the chapter.

2.2 The importance of information in supply chains

The need to share accurate information across the supply chain is of paramount importance. In fact, lack of information or distorted information passed from one end of a supply chain to the other, can create significant problems, including, but not limited to, excessive inventory investment, poor customer service, lost revenues, misguided capacity plans, ineffective transportation, and missed production schedules. Distorted information throughout the supply chain is a common result of what logistics executives have termed the *bullwhip effect.* This phenomenon is characterized by a series of particularly erratic shifts in ordering up and down the supply chain, often with more demand variability at the distributors' and suppliers' linkages than at the retail level (Lee *et al.*, 1997). Four causes of this phenomenon were identified:

1. Demand forecast updating
2. Order batching
3. Price fluctuations
4. Rationing within the supply chain.

Automotive firms such as Ford and Toyota have experienced a similar phenomenon. Even a slight to moderate demand uncertainty and variability becomes magnified when viewed through the eyes of managers at each link in the supply chain. If each manager makes ordering and inventory decisions with an eye to the firm's own interest above those of the change, stockpiling may be simultaneously occurring at as many as seven or eight places across the supply chain, leading in some cases to as many as 100 days of inventory – waiting (Lee *et al.*, 1997).

2.3 Supply chain organizational dynamics

Two major interorganizational dynamics come into play when addressing information sharing across the supply chain: risk and power. All enterprises participating in a supply chain accept a specific role to perform. They also share a common belief that they and all the other supply chain participants will be better off because of their collaborative efforts. Each member specializes in the function or area that best aligns with its competencies. Risk occurs in that each firm must now rely on other supply chain members as well as its own efforts to determine the success of the supply chain. Those members most dependent on the supply chain with the most at stake may take more active roles and assume greater responsibility for fostering cooperation, including the information-sharing efforts, throughout the supply chain.

Power within the supply chain is a central issue, one that in today's marketplace centers on information sharing. Although not universal to all industries, there has been a general shift of power from manufacturers to retailers over the last two decades, which has resulted from a combination of factors. One is the trend toward consolidation at the retail level within the supply chain. Gone are the days of locally owned, independent dealerships in each town. In the interest of capitalizing on the benefits of economies of scale, giant retail conglomerates operate as part of nationwide supply chains.

In fact, relatively few of the thousands of retailers operating in the United States control the majority of dollars in this industry. Clearly, this consolidation impacts the entire supply chain. Although there were reportedly (Plunkett, 2007) 22,495–26,700 (Global Insight 2007) dealerships in the US in 2006, the 100 largest groups sold 17.7% of all new vehicles. Fewer and fewer firms control access to consumer trading areas. The number one public group AutoNation reportedly has 257 dealerships. Many larger organizations are looking outside their own domestic regions to expand their retail business globally (Global Insight, 2007).

Perhaps more importantly, retailers sit in a very important position in terms of information access for the supply chain. For several reasons, major retailers have risen to this position of prominence through sheer size and sales volume, and most importantly, their position within the supply chain right next to the final consumer. Providing current information on consumers' preferences and purchasing behaviors to OEMs can improve products and services. Retailers can capitalize on this position through responsive services to the consumer. Improved communications between the dealer and the OEM could facilitate the growing consumer interest in customizing a new vehicle, optimizing factory options. This combination of factors has put retailers in a very powerful position within the supply chain.

2.4 Creating information visibility in supply chains

2.4.1 What is information visibility?[1]

Information visibility within the supply chain is the process of sharing critical data required to manage the flow of products, services, and information in real time between suppliers and customers. If information is available but cannot be accessed by the parties most able to react to a given situation, its value degrades exponentially.

Increasing information visibility between supply chain participants can help all parties reach their overall goal of increased stockholder value through revenue growth, asset utilization and cost reduction. These multiple levels of trading partners need to share forecasts, manage inventories, schedule labor, optimize deliveries, and in so doing reduce costs, improve productivity, and create greater value for the final customer in the chain. Software for Business Process Optimization (BPO) and Collaborative Planning, Forecasting and Replenishment (CPFR) are evolving to help companies collaboratively forecast and plan amongst partners, manage customer relations, and improve product

[1]This section is based on a benchmarking report developed by the following associates of the Supply Chain Resource Consortium: Steven Edwards, Meenakshi Lakshman, and a number of undergraduate students at North Carolina State University.

life cycles and maintenance. Traditional supply chains are rapidly evolving into 'dynamic trading networks' (Cole *et al.*, 1999) composed of groups of independent business units sharing planning and execution information to satisfy demand with an immediate, coordinated response.

Some of the considerations that must be planned for in implementing an information visibility system include the size of the supply base and customer base with which to share information, the criteria for implementation, the content of information shared and the technology used to share it. Clarifying these issues will help to ensure that all participants have access to the information required to effectively control the flow of materials, manage the level of inventory, fulfill service level agreements and meet quality standards as agreed upon in the relationship performance metrics.

Perhaps no other company has been as successful in implementing information visibility as a competitive strategy than Dell Computer. Dell has fulfilled its commitments to customers through the company's direct model, in which it holds only hours of inventory yet promises customers lead-times of five days. Component suppliers who wish to do business with Dell have to hold some level of inventory, since their cycle times are typically much longer than Dell's (Lewis, 2001). Dell has developed a business model that features a lean, build-to-order manufacturing operation. By utilizing the web, Dell provides its supplier with forecasting information and receives information about the supplier's ability to meet the forecasts. Communication between engineering changes, component availability, capacity, and other data between Dell and its suppliers flows both ways, in addition to forecasting and inventory data. Dell is also able to review suppliers and place web-based orders into their factories in hours. Dell's build to order web-based customer model has become the benchmark for other organizations such as General Motors and Ford who are seeking to create 'build-to-order' models using the web as the platform for taking customer orders.

2.4.2 Benefits of information visibility

Information regarding forecasts, changes in production schedule levels and on-going supply chain performance metrics needs to be conveyed by customers to suppliers on a regular basis. Information flows from suppliers to customer can include current order lead times, capacity levels, order status and inventory levels. Receiving and conveying the correct information will ensure that the suppliers are aware of what needs to be produced while at the same time, the buying firm is sure that it is possible to receive ordered quantities on time, every time. The most important benefit of a visibility system is not that the system is able to correct a supply chain problem, but that it allows people to become aware of problems earlier, and thus take corrective actions more quickly than they otherwise would. The benefits of information visibility include reduced lead times, improved constraint management, better decision making, lower costs, and increased profits. Although problems such as shortages, changes in customer orders, engineering changes, obsolete inventory, and equipment failures can still occur with a visibility system in place, the effects of these problems are less than if the participants in the supply chain were not made aware of these problems until a later date.

When implemented properly, a visibility solution results in the following additional benefits that promote improved supply chain performance:

- *Breaks organizational barriers.* Enables sharing of mission-critical information about business activities and interaction on a near real-time basis across the supply chain.
- *Builds visibility into supply chain.* Provides people a real-time snapshot of supply chain performance metrics.
- *Managing by metrics.* Aligns performance metrics with cross-organizational business processes and assigns ownership of processes and metrics to specific individuals.
- *Reduces the decision cycle process.* Allows an upstream or downstream participant to respond to market or customer demand in hours or days, not weeks or months.
- *Encourages decision-making collaboration.* Facilitates the ability to make decisions collaboratively on the Internet, bringing relevant internal and external stakeholders into the process.
- *Reduces opportunity and problem resolution latency.* Measures and monitors supply chain activities iteratively, allowing people to respond quickly to events as they occur.

Conversely, the dangers of poor execution of supply chain processes include increased lead and cycle times, higher costs, and less informed decision-making.

This case study focuses on an analysis of only those processes that are directly related to the vehicle convertible top material that is manufactured by a large textile manufacturer (A) and is handled further up in the supply chain by a company that treats the fabric (B), one that completes the cut and sew operations (C) and the final OEM assembler in Michigan (D). A broad view of the supply chain map was developed from public information with additional content from industry interviews and a questionnaire sent to each supply chain partner. The focus of the questionnaire was to study the inventory levels, service and responsiveness in the supply chain, and to identify gaps in the supply chain. The data collected from supply chain members was then used to identify:

- Areas of low inventory turns in specific areas of the supply chain.
- Potential areas of improvement across the supply chain members.

- Information requirements to help alleviate the problems detected during analysis.

2.5 Market analysis

Scale and efficiency are extremely important to the automotive industry's large manufacturing firms in order to keep costs down and enhance profit margins. According to *The Economist* (Anon, 2005), 'if their affiliates as well as their wholly-owned companies are counted, the top five company alliances account for 75% of the global market, while adding the next five takes this to 90%, with producers in China, India and Malaysia making up the rest.' 'Suppliers make up 70% of OEM costs, and Tier 1 suppliers passed the cuts down the chain' (Ward's, 2003, p. 63). David Cole of the Center for Automotive Research predicts that as industry consolidation continues, the Tier 1 and Tier 2 suppliers will develop and deliver more complete subsystems to the assembly plants, thus taking on 'significantly higher levels of responsibility' (SAE, 2007, p. 103).

2.5.1 Major global players in automotives

Most major technologies and technical expertise became available in the 1990s from global suppliers which prompted competitive pricing, and emphasized innovative product development to distinguish products. By 2007, the number of textile suppliers has been reduced, either closing operations, declaring bankruptcy in order to reorganize their operations, or being absorbed into larger organizations. Guilford Automotive has reorganized solely as an automotive supplier, particularly in warp knitted headliner fabrics. Milliken & Company remains stable as a privately owned diverse product company.

2.5.2 Structure of supplier value system

In the automobile industry, suppliers to OEMs do not have much power – the automobile manufacturers pass down the general specifications, and it is unlikely that any one supplier could cause the major manufacturers any problems due to the number of current suppliers and potential suppliers who desire to be in the automotive supply chain. The cost pressures of OEMs have intensified, resulting in long-term contracts being opened for new more competitive bidding from global suppliers. Many textile suppliers are reported to have suffered losses as the Big Three have lost market share to Asian companies. Contracts with OEMs, which earned little or no profit for textile companies, are reported to have contributed to the demise of suppliers such as Collins and Aikman. Failure to deliver components can drastically affect production and shut down assembly of vehicles. If a supplier loses a contract then it could take months or years to secure replacement business (Sherefkin, 2005).

The retail buyers, defined as the dealerships, do have some legal power as a group. To date, they have been able to stop the major manufacturers from forming a direct link to the end-consumer, but that may change soon. Again, however, no one dealer has the power to cause major problems for any of the manufacturers.

The automotive industry is very capital-intensive, requiring plants to be built, equipped with machinery, and staffed with people. In general, this makes the threat of new entrants relatively low, as any company wanting to get into this industry would have to expend large amounts of cash before seeing even the slightest return on investment. However, many textile manufacturers possess the capital equipment necessary to manufacture products for the automotive industry by realigning their products/markets. Companies such as Interface, Glen Raven, Freudenberg and IAC Group are positioned to establish themselves further in the automotive textiles market. With the demise of global textile trade quotas in 2005, additional suppliers from Asia entered the North American market. Japanese textile suppliers such as Seiren, Kawashima and Suminoe and Europeans, Eybl and AundE, are not only shipping into international markets but some are also establishing manufacturing plants in regional markets as lead times for developing and producing products become compressed and the costs of transportation increase (Powell, 2004). With approximately 23 textile suppliers in Europe, not including the subsidiary plants, the opportunities for shifts in the supplier chain are likely (Global Insight, 2007). No one supplier of textile fabrics has dominated the market for any length of time, and to remain competitive in the global supply chain, every player must maximize their value to the supply chain.

Experts forecast that the number of cars and light trucks on global roads will reach 1 billion by 2020 (Plunkett, 2007). In the year 2000, the US automobile industry again set an all-time record high in sales of automobiles and light trucks by reaching 17.4 million units. This rise was attributed to the US economy, higher personal income growth, and enhanced consumer confidence. However, the US economy has recently begun to soften, with North American production of cars and light trucks in 2006 falling to 16,374,670 units. Table 2.1 reflects the recent sales by region and forecasted production. The forecast for 2007 for US light vehicle sales will be the lowest in nine years. According to *Automotive News*, European production was 20,811,446 units in 2006 while Asia-Pacific production was 24,828,397. Total global production of cars and light trucks in 2006 was 67,723,891 units (*Automotive News*, 2007). Global production is forecasted to reach 70 million by 2010 (Bhagwat, 2004) and an extended forecast is reflected in Table 2.1. Global competition at all levels will be forced to drive costs down and focus on specific areas of competition.

Table 2.1 Automotive forecast production

Forecast 2010	Region	Sales	Production
	Europe	17,934,370	20,694,029
	North America	20,699,218	17,275,662
	India	1,860,699	2,277,656
**	China	8,831,707	8,834,815
	Japan	6,108,709	10,639,100
Total		55,434,703	59,721,263

**2002 China production was 3 million, and in 2005 5.06 million, 2002 China Sales were 2.7 million and in 2005 4.99 million
Source: Global Insight, 2007, Automotive News Data Center, J.D. Power Automotive Forecasting, and Mexican Automotive Manufacturers Association, Japan Automobile Importers Association, Japan Automobile Dealers.

The US automobile manufacturers have continued to emphasize profitability over market share. However, profit margins are under pressure for both General Motors and Ford. Any significant reduction in production volume will put these two companies' profits under severe pressure. Currently, the US market for new passenger cars and light trucks is quickly approaching saturation. There's little indication that, on a long-term basis, annual growth will be over 1–2%. As a result, the major automakers are turning to developing countries for additional growth. These countries include China, Malaysia, Indonesia, and India. In 2006, China became the world's second largest market for new vehicles, surpassing Japan and exceeded only by the United States, registering sales of 7.22 million units, up 25.13% year-on-year, according to the China Association of Automobile Manufacturers (http://china.org, 2007). Russia ranks number 8 and India number 10 in global vehicle sales market share in 2006 including passenger and commercial vehicles (*Automotive News*, 2007). While some of these are rather large markets, the average incomes are low. This could prevent sales penetration in the short run. Nevertheless, US penetration into these international markets will likely be lower due to competition from European and Asian automotive manufacturers.

Today the growth of markets worldwide and the growth of production in developing economies like China and India forecast major opportunities for global suppliers and consumers. The mounting economic pressures on the supply chain will demand significant investments and mature strategies to succeed in global competition. The single biggest expected change is the growth of the Internet in researching and purchasing material goods. Consumers are increasingly using information from the Internet to determine such things as features, specifications, styles, and designs of different makes and models. It is reported that 62% of new car buyers use the Internet during purchase (AAIA, 2007). In addition, consumers are increasingly demanding more and more flexibility and options in what they can purchase. The industry may soon be able to implement the process of mass customization, thereby circumventing the middleman and dealing directly with the purchaser. This is an effort to reduce the huge amount of inventory that exists in the automotive supply chain (see Fig. 2.1).

2.5.3 Cut and sew operations

The cut and sew industry is considered fragmented, with a great number of competitors providing this service. The level of key computer integration and automation distinguishes the key competitors from the rest of the competition. Some competitors rely almost exclusively on manual labor while others have computer-controlled pattern cutting to allow for exacting specifications. It is expected that since this is a Tier 2 supplier to the automakers, both the mentioned cut and sew operations would fall into the latter (automated) strategic group. The automotive components supplier business (cut and sew) is also fairly fragmented. This industry is also mature with some consolidation amongst competitors. These participants are competing fiercely and directly with one another in many of the markets, including apparel and upholstery, serviced by this industry. Different competitors compete directly on a cost, quality and flexibility basis or carve out niche markets such as convertible tops in which they are differentiated by their technology and 'turnkey' systems. Rivalry among participants is high, but this soft top is a niche market with few players. In fact, it is not unusual for a Tier 2 or 3 company in this specialized convertible top market to have a market share of 60% or greater.

2.5.4 Just In time component suppliers

The automobile manufacturers possess significant power due to their small number in the OEM sector and relative size compared to the remaining tiers in the supply chain. The automobile assemblers are not likely to backward integrate into this niche market, as it would not fit their core business. On the other hand, companies within the various tiers are a credible threat for forward integration. For example, Collins & Aikman became a major supplier of textile products to the transportation industry with over 60% of its business involved in the automotive industry by the mid 1990s. At that time a strategic decision was made to become a major supplier to the first and second tier suppliers in the automotive supply chain. This was accomplished through mergers and acquisitions of companies in other supply chains serving the automotive industry. Collins & Aikman acquired Textron's TAC-Trim unit, nearly doubling sales. However, this rapid growth and the unsuccessful attempts on the part of stakeholders to solve profitability issues may have contributed to the demise of C&A today (Barkholz, 2006). By 2006, Collins & Aikman's

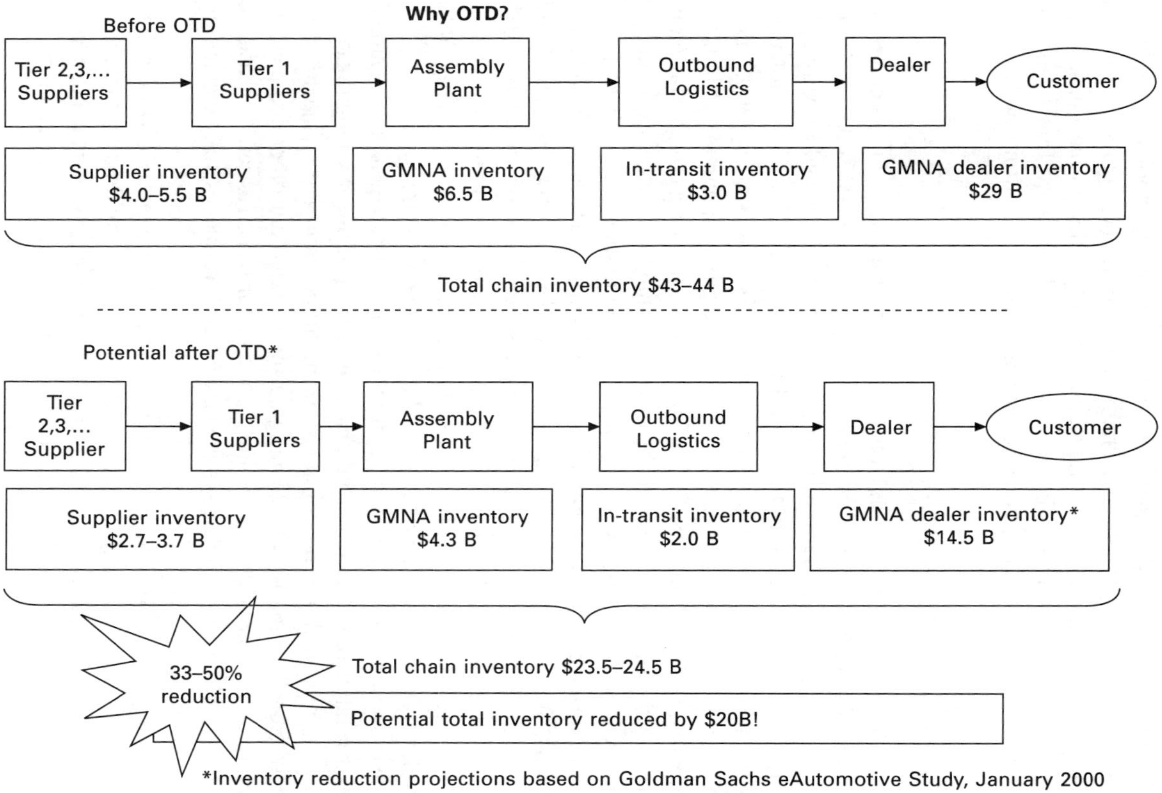

2.1 Impact of mass customization of autos on automotive supply chain.

automotive textile business was closed, and the automotive flooring and acoustic components businesses were available for sale within the bankruptcy court (Anon., 2007, p. 2)

Forward integration does, however, have limitations, since companies may be less of a threat to the completed convertible top automotive component market due to the complexity of mechanical design for these parts. Substitutes for these products include vinyl tops and conventional steel hardtops. A car buyer interested in a convertible vehicle has likely decided that neither of these is a legitimate option, keeping the threat of substitution low.

2.5.5 Competitive advantage

Demand for convertible top products tends to be proportional to fluctuations in demand for automobiles/light trucks in general (Hoovers, 2003). 'Overall sales volume may now be settling at a higher plateau. In 1999 it surged to 16.9 million units, reaching 17.35 million units in 2000. With fast-pace production techniques available to the automotive industry, by 2005 there were over 57 million new cars produced globally per annum' (Plunkett, Automotive Industry Statistics, 2006). These results suggest a possible breakout from the old range into a permanently higher annual sales rate, even though cyclical variations are likely to remain. Leading the increase has been rising sales of non-passenger car vehicles, such as light pickup trucks, SUVs, and minivans. While passenger cars sales declined by 6.1% between 2002 and 2003, light truck sales rose 3.6%, bringing market share for trucks to 54.3% in 2003, up from 48.5% in 1998. Plunkett Research reported that in 2005 total new car sales were 44% of the US market with trucks being 56% (includes light and medium/heavy trucks) (Plunkett, 2007).

The most important trends in the automotive industry are generally centered on two related developments: intensifying competition and globalization. Due to increased domestic competition, manufacturers are trying to enter foreign and niche markets by leveraging their brand and other competitive advantages. Innovative design and innovative features such as safety measures, sensors and global positioning systems are crucial for successfully expanding the niche customer base.

The suppliers of original vehicle parts and replacement parts (also called aftermarket parts) are being forced by automakers to provide parts that are of better quality and last longer than ever before. While good for the consumer and carmakers, this trend has weakened profitability for manufacturers and sellers.

Faced with retail pricing pressure (from both dealer showrooms and aftermarkets), the Big Three in recent years have demanded substantial cost reductions from their outside parts suppliers. Suppliers, in turn, have responded with aggressive programs to enhance productivity and improve efficiency, learning to function in an era of price freezes and reductions. As automakers pare their rosters of suppliers, they increasingly favor companies that can deliver modules (a combination of attached parts) and systems (a functional combination of modules), rather than individual parts. Their ever-increasing demands are leading to consolidation among automotive parts suppliers, for whom mergers and acquisitions are ways to cut costs and provide more modules and systems.

2.6 Results of supply chain mapping: current practices and linkages

After collecting the data from an actual automotive supply chain, it was organized into three key areas: operations, orders, and logistics. The data in each of these areas was used to determine where each company stood with respect to the entire supply chain.

2.6.1 Operations

The first area is operations, which contains information such as cycle time (in days), the forecast communicated to the upstream partner, batch size, and the number of days supply in inventory. As can be seen in Table 2.2, the current inventory in the system rises by a large margin at the partners before D. As well, the communicated forecast from C down is about three weeks, where D gives 20 weeks to C. Inconsistencies in batch size exist between D and C (downstream), and A and B (upstream). This could also be a cause for the high inventory in the system. Finally, the cycle time in days increases dramatically as you move up the chain, from three days at D to 40 days at A. This, combined with the lack of an accurate forecast, could be contributing to a bullwhip effect throughout the chain.

2.6.2 Orders

Orders comprise days of lead time, mode of receipt, and how often it sends an order to the next partner in the chain. The mode of transmission and type

Table 2.2 Material management in automotive supply chain

	Cycle time (days)	Forecast communicated	Batch size	Inventory (days)
D	3	20 weeks	Lot-for-lot	1.3
C	13	3 weeks	Lot-for-lot	59
B	14	20 weeks rolling, 3–4 week lock	4,000 yards	42
A	40		5,000 yards	53.2

of information exchanged varies between the various supply chain partners, as shown in Table 2.3. While EDI (electronic data interchange) is used between the majority of partners, phone/fax is used between A and B. This could potentially cause problems in terms of order entry and confirmation, and also forecast entry and confirmation. While not a primary cause, this, along with the frequency of information exchange, could be a minor factor in the excess inventory in the system that was shown above.

2.6.3 Logistics

The final area is logistics – how the material gets from supplier to customer. Key points in this transaction (shown in Table 2.4) are transit time (and batch size), how often an order requires premium freight, and how often shipments are sent.

The transit time, while minor in most cases, really balloons at C, with five days internal and five days external shipping. This is largely due to their cut and sew operation being located outside the US. This decoupling of processes can cause more inventories to be held in the system – indeed, it seems as if approximately a month's worth of inventory is being held at both B and C to account for the international transit (redundant inventory). As well, the transit batch sizes are very different – 4,000 yards to B every week, 4,000 yards to C every 25 days, but only 55 tops (or approximately 110 yards) to D every day.

One key point to note in this data is the lack of premium freight. While, at first glance, this appears to be a good thing, it is the result of the excess inventory that is present throughout the system. It seems unusual, especially

Table 2.3 Automotive supply chain ordering patterns

	Lead time (days)	Order receipt mode	Frequency
D	–	–	1 week
C	7	EDI	3 weeks
B	21	EDI	4 weeks
A	28	Phone/fax	–

Table 2.4 Logistics of material flow in supply chain

	Transit time (days)	Size	Premium freight	Shipment frequency
D	1	–	<1%	Daily
C	5 (internal) 5 (external)	55 tops	<1%	25 days
B	2	4,000 yards	<1%	Weekly
A	3 (internal)	4,000 yards	<1%	–

in a supply chain of this size, with the great geographic disparity, to have a consistent less than 1% of all shipments be premium freight. While this is not the cause of a problem, it definitely seems to point to too much inventory in the system.

2.6.4 Ideal management practices

Across an ideal supply chain, operational parameters, information management and logistical issues would be balanced to optimize the entire chain's total value proposition. This would provide equitable economic value added for each of the chain participants. Implementation would involve minimizing forecast variance, seamlessly transferring accurate forecast and order data on a real-time basis and maintaining optimal inventory levels at strategic points along the chain.

Ideal operations

The ideal management system across an entire supply chain integrates functions between parties at each of the interfaces between partners. These typically are information transfers (forecasts and orders), inventory management and logistics. In turn, optimal management of these allows for all parties to optimize purchasing, inbound logistics, inbound quality control, receiving and storage, inventory control, order processing, production planning and scheduling, warehousing, outbound transportation and customer service within their own organization. Each member of the chain may have the opportunity to trade inventory management for information management.

Ideal information flow

Ideal information flow provides accurate, reliable electronic data interchange with high frequency and minimal personal intervention. This is optimal because it provides real-time forecast and planning data that provides extended visibility and stability to the suppliers' planning and production efforts. Orders that are transmitted with reduced personal intervention are less likely to contain errors that can disrupt information and operational flows. Likewise, visibility into supplier inventory and in process commitments can give the purchaser some insight into the flexibility of the supplier to meet a special order or need, should one arise.

Ideal logistics

Negotiation, dialogue and agreement would occur to reduce logistical bottlenecks that may inhibit material flow between parties. Strategic placement of inventory and optimization of freight operations, minimize the threat of stock outages or excessive inventory and obsolescence. The benefits of just in time delivery and manufacturing (low to no inventory carrying costs) might be obtained through cooperative efforts between supply chain partners. By analyzing the entire chain from a logistical standpoint, it may become evident that there has not been any optimization of the strategic placement and flow of material between members of the supply chains and significant savings may be available through a cooperative effort.

An entire supply chain overview provides an opportunity to optimize material and information visibility, availability and reliability within and between each supply chain member. Through ideal flow of information, cooperation regarding inventory placement and volumes, and optimization of logistical issues, each member of the supply chain would benefit. Each participant would have the best information available to establish optimal batch sizes to serve customer needs and plan operations to optimize use of assets and potentially improve cash flow. The entire flow of the supply chain is shown in Figs 2.2–2.5.

Some of the potential reasons for this inventory based on analysis of the supply chain are shown in Fig. 2.6. Further, the amount of inventory in the chain is illustrated in Fig. 2.7. The classic pattern of the bullwhip effect is evident, as the final company in the supply chain (the textile manufacturer)

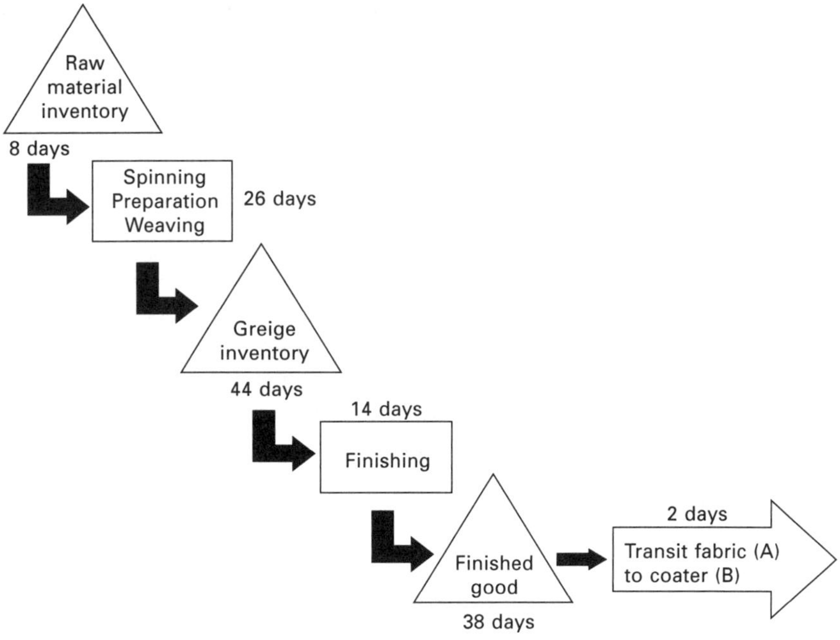

2.2 Supply chain map, fabric manufacturer to coater.

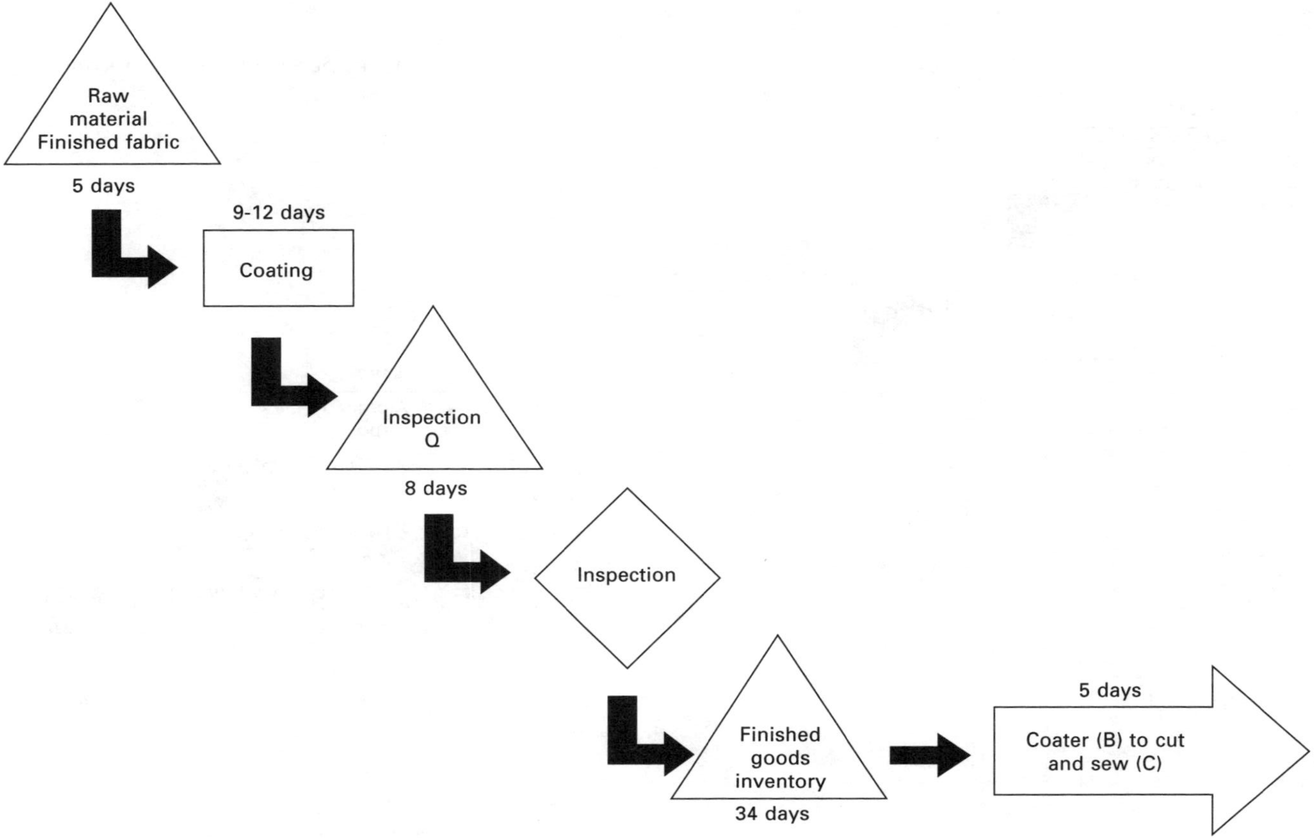

2.3 Supply chain map, coater to cut and sew.

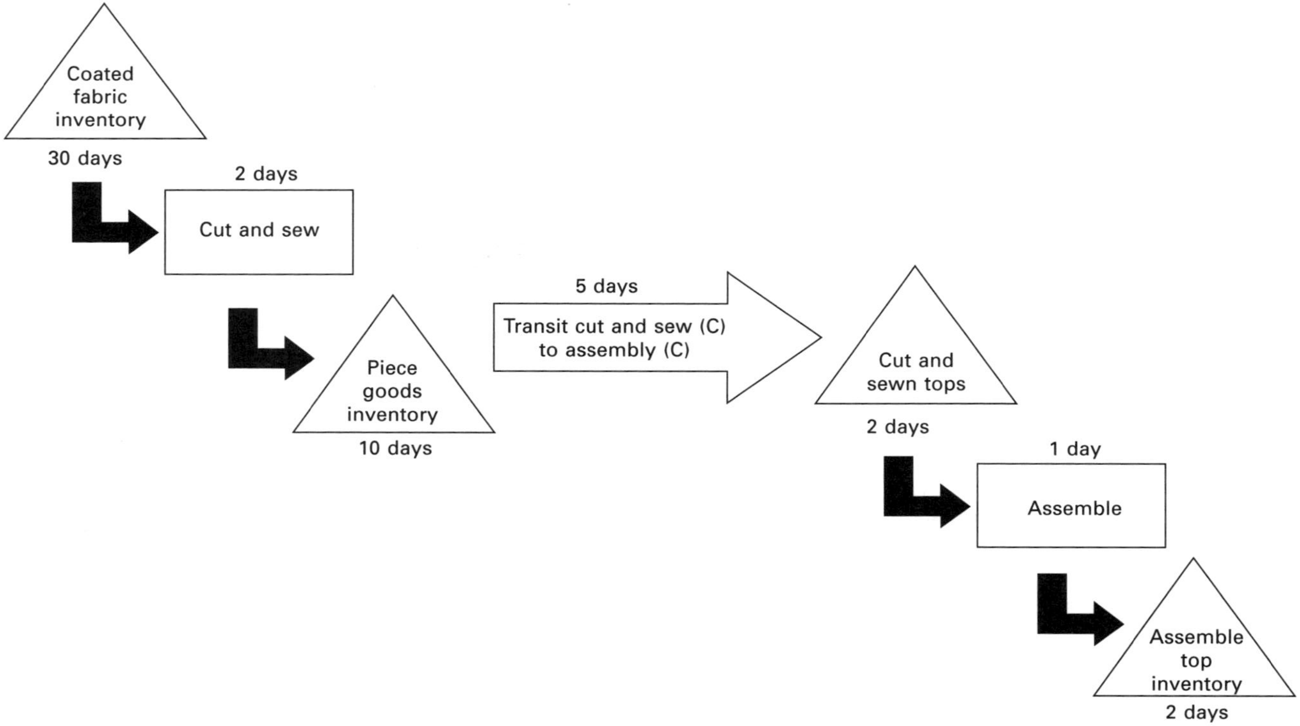

2.4 Supply chain map, cut and sew to assembly.

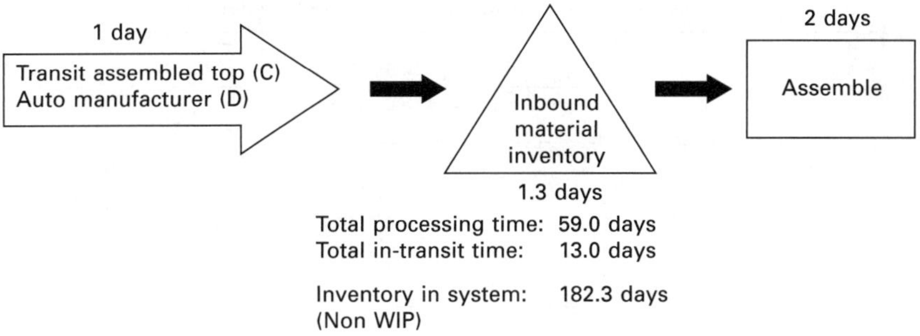

2.5 Supply chain map, assembly to installation.

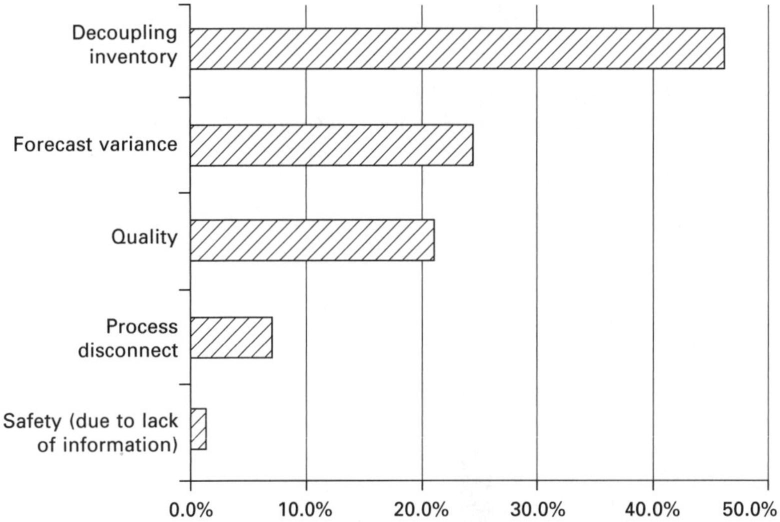

2.6 Reasons for inventory in the automobile supply chain studies.

is left with the greatest level of inventory to hold, due to poor information visibility.

2.7 Tactical recommendations for supply chain mapping

When looking at any supply chain situation, one of the first areas for improvement is optimization of inventory levels. The proposed optimized supply chain in Fig. 2.8 addresses these needs directly. The first recommendation is based on the removal of redundant and excessive inventory. As mentioned above, the finished goods inventory held by B and the materials

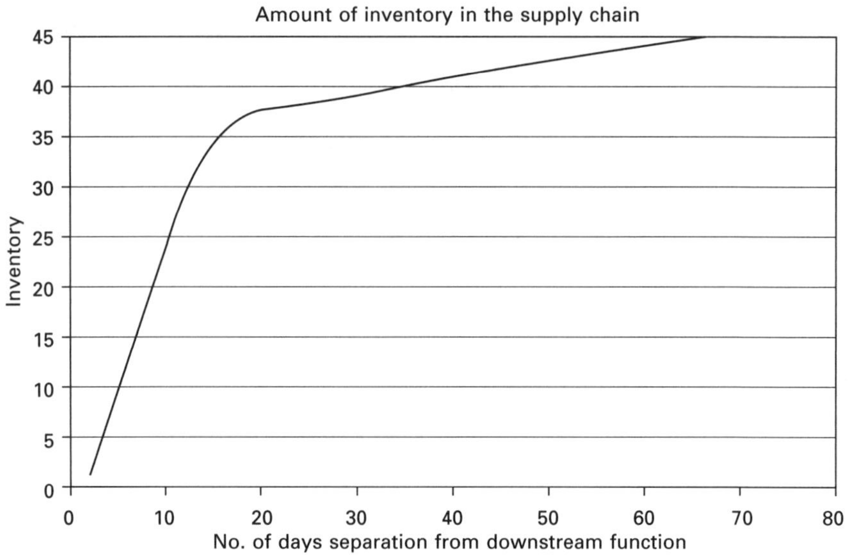

2.7 Bullwhip effect in automotive supply chain.

2.8 Potential inventory reductions by type.

inventory held by C are redundant. The two inventories occupy the same inventory point on the supply chain, separated by transportation time. It is our recommendation that B should eliminate its finished goods inventory. By placing its cut and sew plant offshore, C increased the uncertainty and transit times in the entire supply chain. Therefore, it should be the one to incur the inventory holding costs. This inventory at B could be 'replaced' with information, which could be provided by increased visibility of the forecast and improved information flows.

The second recommendation is that the forecast visibility should be improved across the entire supply chain. D generates a 20-week forecast that is shared with C. However, C only shares a three-week forecast to B and ultimately A. If B and A were afforded the original information, a more effective manufacturing process and inventory management system would be possible. Likewise, providing access to the one-week sequence schedule of D would decrease uncertainty at B. Increasing the forecast visibility would allow for more effective and efficient planning schedules that would reflect demand more closely, thus allowing overall reduction of inventory. All of the recommendations mentioned to this point would also improve the area of flexibility for the supply chain partners, by allowing for lower carrying costs, thus freeing capital. Ultimately, the area of customer service would be improved as well, due to each member of the chain (under D) being 'leaner' and more flexible. This would allow responsiveness to improve among the members.

Another area identified for improvement is information flow. Currently, A receives orders via fax from B. However, an EDI or web-based ordering system would prove much more effective. First, the orders would be much easier to track. The possibility of human error is greatly decreased as the ordering system becomes more and more automated. Next, an order placed by fax has a chance of being misplaced, especially if the fax machine utilized serves other different purposes. An electronic means of ordering would provide a 'dedicated' place for orders to be received. Finally, there is the issue of speed. With Internet connectivity growing faster each day, the time saved by ordering electronically instead of filling out and faxing a form could turn into a significant savings opportunity. Improved information flows would not only allow for faster reaction to orders, improving customer service and responsiveness, but would also provide real time information about demand.

2.7.1 Insourcing/outsourcing consideration in the value system

A final area requiring further study involves significantly reducing the size of the supply chain as a whole. It would improve the entire supply chain if the international plant of C's cut and sew operation could be eliminated. A thorough cost/benefit analysis would need to be performed. However, the time in transit alone serves as justification for exploring alternatives. If C is willing, the operation should be relocated to the US, preferably in a location closer to B's or C's assembly plant. On the other hand, another alternative would be for B to forward integrate into this area, shipping the material directly to C's assembly operation. Further, a second option would be to outsource the cut and sew function to another company, preferably more conveniently located to the entire supply chain.

2.8 Conclusions

There are several essential requirements to going forward and realizing the recommended improvements in the automotive supply chain. The first step is an improved mechanism for sharing and receiving information. Currently, information is shared on a periodic basis ranging from weekly to monthly. This inconsistent information sharing amplifies the bull-whip effect in the supply chain. The validity of the information is in question as well due to intermediate suppliers hedging demand to insure sufficient materials are on-hand. A solution to this inefficient visibility system is for OEM assemblers to provide visibility in depth throughout the supply chain. By penetrating deep into the supply base, variability and uncertainty can be reduced significantly by eliminating demand hedging activities and delays in information transfer between the different tiers of suppliers. This system also allows assemblers to control the type of information that is shared and reduces the likelihood of data entry errors.

Supply chain evaluation and control should be incorporated into the system to measure current performance and the effect of supply chain initiatives. A series of shared metrics that accurately reflect the drivers of value and cost in the system is required to accurately gauge performance. Metrics should be non-financial in nature to allow for measures that accurately reflect supply chain performance while maintaining firm privacy. A series of recommended supply chain performance metrics that ultimately lead to increased value and lower total cost of ownership in this supply chain are shown in Table 2.5.

Table 2.5 Recommended supply chain performance metrics

Value area	Metric
Customer service	On time delivery Correct shipments Customer inquiry response time EDI transactions
Flexibility/reliability	Value added and non-value added time Inventory days Response to change in demand
Costs	Distribution as a percentage of sales

2.9 Sources of further information and advice

2.9.1 Supply chain resources

Journals

Academy of Management Review
Management Science
International Journal of Operations & Production Management
Production and Operations Management
British Journal of Management
Sloan Management Review
Harvard Business Review
Journal of Marketing
Journal of Management Studies
Journal of Business Logistics
IIE Transactions
International Journal of Production Research
Supply Chain Resource Cooperative (http://scm.ncsu.edu)
The Economist Intelligence – Automotive Forecast World: Economic Outlook

Books

Title: *Supply Chain Management*
ISSN: 1359-8546
Publisher Information: Emerald Group Publishing Limited
60/62 Toller Lane, Bradford, West Yorkshire BD8 9BY, United Kingdom

Title: *Journal of Business Logistics*
ISSN: 0735-3766
Publisher Information: Council of Logistics Management
Suite 200, 2805 Butterfield Road, Oak Brook, Illinois 60523, USA.

2.9.2 Automotive textile resources

Journal of Textile and Apparel, Textile Management www.tx.ncsu.edu/jtatm
MobileTex – Textile Media Services
Office for the Study of Automotive Transportation, University of Michigan www.osat.umich.edu
Transportation Research Institute, Ann Arbor, Michigan
Center for Automotive Research http://www.cargroup.org/
Wards Auto – www.wardsauto.com
www.just-auto.com

2.9.3 Industry and market research

J.D. Power www.jdpower.com
CSM www.csmresearch.com
US Dept of Commerce – Office of Automotive Affairs www.commerce.gov
US Department of Transportation www.dot.gov
US Bureau of Census www.census.gov
Bureau of Labor Statistics www.bls.gov
Alliance of Automobile Manufacturers www.autoalliance.org
International Monetary Fund – World Economic Outlook www.imf.org
Transportation Reseach Institute, Automotive Analysis Division, University of Michigan, www.aad.umich.edu
Center for Automotive Research, College of Engineering, The Ohio State University http://car.eng.ohio-state.edu

2.9.4 Global players in the supply chain

International Auto Components Group – North America
Safety Components International/International Textile Group www.safetycomponents.com www.itg-global.com
Lear www.lear.com
Johnson Controls www.johnsoncontrols.com
Magna International/Intier www.intier.com www.magna.com
Faurecia www.faurecia.com
Tachi-S www.tachi-s.co.jp
Toyota Boshoku www.toyota-boshoku.co.jp
Eybl International www.eybl-international.com
Milliken www.milliken.com
  Viktor Achter Ltd
Fezko www.fezko-as.cz
Fibertex www.fibertex.com
Guilford www.guilfordproducts.com
Aunde (Aunde Achter & Ebels) of Germany www.aunde.com
  Gamma Car Fabrics
  DeWitte Lietaer
Autoliv www.autoliv.com
Seiren www.seiren.com
Kawashima www.kawashimaselkon.co.jp/
Kuraray http://www.kuraray.co.jp http://www.clarino-am.com
Suminoe www.suminoe.co.jp.
Tatsumura Textile Co., Ltd. 29, Morimachi, Mibu, Nakagyo-ku, Kyoto 604-8862, Japan
TRW www.trwauto.com

Borgstena www.borgstena.com
DuPont Automotive www2.dupont.com/Automotive/en_US/
Freudenberg www.nonwovens-group.com
Takata Highlands www.takata.com www.highlandindustries.com
Grupo Copo www.grupocopo.com
International Auto Components Group www.iacgroup.com

2.10 References

AAIA, 2007. *2005/2006 Aftermarket Fact book*, Automotive Aftermarket Industry Association, www.aftermarket.org. retrieved, June 21, 2007.

Anon, 2005. 'Extinction of the predator', *The Economist*, Special Report on The Global Car Industry, Vol. 376, Issue 8443, September 10, 2005, pp. 63–65.

Anon, 2007. 'Collins & Aikman president quits ahead of sale process', *MobileTex*, March 2007, p. 2.

Automotive News Europe, 2007.

Barkholz, D. 2006. Creditors question Big 3's part in C&A bankruptcy slide, *Automotive News*, August.

Bhagwat, V. 2004. Automotive Textiles, *Asian Textile Journal*, April, pp. 55–61.

Cagan, J. and Vogel, C. 2002. *Creating Breakthrough Products,* Chapter 9, Automotive Design, Prentice Hall, Upper Saddle River, NJ.

Cole, S.J., Woodring, S.D., Chun, H. and Gatoff, J. 1999. 'Dynamic Trading Networks', *The Forrester Report*, January.

Global Insight, *Automotive News,* 2007.

Hoovers Automotive Industry Overview, 2003. www.hoover.com/industry;auto

Lee, L. Hau, Padmanabhan, V and Whang, S. 'The Bullwhip Effect in Supply Chains', *Sloan Management Review*, Spring, pp. 93–102.

Lewis, N. 2001. 'Valuechain.Dell.Com Provides Pipeline to Info Exchange.' Dell Portal Adds 'Value'. CMP Media Inc., 2001.

Nieuwenhuis, P. and Wells, P. 2003. *The automotive industry and the environment: A technical, business and social future,* CRC Press, Woodhead Publishing Limited, Cambridge, England.

Plunkett, Automotive Industry Statistics, 2006. www.plunkettresearch.com, retrieved June 18, 2007.

Plunkett Research Ltd. 2007. Automobile Industry Statistics, and Automotive Industry Trends, www.plunkettresearch.com, retrieved May 24, 2007.

Powell, N. B. 2004. Design Driven: The Development of New Materials in Automotive Interiors, *Journal of Textiles Apparel and Technology Management, Transportation Textiles Issue*, Vol. 3, Issue 4, www.tx.ncsu.edu/jtatm.

Sherefkin, R. 2005. Cloth maker closes; carmakers scramble, *Automotive News*, January 31, p. 4.

Society of Automotive Engineers, 2003. *Sustainability Mobility Project*, World Congress, Detroit, MI.

Society of Automotive Engineers, 2007 www.sae.org.

Ward's Automotive, 2003. WAW 24th Annual Supplier Survey. Southfield, Michigan *Ward's Automotive Reports*, April, 2007.

Wilson, A. 2007. Automotive Textiles – The global supply chain to OEMs, Textile Media Services Ltd, 1st edition. Keighley, UK. May.

http://china.org, 'China becomes 2nd Largest Market for New Vehicles', http://china.org.cn/english/BAT/195815.htm, retrieved June 24, 2007.

3 Woven and knitted fabrics used in automotive interiors

T STEGMAIER, J MAVELY, M SCHWEINS, V VON ARNIM, G SCHMEER-LIOE, P SCHNEIDER, H FINCKH and H PLANCK, ITV – Institute of Textile Research and Process Engineering, Germany

Abstract: After an introduction to the demands of automotive textiles, an overview is given of the different types of textile components in car interiors and of market data of textiles in the automotive production. Details are given of the construction of woven, flat and 3-D knitted fabrics and of the finishing process with lamination to multilayer systems.

For testing, the different and most important standards are mentioned as well as special measuring devices for the haptic, soil release or self cleaning and electrostatic charging.

As for future trends, the R&D in fibre polymers and in the simulation of textile constructions are considered.

Key words: automotive textiles, texile components, market data, textile construction, woven fabric, knitted fabric, 3-D knitted fabric, finishing, lamination, testing, standards, haptic, soil release, self cleaning, electrostatic charging, FEM, simulation.

3.1 Introduction

Textiles play an important role in the growing worldwide vehicle market. Owing to the various applications in which they are used, they have to meet a comprehensive list of requirements with regard to their visual, comfort, security and fastness qualities. Therefore they have to pass corresponding test procedures to determine their relevant mechanical, physical and chemical properties.

The finishing of textiles, using processes like weaving, knitting and warp knitting, plays an important role in their mechanical and surface-related properties. The use of spacer textiles is attractive as they have the advantage of improving air permeability and humidity in a seat's construction. To produce the final product, a textile material often has to undergo processes such as lamination, coating, deep drawing and make-up. In future, new fibre developments will extend both the resources available and areas of application. These developments will be supported by new numerical simulation tools for their construction, processing, and application in vehicles.

3.2 Various types of textile component in car interiors

3.2.1 Basic requirements

Textile components in automotive applications have to provide a good visual impression, as well as comfort and safety. Therefore they have to fulfil complex demands:

- colour, light and appearance have to be considered
- for comfort, their bio-physiological properties have to be adapted
- textile properties that have to be considered include:
  - haptic, temperature behaviour (cold-hot)
  - contact with other materials like clothing, human skin
  - resistance to ageing
- for noise in the interior, sound absorption has to be integrated.

The application areas for textiles in automotives are extensive, and include:

- seat coverings in combination with trim panels such as side and back panels
- door and side coverings with airbags
- roof rack, pillar lining, sun shield, headliner, boot trim
- floor area, dashboard trims
- security elements such as safety belts, airbags
- filter materials (air, oil, petrol, interior)
- external soft top, wind deflector for convertibles
- reinforcement in fan belts, tyres, tubes
- sound absorption in the engine compartment.

Up to 10 m^2 textiles, 160 cm in width (without the headliner) are needed for an automotive interior [1].

Apart from design, textiles have to fulfil a certain list of technical requirements, which strongly depend on the application. One can divide them into the following groups:

- Processing properties such as:
  - sewability
  - sewing strength
  - seam slippage resistance
  - stiffness
  - elongation
  - tensile strength.
- Service properties such as:
  - mechanical behaviour (strength, elongation, bending and folding, tear propagation strength, dimension stability, pressure resistance)
  - ageing behaviour (heat/cold, temperature, humidity, light) [2]

- colour fastness (friction, light, chemicals, sweating)
- friction behaviour (brushing-up, abrasion, pilling, fibre migration)
- bio-physiological demands (air permeability, heat and humidity transport in upholstery)
- emission behaviour (toxic elements, smelling, fogging, total emission) [3]
- contamination and cleaning behaviour
- electrostatic behaviour
- flammability
- optical, haptic, design.

Depending on the application, some additional specific requirements may have to be considered.

New designs in automotive construction are increasingly influenced by the legal regulations on emissions and those on the waste from old vehicles, which require the use of materials that can be recycled. These influence nearly all textile developments in the automotive sector.

3.2.2 Market data

Quite a large number of different textiles are used in the production of a car [4]. In the global vehicle market, about 65.9 million vehicles are produced each year. The forecast is for 56 million cars and 22 million light trucks, a total of 78 million vehicles in 2011 [5]. Some 5 to 8 kg (15 to 18 m^2) of textiles are used in a car interior. This adds up to about 240 million square metres of textiles used for car interiors in Europe and 91 million square metres of carpet and boot linings. The EU is the world's largest automotive manufacturer, producing about 16 million new passenger cars and 1.8 million light commercial vehicles in 2006.

The different textile usage for major components is:

- 6–8 m^2 – seating
- 5–6 m^2 – floor
- 2 m^2 – headliner
- 3–3.5 m^2 – boot/cargo
- 0.6–0.8 m^2 – door panels
- 0.4–0.5 m^2 – pillar cover
- 0.7–0.8 m^2 – parcel shelf [4].

These textiles are made from woven, knitted, warp knitted, tufted and laminated fabrics, and nonwoven, electrostatic flocked surfaces [6].

Fibres are chosen with regard to the demands for fastness, comfort and price.

3.3 Overview of main types of textiles used

Different textile materials are used depending on the part of the automotive in which they are to be used [7]:

- A-, B-, C- pillar, headlining: knitted fabrics
- door panel: upholstery material
- floor: carpet
- seat: upholstery material
- boot: carpet.

It is interesting to note how much fabric is actually used for the seat and door covers, including the headrests, compared to the fabric used in the headliners and parcel shelf covers. It has been found that around 88% of the fabrics used in the automotive industry are used in seating and door covers.
Polymer materials are used as follows:

- polyester (mainly for upholstery)
- polyamide (mainly for carpet, and on a limited scale for upholstery)
- polypropylene (for carpet, and on a limited scale for upholstery)
- polyurethane (as an additive fibre for technical applications, e.g. elasthan)
- polyacrylic (for convertible roofs)
- wool or cotton (as additional fibres).

A large proportion of these contain spun-dyed polyester fibres with about 40 000 tons of polyester yarns used in Europe. As well as man-made fibres, natural fibres such as hemp, flax, wood fibres and cotton are used in composites, with each vehicle containing about 16 kg. Flame-retardant properties are provided by blending different fibres.

Therefore, polyester dominates the raw materials used in fabrics in the automotive industry, its share of the total materials used being about 78%. Of this percentage, 59.4% is air-textured PES yarn (ATY), followed by PES friction-textured yarn (FTY) at 30.6%.

So more than 70% of automotive textiles are made from textured yarns, of which the major portion, approximately 60%, is air-jet textured (Taslan) and the remaining 40% is false-twist textured. The proportion of air-textured yarns used has risen over the last few years because of the increase in the use of woven fabrics, where the overall count at 550 dtex has become finer. The quality of the yarns used has improved, so the desired abrasion resistance properties can be achieved, even using a finer count. Also, using air-texturising, a natural-looking appearance can be more easily achieved, when several components are fed into the jet with different levels of overfeed to achieve a special colour blend.

In 2005, the worldwide production of automotive textiles [8] was calculated to be:

- fabrics, knits 214 000 tons
- nonwovens 114 000 tons
- composites 976 000 tons
- others 179 000 tons

with an astonishingly high increase in China.

3.4 Woven and knitted fabrics

3.4.1 Woven fabrics

Some facts and figures can be found at www.inteletex.com: woven fabrics dominate this sector with 51% of interior automotive fabrics produced by this method. These fabrics are mainly dobby weaves, with 32% of the share, followed by Jacquard weaves with 18% but this share is increasing. Very little woven plush is used (0.45%). All this fabric is used for the seat and door covers. Jacquard- and dobby-designed fabrics are used mainly in the seat area and the dobby woven plain fabric is found in the seat wings, bolsters and rear covers of the front seats.

3.4.2 Flat knits

Circular knits are the second largest fabric group, having a 19% share of this particular market. Leading the way are circular knitted velour fabrics with 15% and circular flat fabrics (single or double jersey) with nearly 4%. Circular velour is used on the seat and door covers, with flat circular knits used mainly for the rear covers of the front seats and headliners.

Meanwhile warp knitted fabrics are the third most popular group with a 12% market share. These include the Tricot family of knits (6%), brushed or structured, with Tricot pile sinker fabrics having a 4% share. Raschel knits represent just 0.79%.

3.4.3 Spacer knits

Spacer textiles in upholstery are made from warp knitted (double-Raschel machine) or knitted (circular knitting machine [9]) fabrics [10]. They are made from two separate textile layers which are connected by stiff spacer yarns.

Monofilaments (single) are mainly used for spacer yarns. The height of the structure is governed by the distance of the two needle bars of the Raschel machine and by the distance between the cylinder and rib discs in the circular knitting machine. The force that is necessary to keep the two textile layers apart depends on the material, thickness and structural integration of the spacer yarns (monofilaments) into the basic layers.

Spacer structures have an elastic pressure behaviour: by pressing on the surface they are compressed. When the pressure is released, they relax in an elastic way. The space between the two layers is an air-filled cavity from which the air is removed during compression, and into which air is sucked during decompression. Unlike foam constructions, these textile constructions are able to breathe.

An increasing interest is being shown in spacer structures and their use in different applications. Nearly all the available textile and non-textile yarns can be used for their production. Due to their construction, spacer structures have the following advantages:

- light weight, soft touch, skin friendly
- breathability (high air permeability, defined climatic conditions)
- humidity transport and humidity absorption
- defined elasticity in length and breadth
- thermo regulating
- fastness to washing
- good reaction to long-term loading
- sound absorbing
- filtering
- sewability
- resistant to ageing.

Spacer textiles on a base structure of knitted textiles can be used to advantage in applications with climatisation requirements. In different project studies, several vehicle seats were developed that regulated the climate in the seat according to the time of the season and the temperature. Since 1990, such ventilation systems for seat climatisation have been in mass production for the back of the seat. The construction consists of a combination of ventilation and breathable seat layers. When the driver perspires, humidity is able to escape between the back and seat upholstery through the textile into the inner layer of the seat, where a circulating air flow removes the humidity. The laminate used has a protection layer and a very thin Gore-Tex membrane. The water vapour crosses through this laminate into the deeper air channels of a spacer textile connected to the air ventilation system, which transports the humid air outside.

Spacer structures offer further advantages to applications:

- Due to their broad flexibility with regard to mesh structure, it is possible to choose the inner and outer material, together with spacer yarns of different material, fibre type, density, structure and space combinations.
- The back side can be adapted separately to the front side to obtain good properties of lamination, bending radius, elongation and stiffness.

There are significant differences between knitting and warp knitting, which can be used to advantage – depending on the application and production conditions:

- The main advantage of spacer knits is the free design of Jacquard patterning, which is also possible in warp knitting, but with more limitations.
- The elongation of knits is often higher than that of warp knits.
- Spacer knits are of interest for small and medium-sized production runs due to their high flexibility in circular knitting machines. No work-intensive warp beams have to be prepared for the production of spacer knits. It is possible to produce them with only a few cones to fulfil the total technical requirements.
- The front surface of spacer knits can be designed by electronic patterning, free from limitations, optimally fulfilling visual requirements and technical demands. The electronic needle allows large and small patterns to be produced in the same way.

The advantages of flexible production and free patterning are becoming of more and more interest. Machine and construction development is taking place at ITV Denkendorf.

3.5 Finishing of textiles

Great demands are made on the properties of textiles for the interiors of vehicles. Many of these properties, as well as special functions and their permanence, are achieved by the finishing of textiles and must be considered separately. Important properties include the following: low susceptibility to soiling, excellent cleanability, no or only very low emissions of temporary organic bonds which smell or condensate (fogging), high resistance to abrasion, good pilling behaviour, excellent fastness to rubbing, colour, welding, good humidity transport behaviour (seat climate), high degree of light fastness, high resistance to flammability, and little tendency to produce electrostatic charge.

The most problematic types of spots and soiling on textiles are based on aqueous and/or oily materials. Chemical compounds, therefore, are used as protection against them, thus preventing textiles from being wetted by oily materials (oleophobicity). Textile finishing, additionally, affects water repellence behaviour (hydrophobicity) while also reducing dry soiling. Textiles finished as described above show a very low degree of surface energy. The forces of adhesion between a textile surface finished with oil-repellent agents and oily liquids are very low so that the forces of cohesion predominate, thus preventing the textile surface from being wetted.

The compounds mainly used are fluorocarbons, which are based on a structural component composed of acrylic acid or polyurethane and perfluorinated alkyl radicals. The polysiloxane bonds represent a further product group which, however, cannot affect the repellence of oily bonds.

Bacteria, fungi, mildew, yeasts, and algae can be killed by means of an antimicrobial finish on textiles. However, its contact with the skin of the driver and passengers, in those particular areas of automotives where it is unavoidable, has a detrimental effect because it also has a cell-toxicological effect on the aforementioned elements in human cells. Long-term stability is restricted because of the continuous release of cell toxins required to achieve the desired effect.

Many of the formerly used antimicrobial agents are no longer used for textile finishing in most industrial countries because of their enormous toxic effect and/or environmental impact. Phenols and organometallic bonds of quicksilver, for example, are included in those materials. Many antimicrobial agents show poor biological degradation behaviour. Further side-effects involve the danger of the emergence of resistant bacteria and fungi as well as of allergies.

Non-ionic antimicrobial agents – also designated as heterocyclic compounds and benzimidazole derivates – are used instead. Their advantage is that they are free from heavy metals, phenols as well as halogens, and thus cause fewer problems for wastewater. Salicylanides – mostly halogenated and barely water-soluble – are used for both bacteria and fungi. Their permanence to washing is improved by combining them with resin pre-condensates. Undecyclic acid derivates – especially alkylamines – are considered to be particularly skin-tolerant for humans.

The finishing of textiles requires – independent of the material system used – the following:

- The textile's visual impression (structure, colour) and feel must be preserved.
- The finishing agent must be permanent with regard to washing and use.
- Downstream processing such as laminating, gluing, bonding or sewing must not be adversely affected.
- Fogging values, emission behaviour, fastness, mechanical-physical properties as well as antistatic behaviour must not be impaired.

3.6 Processing multilayer systems

The textile properties of these fabrics could be greatly affected by subsequent processes such as coating and laminating (e.g., flame-laminating of pdyurethane foams). Problems between the lamination partners have to be taken into account with regard to evenness, adhesion, emissions and resistances [11].

Knitted fabrics are mainly used for rear window shelves or headliners. For these applications, foams should also be replaced by non-wovens. In particular the acoustic properties have to be taken into account.

Upholstery fabrics are mainly made of polyester, sometimes with a small wool component. For upholstery, the use of flocked yarns or knitted fabrics is declining, although combinations of different fabrics or textured fabrics have become common [12] (e.g., floatings).

Leather and leatherette are fashionable for use in automotive interiors. An unsolved problem for leatherette is its poor light- and colour-fastness. Nevertheless, because of its price advantage compared to leather, it is in demand.

Direct processes for the assembly of composites made from renewable raw materials (flax, hemp) are improving rapidly. In these the adhesive, in the form of hot-melt fibres, is mixed into the non-woven material and the composite is formed and cured by heat in a press [13], thus saving some laborious production steps.

Using back injection for fabrics, inherently stable component parts could be generated. It is most important to have a close and durable bonding to the three-dimensional mould, which mostly consists of plastic materials. For back injection as well as for deep-drawing, there are special requirements for strength and extension (sometimes more than 40% is required), with an even and tight surface and heat resistance. The penetration of liquid plastic through the fabric during back injection could be avoided by using an additional layer, e.g. a bonded non-woven [14].

Supporting components made of fibre-reinforced composites, e.g. axle carriers for buses, are being produced. The benefits of these parts are:

- lightweight construction
- integration of further functions which is not possible if using metal
- the natural frequency decreases by about 10 Hz, which increases travelling comfort.

A new development in the field of lightweight constructions is the combination of two layers of steel sheet with a textile layer in between, bonded together by adhesive [15]. Using this composite, high stiffness and very good deep-drawing property in addition to body sound damping could be combined for the first time.

3.7 Test procedures and special properties

3.7.1 Standards

To characterise textile materials for the automotive industry and ensure uniformity for each delivery, besides the numerous national and international test standards, standards exist which are produced by the automotive manufacturers themselves, the so-called 'Technical delivery conditions' (TC).

They describe characteristic properties such as material composition, construction and physical properties, as well as application-oriented characteristics with their nominal values. They also contain test specifications (TS), which are relevant for the determination of the characteristics. These test specifications are developed and supervised through the members of the automotive association (in Germany this is the VDA). This ensures that the high requirements and obligations of textile components for the automobile are guaranteed.

Tables 3.1 and 3.2 exemplify the requirement profiles that should be met for a bonnet fabric.

The test methods for the burning behaviour of a car interior are [16]:

- 95/28/ES (European Directive coaches) oMVSS 302 and
- Producer's specification.

Until now, haptic has been the next most important impression for humans (after visual impression, smell and noise). The evaluation of the haptic is still difficult to reproduce, because most of the tests are based on people's impressions [17]. The Kawabata test procedure is quite comprehensive and needs great experience in evaluation. So there are intensive trials to reduce the test demands to a small number of methods and to calculate a summary number of the individual results. In the industry, the well-known measuring unit – the FAST System – characterises the haptic by the three properties of thickness, bending and compressibility.

ITV Denkendorf proposes an alternative technique using the Pulling-Through Method. During the test, developed by ITV, a sample strip is pulled through a testing ring that is small in comparison to the sample's width, with a specified constant speed (see Fig. 3.1), and is formed by it. The forming resistance, which is influenced by the friction between sample and ring, is measured as a pulling-through resistance in the form of a distance-dependent

Table 3.1 Identifying characteristics

Characteristics	Standard/test specification	Test conditions
Mass per unit area	DIN EN 12127	
Thickness	DIN EN ISO 5084	
Tensile strength	DIN EN ISO 13934-1	• original • after water storage • after thermal storage
Tensile elongation	DIN EN ISO 13934-1	• original • after water storage • after thermal storage
Air permeability	DIN EN ISO 9237	• original
Tear growth strength	DIN EN ISO 13937-2	• original • after water storage • after thermal storage

Table 3.2 Application-oriented characteristics

Characteristics	Standard/test specification	Test conditions
Adhesion	DIN 53357	• original • after thermal storage
Dimensional stability	DIN 53892/Test specification	• original • after thermal storage
Stitch tear resistance	DIN 53331	• original
Water penetration; hydrostatic pressure	DIN EN ISO 4892-2	• original • after thermal storage
Fabric/seams	DIN EN 20811	• after aging
Abrasion resistance	DIN EN ISO 12947-4	• original
Colour fastness to rubbing	DIN EN ISO 105-X12	• original
Repeated flexure test	DIN 53359	• original • after thermal storage
Stiffness under flexure	DIN 53121	• original • after thermal storage
Break-through force	DIN EN 13567	• original
Scratch resistance	Test specification	• original • after thermal storage
Rot resistance	Test specification	• original
Static/dynamic elongation	Test specification	• original • after thermal storage
Water repellence test	DIN EN ISO 24920	• original • after thermal storage
Resistance to media	Test specification	• original
Light fastness	VDA 75202	• original
Heat/light fastness	VDA 75202	• original
Aging behaviour	DIN EN ISO 4892-3	• original
Fogging behaviour	DIN 75201	• original
Odour under temperature	VDA 270	• original
Flammability	DIN 75200	• original
Soiling and cleaning behaviour	Test specification	• original
Mould culture	Test specification	• original

force. This strength reflects those measurement categories which essentially influence tactile properties, such as tensile force, dynamic friction, bending and shear force. The diameter of the testing ring d_{MR} is adapted to the mass per unit area of the particular sample, as defined by the following relation:

$$d_{MR} = PV \cdot \sqrt{mF}\,(1/100)$$

$$= \frac{p_N - p_{RX}}{p_N} \cdot 100\,\sqrt{mF}\;1/100$$

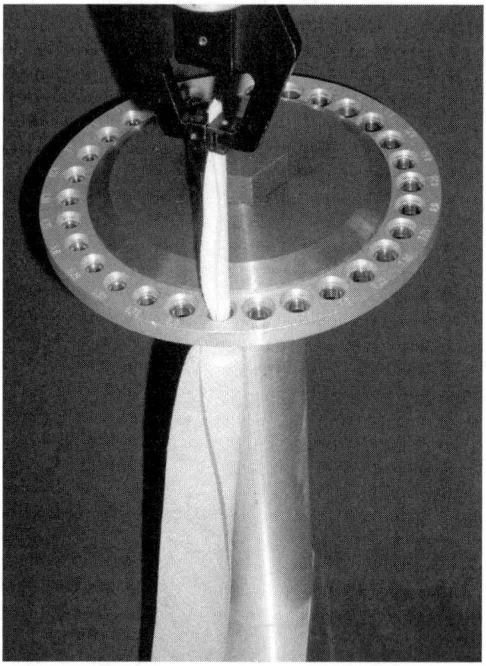

3.1 Pulling-through resistance device with different ring diameters.

where d_{MR} = diameter of the testing ring in mm, PV = pore volume of the sample in %, mF = weight per unit area in g/m^2. This relation applies for a sample width of 50 mm.

The pulling-through resistance is easily measured as a force-distance signal, when the testing ring is fixed horizontally in a tensile testing machine and the sample is pulled through the ring under a combined tension and compression.

The following test conditions were used:

sample width:	50 mm
deforming rate:	500 mm/min
distance between testing ring and clamping arrangement:	100 mm

The sample is folded twice in the clamping range so that a sample width of 6.25 mm results. The measurements were carried out in warp and weft directions.

3.7.2 Methods to test soil release

Numerous methods are used in the automotive sector for the evaluation of soil release. They differ with respect to:

- the type of soiling: dry soiling or wet soiling
- the soil substance used (sand, smut, oil, fat, butter, ketchup, red wine)
- the method of applying the soil (Martindale, shaking in a bottle or other appliance)
- after cleaning: suck off or wash off (soil-release or oily-stain-release methods)
- type of cleaner (foam cleaner, tensides, stain removal)
- the rating (photographic standards, visual rating, relative colour strength, grey standard).

3.7.3 Electrostatic charging in seat upholstery

Textile materials and products made from them are electrostatically active and become charged under certain conditions, for example from rubbing against other textiles or objects, separation, gas flows or ultraviolet radiation. In most cases this static charge is not only a problem during manufacturing and processing but also an annoyance when the product is being used. In a car, static charges of either the upholstery fabrics or of the car passengers themselves may not only affect passengers' comfort but may also endanger sensitive electronic components or even cause an explosion during refuelling. Accidents – in some cases during refuelling – always happen at a relative humidity below 50%.

Electrostatic charging of passengers in a car by the upholstery is one of several sources of electrostatic charging in the overall passenger/car system which has to be considered. The unpleasant charge exchange when leaving the car, caused by touching the bodywork for example, depends on the electrical resistance of the 'person/clothing' system in which the electrical resistance of shoes plays an important role.

The problem of electrostatic charge when leaving a car can, in principle, be solved. Textile-related solutions exist:

- by using conductive fibres and yarns in the upholstery fabrics used for car seats
- by giving upholstery fabrics a permanent antistatic finish
- by inserting electrically conductive layers between the upholstery fabric and the foam.

Attempts to find practicable solutions must, however, take into consideration the overall system within which the upholstery fabric will be used. This must include the stresses resulting from use, such as compression, friction, moisture, contamination and the design of the seat – especially its electrical grounding. Even interactions with the environment and environmental influences, for example contamination problems in the context of long-term behaviour, are factors exerting a decisive influence.

The development of practicable and economically efficient solutions has so far been difficult since previous laboratory test methods have been restricted to investigating the electrostatic properties of the upholstery fabric alone. An evaluation of the effect of 'antistatic measures' on upholstery fabrics will not, however, be possible unless the entire system of 'seat and environment' is examined. To this end suitable testing and measuring methods are being established at ITV Denkendorf. These not only include material testing but also the development of a simulation method which comes close to real conditions of use:

- Investigation of the dynamic charging and discharging of textile samples by means of test equipment available or developed by ITV.
- Analyses of combinations of materials and of entire systems by realistic simulation testing in an ITV upholstery testing device.
- Measurements of the seat/person system to validate the simulation tests.
- Assessment of long-term durability on the basis of the change in electrostatic properties which result from ageing.
- Evaluation of the influence of contamination properties and other environmental properties.

The investigations and developments should consider the following interrelationships:

- Correlations between the electrostatic charging and discharging of persons leaving a car and the physical design of textile upholstery systems.
- Effect of the textile construction of the seat as a total system on electrostatic charging and discharging.
- Effect of electrically conductive elements in upholstery systems on contamination properties.
- Determination of constructive requirements in design for charge dissipation or charge distribution within the seat system.
- Influences of temperature and relative humidity.

On the basis of these results, new textile constructions and also new total systems are being developed. Figure 3.2 shows three typical measurement reports of simulation testing: In each case the electrostatic charge of the upholstery fabric and 'friction body' – simulating a car passenger – are recorded. Here the seat is grounded but the 'friction body' is not. The first diagram shows a standard upholstery fabric of relatively high charge level. The second diagram illustrates the charge behaviour for an upholstery fabric with conductive components. The electrostatic charge of the upholstery fabric is reduced by a factor of 10 although this reduction is not so great at the separation of the material systems – which simulates the passenger leaving their seat. Practical experiences confirm these lab values. The last diagram shows the charge behaviour for an upholstery fabric with an antistatic finish.

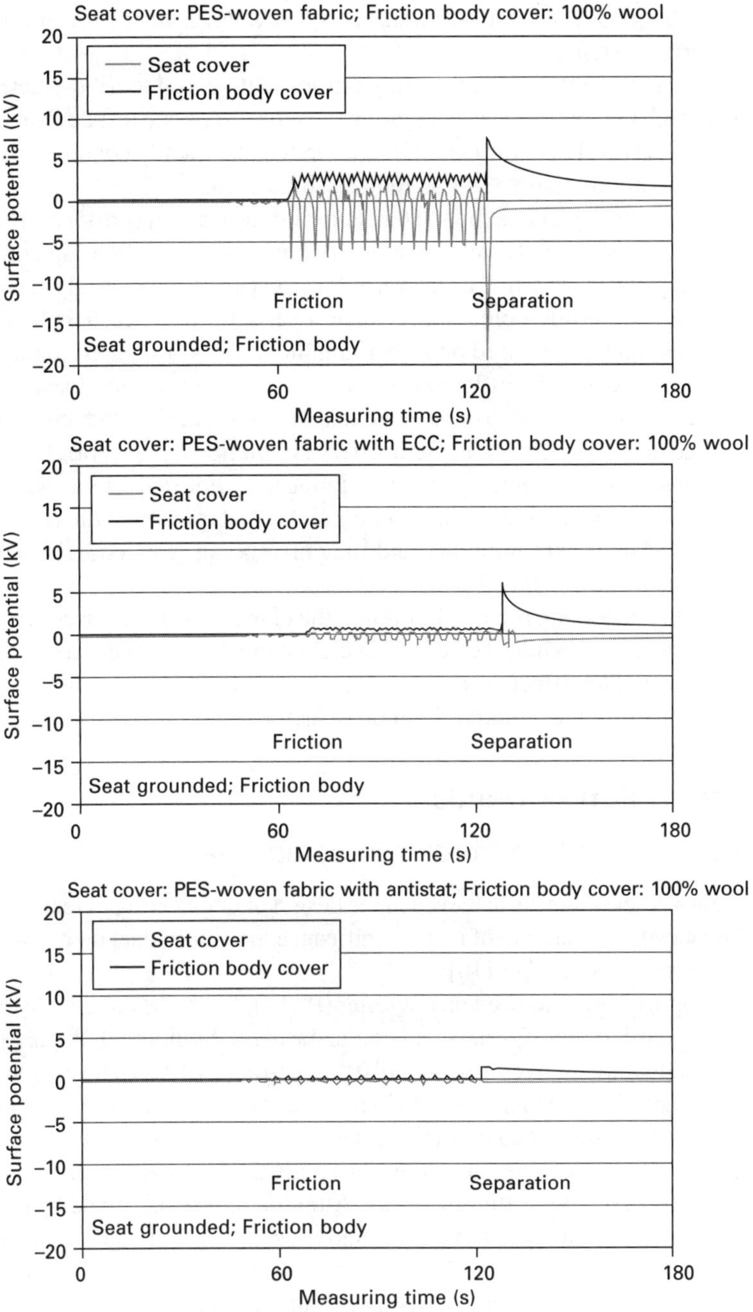

3.2 Test records for simulation testing on the ITV upholstery testing rig: electrostatic charging of different upholstery fabrics (PES) and friction body cover (WO); seat upholstery grounded, friction body not grounded. ECC: conductive components.

The charging can also be reduced to a low level at the separation of the material systems.

Investigations of the upholstery fabrics with regard to tribo-electric charge and discharge properties when grounded – on a separate test device – correlate with these results. In the case of incorporated electrically conductive elements, the maximum charge level is reduced by a factor of 2–3.

The friction partners (upholstery and clothing) have a considerable influence on the charge level as well as on the measured polarity. Wool in combination with polyester, which is usually used for seat covers, generally shows a high charge level while colours also have a certain influence. Distinctly higher charges can be measured on separating the textiles compared to those for the friction process. The influence of grounding the upholstery doesn't appear to be significant since none of the materials shows a specific conductive behaviour.

The measurements of the upholstery simulation device correlate well with trials using people getting up from upholstery. For real-life tests, apart from the influence of clothes, the state of the test person with regard to, for example, their skin humidity and the corresponding electrical conductivity, also has to be considered.

With decreasing relative humidity, the charge level increases considerably – as expected – while the conductive elements (e.g., finish) also have strong influence. The effect of antistatic agents on hygroscopic bases is strongly influenced by the 'relative humidity' factor.

3.8 Future trends

3.8.1 Trends in fibres and composites

Improvements may be achieved in the design or engineering of textile materials. Enhanced materials are of more significance than individual new developments that hit the front page [18].

Approved fibres based on polyester (PET) [19] can be continually improved with regard to light-fastness and colour-fastness. Spun-dyed fibres, especially, have advantages over yarn-dyed or piece-dyed fibres. The extraordinarily high light- and weather-fastness (in particular UV-fastness) of polyacrylonitrile (PAN) is utilised for convertible soft tops. Airbags are still made of polyamide (PA), because polyamide yarns have advantages over polyester yarns in terms of their low initial modulus. Attempts are being made to replace the 'weak point seam' by alternative designs in weaving. For composites consisting of a singular type, spun-dyed polypropylene (PP) is used [20], which is mixed in composites or carpets as filling material.

For renewable materials, attempts are being made to increase market shares by creating more homogeneous fibre qualities in terms of equalising seasonal oscillations and by adjusting the fibre fineness through the retting process. For hemp, new disintegration methods, e.g. the 'impact method' [20], or fermentative methods [21] provide a higher fibre quality.

New applications for using carbon fibres for seat heating are opening up. Their advantages over common copper strands are their flexibility, corrosion resistance and the punching quality of the warp knitting. The yarn count ranges from 67tex (1000 filaments) up to 800tex (12 000 filaments) [22]. In particular, staple fibre yarns, flat yarns as well as elastic yarns (air-bulked, knit-de-knit and false twist) [23] are in use.

Apart from woven and knitted fabrics with a Jacquard effect that are used for their attractive designs, it is mainly non-wovens that are increasingly being used. Great efforts are being made to replace polyurethane foams as upholstery backing and solutions using bonded fibres [24] and needle-punched structures [25] have been found.

In summary, it is likely that new raw materials will be used in automotive textiles. These new fibres and filaments will have higher strength against abrasion, more elongation for easier fabric moulding (like PBT), less flammability (like CS or FR), more antistatic properties and more bacterial resistance. The drive towards lightweight/fuel-efficient vehicles could also mean more use of textiles in the engine compartment [26], and increased use of raw materials such as Nomex.

It is likely that there will also be an increase in the use of knitted fabrics in automotive textiles subject to new fabric developments for seating like the Raschel knitted spacer fabrics.

Self-lightening textiles, which are in the development stage in different labs, will have a great future. There will be additional opportunities for the use of textiles in combination with electronics, the so-called smart textiles, integrating sensors or actuators in textile constructions.

3.8.2 Finite element computations of textile products

Numerical simulation using the finite element method (FEM) is widely used in the automotive industry to compute car deformation, energy absorption, safety belts, inflating of airbags and their interaction with people. Because of the complex mechanical behaviour of textiles, FEM will gain more and more importance. Intensive research work, increased computational power, new modelling techniques and costly material models will enable us to compute deformations, draping, drawing, stabbing, ballistic impacts and saws.

The aim of computing is to find the optimal product properties with reference to the application. This is a quite complex and time-consuming task because of the large number of yarn types and textile constructions. Beside the mechanical behaviour of textiles (stiffness, stress-strain behaviour, bending rigidity, stabbing, tear resistance, drapability), many other properties such as thermal behaviour or visual appearance play an important role too.

Therefore a great deal of knowledge is required, not only of material behaviour, physics and testing methods but also of textile producing techniques and finishing treatments.

Numerical simulation using the finite element method (FEM) can be a very helpful tool. ITV uses a new micro-modelling technique which is unlike other modelling techniques using membrane elements, as in the simulation of airbag inflation. Here all threads, and to a certain degree even filaments, are included in the fabric model [27, 28]. All threads/filaments can move in the simulation under friction conditions and are checked for collision during the complete simulation. The single-thread modelling technique has made it possible to simulate monofilament behaviour in woven fabric with reference to bending, strain and breaking properties. This has been shown by simulating the penetrating resistance of wire fabric of fencing masks. Further developments have also led to yarn models which can represent multifilament yarn behaviour in textile fabrics, which now opens up a wide range of possibilities for using FEM in textile applications including brand new FE simulations of the textile manufacturing process itself. Examples are the process of weaving a fabric of aramid multifilament yarns, braiding cords and knitting loops.

Micro-models of fabrics including yarns and filaments have another huge advantage over other common fabric models, which is that they look very

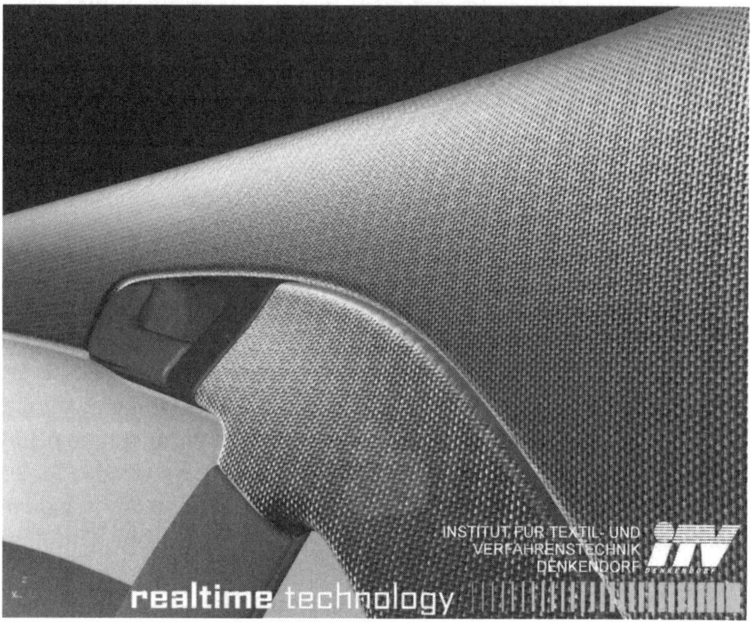

3.3 Real-time rendering of a woven fabric on a car door based on a computed micro-model of textile fabric showing realistic reflections and shadows.

realistic. The visual appearance of textile fabrics is a result of yarn and binding structure, density, yarn/filament colour, reflection and shadows. As ITV's new micro-models consider all these important features, turning the 3D-fabric model under any desired light conditions shows the reflections, shadows and colours as they would look in reality. In addition, these complex fabric models can also be loaded with tensile, draping or shear stresses, so simulations can show how the fabrics would look when draped on arbitrary complex formed parts, as shown in Fig. 3.3. This enables the designer to evaluate any change in textile appearance and leads to a new kind of virtual prototyping facility for textile products. In the near future this will play an important role in textile design, especially in the automotive industry (e.g., car interior parts like seats, side panels, etc.).

3.9 References

1. Tuchfabrik Willy Schmitz: *Rundum-Glücklich-Sevice vom Mini bis zur E-Klasse*, BW Technics 3/2001.
2. Schmidtmann, Jens; Bredereck, Karl: *Alterungsverhalten von Automobiltextilien*, Paper on the Man Made Fiber Congress, Dornbirn, Austria, September 2001.
3. *Entwicklung neuartiger Werkstoffe mit kontrolliertem Sorptionsverhalten für die Automobilinnenausstattung*, Abschlußbericht Universität Stuttgart (2000).
4. Stine, Jeff: *Nonwovens in Automotive Applications*, Paper on the Man Made Fiber Congress, Dornbirn, Austria, 19–21 September 2007.
5. Henrio, Francoise: *Multi-sensorial Fibres and Coverings for Car Interiors*, Paper on the Man Made Fiber Congress, Dornbirn, Austria, 19–21 September 2007.
6. Bianchi, Pierre: *Beflockte Oberflächen für die Automobilindustrie*, Paper on the Man Made Fiber Congress, Dornbirn, Austria, September 2001.
7. Lutz, Stefan: *Technical Requirements for Automotive Textiles*, Paper on the Man Made Fiber Congress, Dornbirn, Austria, 21–23 September 2005.
8. Stegmaier, Thomas; Erth, Holger: *Marktdaten und neue Entwicklungen auf dem Gebiet der Textilien für mobile Anwendungen*, 14. International TechTextil-Symposium for Technical Textiles, Nonwovens and Reinforced Materials, Frankfurt am Main, Germany, 11–14 June 2007.
9. Kunde, K.: *Hightech-Textilien rundgestrickt*, Maschen-Industrie 12 (2001).
10. Moser, Werner: *Innovatives Abstandsgestrick für technische Anwendungen*, Paper on the Man Made Fiber Congress, Dornbirn, Austria, 20–22 September 2006.
11. Kunz, Alexander: *Beeinflussung von Eigenschaften von Textilien beim Kaschieren*, Paper on the Man Made Fiber Congress, Dornbirn, Austria, September 2001.
12. *Routenplaner Automobiltextilien*, kettenwirk-praxis 4 (2001), Obertshausen.
13. Reußmann, Mieck: *Direktverfahren auf dem Vormarsch*, BW Technics 4/2001.
14. Walter, Peter: *Einfluss von Hinterspritzen, Hinterpressen und Kaschieren auf textile Oberflächen*, Paper on the Man Made Fiber Congress, Dornbirn, Austria, September 2001.
15. Stegmaier, Thomas; Yilmaz, Dagmar; Hager, Tom; Schneider, Petra; Planck, Heinrich: *Steel and textiles: a new combination of lightweight materials*, 14. International TechTextil-Symposium for Technical Textiles, Nonwovens and Reinforced Materials, Frankfurt am Main, Germany, 11–14 June 2007.

16. Hoeyland, Pros Van: *Flammability Test Standards and Criteria*, Paper on the Man Made Fiber Congress, Dornbirn, Austria, 21–23 September 2005.
17. Lutz, Stefan: *Wertigkeit von Polsterstoffen im Automobil*, Paper on the Man Made Fiber Congress, Dornbirn, Austria, 19–21 September 2007.
18. Génin, Antoine: *Entwicklung und Trends in der textilen Innenausstattung von Automobilen*, Paper on the Man Made Fiber Congress, Dornbirn, Austria, September 2001.
19. Morgenstern, Wolfgang: *Spinn- und spulengefärbte Garne in Autotextilien*, Paper on the Man Made Fiber Congress, Dornbirn, Austria, September 2001.
20. Fa. Möllertech: *Der Widerspenstigen Zähmung*, BW Technics 3/2001.
21. Sarsour, Jamal; Linke, Michael; Planck, Heinrich: *Aufschluss von Bastfasern*, Melliand Textilberichte 2007.
22. *Carbonfasern for Sitzheizungen*, BW Technics 4/2001.
23. Trevira: *Garne für die Automobilindustrie*, BW Technics 3/2001.
24. Suzuki, Atsushi: *'ELK' – Elastische Bauelemente aus PES-Fasern*, Paper on the Man Made Fiber Congress, Dornbirn, Austria, September 2001.
25. Schmidt, Hermann: *Anforderung und Erprobung von Abstandsvliesstoff CAIWEB im Automobil*, Paper on the Man Made Fiber Congress, Dornbirn, Austria, September 2001.
26. Stegmaier, Thomas; Schneider, Petra; Finckh, Hermann; Gündisch, Werner: *Leichtbaustrukturen für mobile Anwendungen durch innovative Flocktechniken mit Metallfasern*, Paper on the 16th International Flock Symposium, Dresden, Germany, 19–20. March 2001.
27. Finckh, Hermann: *The virtual way to develop textiles – from numerical simulation of the manufacturing process of woven fabrics to realtime rendering of textile parts*, 5th RTT conference, Munich/Ettal, Germany, 2004.
28. Finckh, Hermann: *Textile micro models as a result of idealized simulation of production processes*, International Symposium on Finite Element Modelling of Textiles and Textile Composites, St. Petersburg, 26–28 September 2007.

4 Nonwovens used in automobiles

S J RUSSELL and M J TIPPER,
University of Leeds, UK

Abstract: Nonwoven fabrics are essential components in modern vehicles with over forty areas of application depending on the vehicle type and technical design specifications. An overview of the main uses and functions of nonwoven fabrics in vehicle interiors, filtration and acoustic insulation is provided with reference to polymer composition and fabric construction. Current and future requirements for nonwoven automotive fabrics are summarised including the need for recyclability at the end of life.

Key words: filtration, headliners, acoustic insulation, bonnet liners, parcel shelf, seat, carpet.

4.1 Introduction

This chapter considers the utilisation of nonwoven fabrics in OEM (Original Equipment Manufacturer) automotive interiors and in automotive filtration. The importance of nonwovens in the automotive industry continues to be fuelled by the ability of such fabrics to be customised and engineered to meet the requirements of modern vehicle assembly operations and technical performance in use. Nonwovens are particularly attractive in the automotive industry because of their ability to:

- be integrated into multi-layer, modular components with other materials including foam
- be incorporated into lightweight and low-density modules
- meet stringent cost-performance targets in what is a highly price-sensitive industry
- be deep-draw moulded into complex shapes at relatively low temperature
- contain recycled raw materials and still meet performance requirements
- be compatible with emerging recycling processes.

The global market for automotive interiors was approximately US$165 billion in 2005 and is expected to grow to an estimated US$210 billion up to 2017 (1). The total weight of textile components including nonwovens as a proportion of the total vehicle weight has gradually increased. Looking to the future, in a medium-sized vehicle (segment D), the weight of textile materials is expected to rise from a current average of about 21 kg to 35 kg by 2020 (1). Of the current 21 kg, about 8 kg is interior textiles, 4.5 kg of which is floorcovering and the remaining 3.5 kg is the headliner and upholstery (1).

Some of the existing automotive applications for nonwoven fabrics are summarised in Table 4.1 (2) and Fig. 4.1. In vehicle interiors, nonwoven facings contribute to the overall aesthetics of the interior while fabrics that are hidden in the construction of automotive parts perform reinforcing, filtration, barrier, sealing or acoustic insulation functions. The growing importance of nonwoven fabric reinforcement in composites and in foam replacement is a further major area of development that is considered elsewhere in this book. The interior use of textile and nonwoven fabrics according to vehicle segment is summarised in Table 4.2 (1). Woven fabrics still predominate in automotive interiors irrespective of vehicle segment. With the exception of Alcantara® or Ultrasuede® as it is known in the USA, utilisation of nonwovens is confined mainly to the lower cost vehicle segments A, B, C, vans, SUVs and MPVs with limited utilisation in the D, E and F vehicle segments. The fibre composition and selection of fabric types also varies amongst vehicle manufacturers in the different global regions. The type and cost of the vehicle influences the choice of fabric; for example, in Europe, group C vehicles and above traditionally have a tufted rather than a needlepunched floorcovering.

In Japan and Europe nonwoven fabrics are more extensively utilised throughout the vehicle than in the NAFTA (North American Free Trade Agreement) countries of the USA, Canada and Mexico. While standard specifications and test methods relating to nonwoven materials used in vehicles have been introduced by organisations such as the VDA (German Association of the Automotive Industry), the SAE, the ASTM, DIN and the AATCC, each vehicle manufacturer operates its own bespoke test specifications relating

Table 4.1 Automotive applications for nonwoven fabrics

Facing fabrics	Backings and fillings	Filter media	Miscellaneous components
Headliner	Headliner	Cabin air	Electrical insulation
Main floorcovering and boot liner	Main floorcovering and boot liner	Engine air	Composite (panel) reinforcement and veils
Seating and rear seat backs	Seating	Transmission	Tyre reinforcement
Bonnet liner	Door and dashboard panels	Oil	Gaskets and seals
Parcel shelf	Acoustic insulation	Brake hose	Catalytic converter
Door panel and trim	Thermal insulation		Battery separator
A, B, C pillar coverings	Adhesive layers		Side impact airbag sleeves
	Fire barriers		

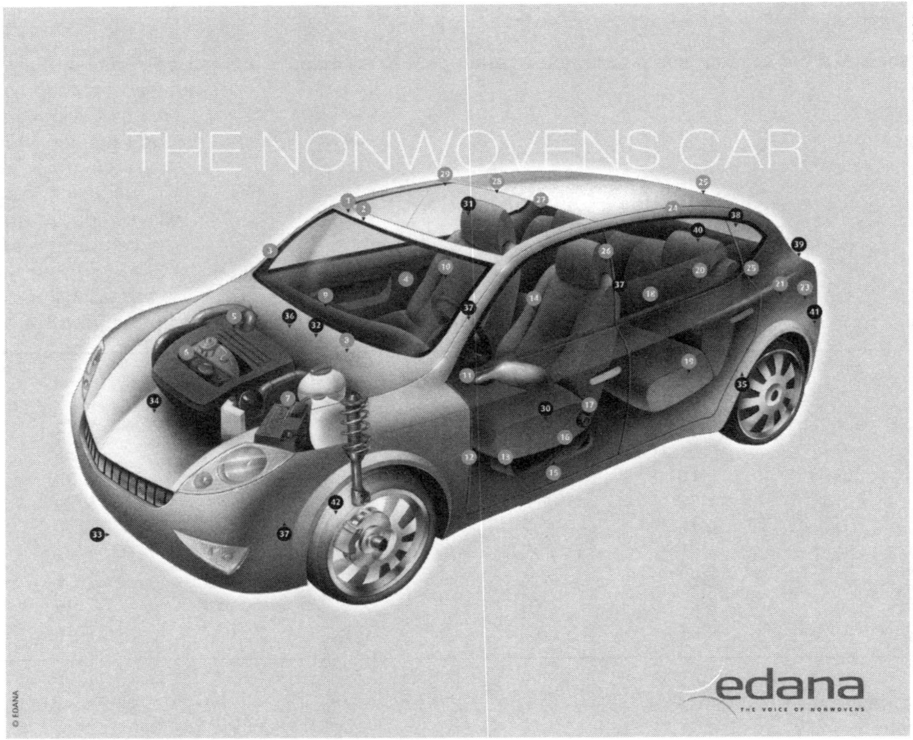

1	Covering material for sun-visors
2	Padding for sun-visors
3	A, B, C, column padding
4	Door trim pads
5	Fuel filters
6	Oil filters
7	Battery separators
8	Cabin air filters
9	Loudspeaker cover
10	Covering for moulded seats
11	Transmission tunnel
12	Carpet & carpet reinforcement
13	Car mats
14	Vinyl backing for seat covers
15	Backing for tufted carpeting
16	Covering for seat belt anchorage
17	Covering for seat belt
18	Decorative fabric
19	Polyurethane coated backing
20	Seat slip agents
21	Boot (trunk) liners
22	Moulded fuel tanks
23	Bodywork parts
24	Window frames
25	Headliner facings
26	Upholstery backing
27	Loudspeaker housing
28	Sunroof
29	Saloon roof
	Acoustic absorber applications
30	Doors
31	Headliner
32	Inner & outer dashboard insulation
33	Under engine shield
34	Moulded bonnet liner
35	Rear wheel arch liner
36	Cowl
37	Pillar trim panels
38	Parcel shelf
39	Trunk trims
40	Rear seat strainer
41	Air extractor
42	Wheel arch liners

4.1 Nonwovens in automotive applications (courtesy of EDANA).

Table 4.2 Interior automotive fabrics by type of fabric construction (%)

Vehicle segment	Woven	Circular knitted	Raschel	Tricot	Leather	Nonwoven		
						Alcantara	Needlepunched	Other Nonwoven
A	45.9	24.1	4.9	0	11.3	0	2.2	11.5
B	48	23.1	2.2	10.2	4.7	1.1	3	7.9
C	49.2	23.9	0.5	14.8	7.7	0	1.6	2
D	49.4	14.6	0.4	12.4	19.4	3.3	0.5	0
E	57.5	7.9	0	7.1	27.5	0	0.2	0
F	55.8	0	0	9.9	33.4	0.9	0	0
G	20.9	2.9	0	3.9	61	9.2	1	0
H	56.7	23.4	9	0	0	0	0	11
Vans	43.3	13.4	2	19.6	10.2	4.6	1.5	5.3
MPVs	56.6	26.3	1.3	6.1	2.4	0	1.7	5
SUVs	31.4	0	0	5.7	58.2	2.5	0	2.2

Where the segments indicate the following vehicle types:
A Segment = Micro-size
B Segment = Mini-size
C Segment = Compact size
D Segment = Medium size
E Segment = Large size
F Segment = Luxury
G Segment = 2-seater sports
H Segment = Box vehicles
Vans = includes compact vans
Note: the data includes parcel shelf, headrests, headliners, bolsters and seat covers but excludes boot linings and interior carpets.

to various aspects of fabric performance. Some of the important technical performance parameters in automotive nonwoven fabrics include: colour uniformity and light fastness (effects of temperature and UV degradation), fogging (e.g. with latex backing), flame resistance, soiling and cleaning, sound absorption, static electrification, mouldability, abrasion resistance and durability.

4.2 Headliners

Headliners are modular parts fitted tightly into the interior roof of a vehicle to provide aesthetics, sound absorption, thermal insulation and cushioning in a light-weight construction. They are semi-rigid modules consisting of at least three layers laminated together: an aesthetic facing fabric, a foam backing and the substrate, which is also referred to as the core. Nonwoven fabrics in headliners take the form of facings, backings or substrates depending on the engineering design of the particular module.

In Europe, warp knitted tricot fabrics account for 50.7% of facing fabric production, while needlepunched and stitchbonded fabrics represent 24.6% and 15.6% respectively (1). Circular knitted fabrics make up the remainder. In the USA, OEMs traditionally produce headliner and pillar trim facing fabrics from tricot warp knitted polyamide fabrics. The choice of facing fabric in headliners is influenced by the prevailing automotive specifications in different countries, which are not uniform. Particularly in Japan and Europe, headliners faced with dope-dyed 100% PET needlepunched fabrics have to a large extent replaced conventional warp knitted fabrics partly because they are readily assembled and fitted into the roof of the vehicle. Fabrics produced from 100% PP are also available. Traditionally, the higher durability standards for vehicle interiors in the USA have meant that knitted and woven facing fabrics have been selected in preference to needlepunched fabrics since to meet these specifications nonwoven fabrics of relatively high fabric weights are required, reducing their cost competitiveness. In low–medium segment vehicles, needlepunched headliner fabrics are common in Japan and Europe; the fabrics have acceptable abrasion resistance and particularly good thermal moulding characteristics, which aids the vehicle assembly process.

Nonwoven facings are made of needlepunched, stitch-bonded or to a more limited extent, hydroentangled fabrics produced from dry-laid (usually carded and crosslapped) spun-dyed PET fibres with a linear density of ≤3.3 dtex and a mean fibre length of 50–60 mm, which compete with traditional warp knitted tricots. Flat or random velour needlepunched facings are selected depending on application. Because of the large uninterrupted visible area provided by the headliner in the vehicle, colour consistency and freedom from faults such as fused fibres or contamination in the fabric is particularly important. An example of a process line arrangement for producing flat

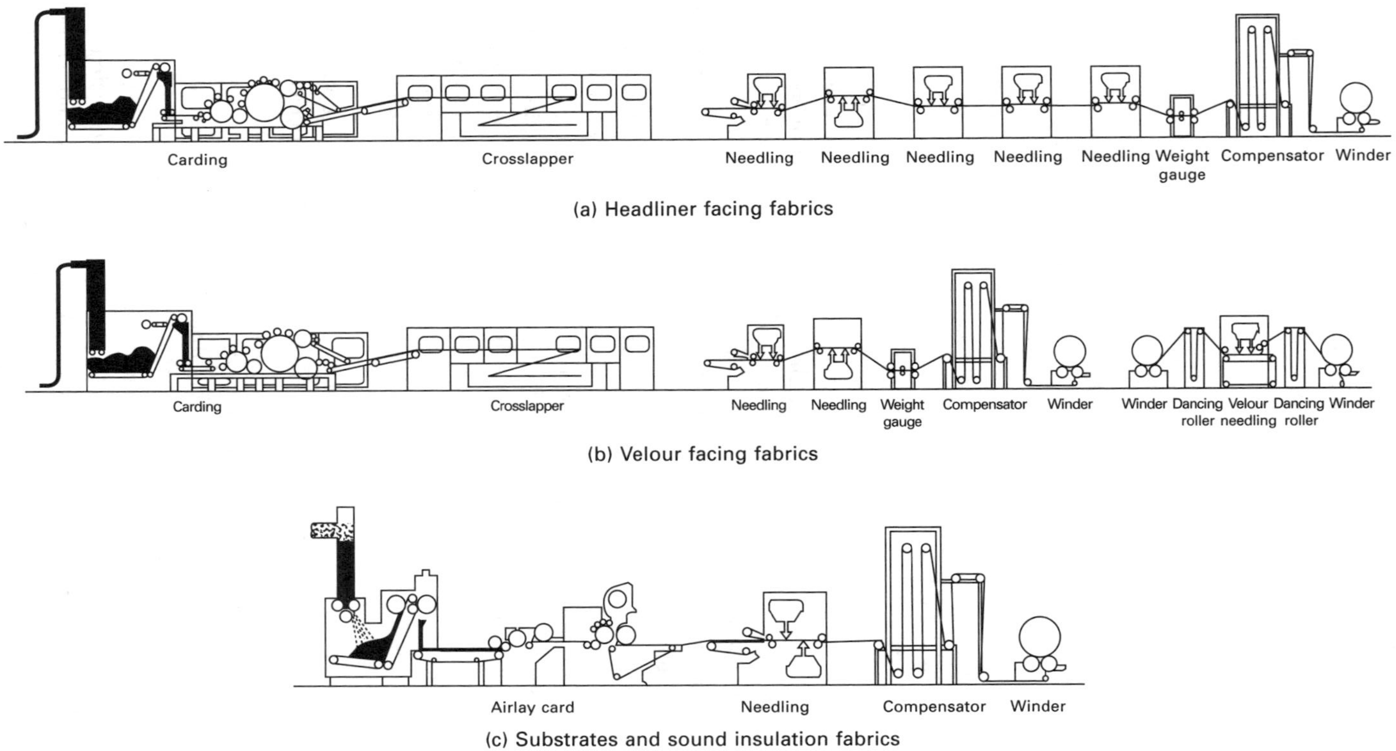

4.2 Production of needlepunched fabrics for automotive products (courtesy of Oerlikon Neumag, Austria).

needlepunched headliner fabric is given in Fig. 4.2(a). Multiple needle looms are arranged in sequence to achieve a high needling density and a dense, flat fabric surface that is uniform in appearance. To increase competitiveness, lighter-weight needlepunched fabrics are produced by decreasing the fibre linear density in the fabric to compensate for the reduction in weight, which has the added benefit of improving softness. There is also an increasing requirement to make headliner facings soil resistant. In addition to needlepunched nonwovens, hydroentangled fabrics that are durable and soft have attracted attention. Evolon® fabrics produced from spunlaid and hydroentangled splittable PA/PET bicomponent fibres have excellent strength and softness and may be printed to give a tailored appearance (3). Other hydroentangled fabrics launched for headliners, visors and trim components include Miratec® fabrics, which can be made with a variety of highly complex three-dimensional structures and are jet-dyeable (4).

The incorporation of semi-rigid polyurethane foam in laminated headliners provides a means of controlling the stiffness of the headliner when the foam is cured, increases sound absorption and provides impact cushioning for the vehicle occupants. However, lamination of foam directly to the facing fabric can negatively affect the mouldability of the laminate, particularly where highly contoured and deeply recessed shapes are required. Recycling the fabric component at the end of life is also complicated in modules containing foam. An alternative approach is to eliminate foam altogether and to bond the facing directly to a high-loft nonwoven fabric that functions as a sound absorber and provides impact cushioning; these components may also be bonded directly to the substrate (core). An advantage to this approach is the possibility of making the entire construction from a homogeneous polymer such as polyester or polypropylene, thereby simplifying the end of life recycling. In Europe, needlepunched PET facing fabrics, with a PET substrate have been developed to promote recycling of the module at the end of life in line with the European End of Life Vehicle (ELV) Directive. Reclaimed fibrous materials may also be used as foam replacement; refiberised mixed garment waste is dry-laid and combined with thermo-curable phenolic resins to form a thermoformable nonwoven layer providing mouldability, dimensional stability and sound insulation properties. In headliners containing foam, polyester spunbonds are utilised for lamination and chemically bonded glass fibre nonwoven fabrics provide reinforcement and stiffness. The glass component of the headliner in a foam laminate may be introduced in various ways including (5):

- Deposition of chopped glass fibres on to a scrim coated in adhesive followed by the application of adhesive to the glass fibre surface to improve adhesion and permit the attachment of a foam board layer. This arrangement of layers may be built up sequentially to produce a thicker structure and is pressure-laminated in an oven.

- Formation of a glass mat (instead of providing loose fibres) which is adhered to a spunbond fabric and another glass mat to produce a tri-layer, chopped glass/spunbond adhesive/chopped glass structure. The construct is pressure-laminated in an oven.

Alternatively, the individual layers of the headliner may be laminated inside the mould itself and simultaneously moulded in a one-step process (6). Glass fibre fabrics are particularly useful in noise reduction but have the disadvantage of a relatively high density compared to polymers, which increases the weight of the module, and there are perceived health and safety concerns in handling during vehicle manufacture. Partly for this reason and the desire to improve environmental sustainability, high modulus natural fibre reinforcements have been evaluated including flax, hemp, sisal and jute as well as blends with recycled fibres. In addition to headliners, this is of course particularly relevant to door panel trims, boot liners and parcel shelf construction. Research by Yan *et al.* (7) revealed that acrylic-copolymer bonded composites exhibited good mechanical properties but there was no significant difference in thermal conductivity compared to polyvinyl-alcohol bonded composites. To replace glass reinforced PU foam headliner cores and increase opportunities for end of life recycling, polyolefin foams have been introduced as well as 100% PET headliner modules. Additionally, the development of nonwoven fibre cores in place of foam provides opportunity to reduce the weight of the module (8).

Requirements in passenger comfort and safety are also influencing headliner design. Developments include the incorporation of side impact airbags (such as the inflatable tubular system) within the headliner unit that deploy downwards when triggered (9, 10), internal storage rails with mounts in the headliner (11) and additional small storage areas fitted within or on to the headliner module. Additionally, there has been a migration of electronic switches and controls to overhead positions mounted in the headliner including one for an internal fire suppressing system (12) as well as audio and telecommunication controls. The integration of LEDs in to the headliner to control ambient lighting has been developed (13) and optical fibres for lighting as well as electrically conductive filaments within fabrics offer potential for the development of an improved vehicle cabin environment.

4.3 Bonnet liners

The bonnet or hood liner and associated fabric linings in the engine compartment have both thermal and sound insulation functions. The bonnet liner provides a protective layer between the top of the engine and the underside of the bonnet, which may be constructed from metal or a fibre reinforced plastic composite depending on the vehicle manufacturer. Bonnet liner modules are composed of a stiffening component such as a glass fabric, foam or a resin-bonded nonwoven fabric composed of reclaimed fibres covered in a nonwoven facing such as a PET spunbond. Bonnet liner fabric construction depends on the vehicle manufacturer. Spunbond, dry-laid chemically bonded and dry-laid needlepunched fabrics are all utilised. A traditional construction is a nonwoven fabric composed of glass fibre combined with a film layer but these have evolved into more sophisticated nonwoven fabrics composed in part of microfibre nonwovens. Nonwovens containing high temperature resistant fibres, for example oxidised PAN, are also available. Modern bonnet liners enable active rather than passive noise reduction limiting the vibration reaching the cabin. Since road noise generated by tyre contact with the surface is a major source of noise entering the cabin, the selection of linings for the wheel wells is also of major importance and can contribute to noise reduction through appropriate tuning. Mouldable and recyclable nonwoven fabrics composed of PP and PET have been developed that are needled to a facing fabric producing a fully integrated acoustic component.

4.4 Boot (trunk) liners

In the boot and luggage compartments, specific applications of nonwoven fabrics include the boot-liner facing fabric, including reinforcements for coated liners, the liner found under the parcel shelf immediately above the boot, the rear-panel inside the boot enclosure and on the back of the fold-down seats. Irrespective of geographical region, nonwovens are predominant in boot liners and luggage compartments. In Europe about 85% of boot liner fabrics are nonwoven, with 10% tufted and 5% being composed of other materials (1). In European vehicle segments A–D flat needled fabrics are important of *ca.* 180–350 gm^{-2} or needled velour fabrics with an average weight of 400 gm^{-2}. A large proportion of boot liners in the USA are also produced from needlepunched fabrics. Facings are commonly staple fibre PET or PP needlepunched fabrics in either flat or velour construction. Cost is one driver for the selection of PP together with the increased opportunities for recycling the trim waste. To improve sound insulation, these are frequently backed with either fabrics produced from waste fibres reclaimed from pulled clothing waste or low-grade bast fibres, e.g. jute, depending on region. Needlepunched underlay fabrics for boot linings are produced from batts composed of recycled fibres obtained from pulled clothing waste. The fabrics are produced by chemical bonding or by thermal bonding when the waste fibres are blended with PP fibre. The PP is either virgin or reclaimed waste fibre. These substrates are faced with either needlepunched PP or PET fabrics that match the rest of the vehicle interior or with durable and impervious polymeric coatings.

4.5 Door and parcel shelf

In the door, the panel trim including the inserts or bolster, underlying reinforcement fabrics and lower facing fabrics may be composed of nonwovens. The facings are composed of flat or random velour needlepunched materials but there is also potential for use of hydroentangled fabrics. Nonwoven interior fabrics are subject to high levels of UV light exposure and this is particularly true in the parcel shelf. For this reason, spun-dyed fibres are preferred. This is true throughout the automotive interior. In the USA, needled fabrics are traditionally produced from PP fibre of 15–18 denier whereas in Europe and Japan finer spun-dyed PET fibre is preferred of 6 denier, or below. This partly reflects the lower abrasion resistance specifications that exist outside the USA (14). Commercially, white hydroentangled fabrics are designed to be joined with facing fabrics or foamed polyurethane to protect from glue or latex penetration in pillar trims, door panels and injection moulded dashboards. Such fabrics are composed of 100% PET, 75% PET/25% viscose or 50% PET/50% PP and are in the range of 0.5–2 mm thick and 50–200 gm^{-2} (15). The reinforcements are normally based on spunbond fabrics. Environmentally sustainable natural fibres are also found in door panel applications. Short-cut bast fibres, such as flax and ramie have been used to reinforce thermoplastic composite structures for some years. More recent developments have included the use of needlepunched hemp (*Cannabis sativa*)/flax fabrics (50/50) to reinforce thermoset compression moulded composite products. Epoxy resin is used as the impregnation medium in the BMW 5 series.

4.6 Seat

Nonwovens are extensively used as backings, bolster fabrics, reinforcements (particularly coating substrates) including needlepunched, hydroentangled and spunbond fabrics. Durability, soil resistance, UV resistance and appearance retention, are obvious requirements in facings for automotive seats, which traditionally has limited the penetration of nonwoven fabrics. Flat woven fabrics are particularly important in European vehicles with woven velours being more popular in Japan and NAFTA countries. Other facings are constructed from leather, double needle bar Raschel, circular knits, tricot and warp knits.

Nonwoven fabrics in seat facings are limited to durable, synthetic suede, microfibre nonwovens, which may be impregnated with polyurethane resin to simulate natural leather. Traditionally, the fabrics are produced from islands in the sea bicomponent fibres that are separated by dissolution of the matrix after needlepunching. Such fabrics represent only a small proportion of the total seating market (*ca*. 2%). One of the best known synthetic suede fabrics used in automotive interiors is Alcantara® (16). The fabrics are used by many of the main vehicle manufacturers, usually but not exclusively for sports and medium to high-end models. The fabrics were originally produced by the Italian ENI Group based on Toray technology patented in the early 1970s. Such fabrics are incorporated not only in seat facings and trimmings but also in door trims, dashboard trims and headliner facing fabrics. Other high quality synthetic leather fabrics, which may or may not be coated with polyurethane include Evolon® and Amaretta®. The recent development of nonwoven roll goods produced from 75–80% reclaimed leather fibres has the potential to find applications in automotive seating applications. The process involves hydroentangling a web of reclaimed natural leather fibres blended with synthetic fibres. The synthetic thermoplastic bicomponent fibres are heated prior to hydroentanglement to fuse and form a supporting network for the leather fibres (17).

In addition to facings, nonwoven substrates coated with pigmented PVC or polyurethane resin (PUR) are used as side or rear panels for leather seats. In backings, considerable research has been conducted on foam-replacement materials composed of PET and other polymeric fibres produced from perpendicular-laid nonwovens, based principally on the STRUTO process. Carded webs containing bicomponent or low-melt fibres are continuously corrugated (the corrugations aligned in the CD) and then through-air bonded to stabilise the structure. In addition to 100% PET constructions, fabrics are produced from reclaimed waste fibres intimately blended with thermoplastics to meet particular price-points. The vertical orientation of fibres relative to the fabric plane increases the lateral compression-resistance and recovery of the fabrics as compared to a carded and cross-lapped fabric and can be engineered by the introduction of a scrim or by adjusting the corrugation amplitude and frequency, the percentage bicomponent fibre content, the fibre type and fibre diameter. Rear seat-back facings include those produced from latex-backed flat and velour needlepunched fabrics composed of spun-dyed PET or PP staple fibres.

4.7 Floorcovering

Automotive facings in floorcoverings are either tufted or needlepunched fabrics produced from spun-dyed PA (mostly tufted), PET or PP fibres (needlepunched) depending on the durability, cost and required specifications, which vary globally between vehicle manufacturers. In respect of tufted fabrics, cut pile is more extensively selected than loop pile. A high proportion of facing fabrics in North American vehicles is tufted, whereas in Japan and Europe a larger proportion is needlepunched, particularly in vehicle segments A to C. Of the interior carpet about 55% is composed of nonwoven fabrics with tufted accounting for 45% (1). In Europe, needled velour fabrics are found in vehicle classes A to C, where cost is paramount, but there is also some penetration in more expensive vehicles (segments D and F) where tufted fabrics have hitherto dominated the main floor areas. The gradual shift towards needlepunched fabric is being driven principally by a desire to reduce cost but also reflects technical progress in achieving improved aesthetic appearance, abrasion resistance, pile stability and cleanability of needled fabrics, which historically have been inferior to tufted fabrics.

Needlepunched facings are made in flat or structured (random velour and rhombic) constructions by carding, crosslapping and needlepunching spun-dyed, crimped PET staple fibres of *ca.* 6.7–17 dtex in Europe with PP staple fibre finding applications in the USA. Abrasion resistance increases with fibre linear density, and in PET velour constructions, different fibre linear densities may be blended together to balance pile stability and aesthetic requirements. To produce velours, the batt is pre-needled and then structured on a second loom to form a surface pile. In the case of rhombic patterns, the velour fabric is needled again on a velour machine. Fig. 4.2(b) is an example of a needled velour fabric production line. Random velours are produced using crown (three barbs, one on each apex) or fine fork needles rather than conventional barbed needles found in flat needlepunching, and the conventional perforated bed-plate is replaced with a brush conveyor to support the pre-needled fabric. As the needles penetrate, a pile surface is formed on the side of the fabric in contact with the brush conveyor. To compete with tufted carpets, the durability of needled fabrics has been progressively improved by increasing the pile stability. A major contribution has been the introduction of double velour needlepunching, where two fabrics are needled together in a second random velour machine to produce a high density velour surface. Increasing the density and mechanical durability of needled velours facilitates a potential reduction in the fabric weight. In vehicle segments A and B there is a trend toward low-weight flat needlepunched fabrics of <350 gm^{-2} in the mainfloor areas with even lighter-weight fabrics of *ca.* 180 gm^{-2} in the boot. Needlepunched velour fabrics range from *ca.* 350 gm^{-2} in the mainfloor of vehicle segments A and B to *ca.* 500–750 gm^{-2} in vehicle segments C and D.

In addition to increasing the fabric density, the introduction of two differently coloured fabrics during double velour needlepunching provides the opportunity to introduce complex design effects in the facing. One example of a needlepunching system capable of producing patterned rib and velour structured fabrics is the Orelikon Neumag Carpet Star; the system also enables a surround and borders on all sides of the fabric. A rapid advance system transports the fabric during the intervals between needle penetrations to produce patterns in the main carpet and in the borders. While structuring looms provide the possibility of producing highly patterned constructions, unpatterned, single colour fabrics (usually black or grey) still predominate.

The type of backcoating on the floorcovering depends on the particular vehicle segment and includes latex backing, thermoplastic PP backing and thermoplastic PE powder backing. When heated, a thermoplastic backcoating on the fabric aids fitting of the floorcovering to the complex three-dimensional shape of the vehicle floor compartment. Conventionally, both flat and velour needlepunched fabrics are latex-backed with, for example, SBR. Figure 4.3 highlights the differences in transverse structure between flat and velour needlepunched fabrics. Compared to tufted fabrics, needlepunched fabrics have a shorter pile and a different surface structure, which affects the appearance and wear resistance of the floorcovering, particularly in relation to pile crushing/recovery and soiling. Recent developments include the introduction of durable, latex-free needlepunched facings produced using binder fibres to facilitate reduced levels of fogging in the cabin, simplify module assembly and recycling at the end of life.

In tufted fabrics, spunbond primary backings are required to secure the tufted yarns in place (Fig. 4.4). During tufting, filament yarns are pushed through the nonwoven backing by the needles to form the tufted pile. The backing facilitates deep moulding of the tufted fabric during formation of the floorcovering module and contributes to tuft-holding and dimensional stability during and after the moulding process. The inherent mouldability of isotropic nonwoven primary backings in tufted carpets is particularly advantageous (Fig. 4.5). Automotive floorcoverings are produced as deep moulded modules and are pre-assembled with the underlying sound insulation ready for integration into the vehicle. Nonwoven backings are *ca.* 100 gm^{-2} thermally bonded spunbond fabrics composed of either PET (e.g., Lutradur®) or thermally bonded sheath-core bicomponent filaments consisting of a PA6 sheath and a PET core (e.g., Colback®).

4.8 Acoustic insulation

The reduction of noise in the cabin is a key objective in vehicle design in both small and large vehicles. A passive or active approach to noise reduction may be adopted or a combination of both. Nonwoven fabrics are extensively utilised in passive noise reduction where the sound absorption coefficient is particularly important at the material and component levels. There are three basic acoustic effects that can be influenced by the choice of nonwoven fabric: sound reflection, sound transmission and sound absorption. Sound transmission and sound absorption are the main factors affecting sound isolation in vehicles and depend on the interaction of the sound wave with the fibres in the nonwoven fabric. While nonwoven fabrics are good sound absorbers, particularly at high frequency, they make less effective sound barriers partly because of their low density. In sound barriers, the transmission loss is particularly important, which is the ratio of the sound energy incident on a

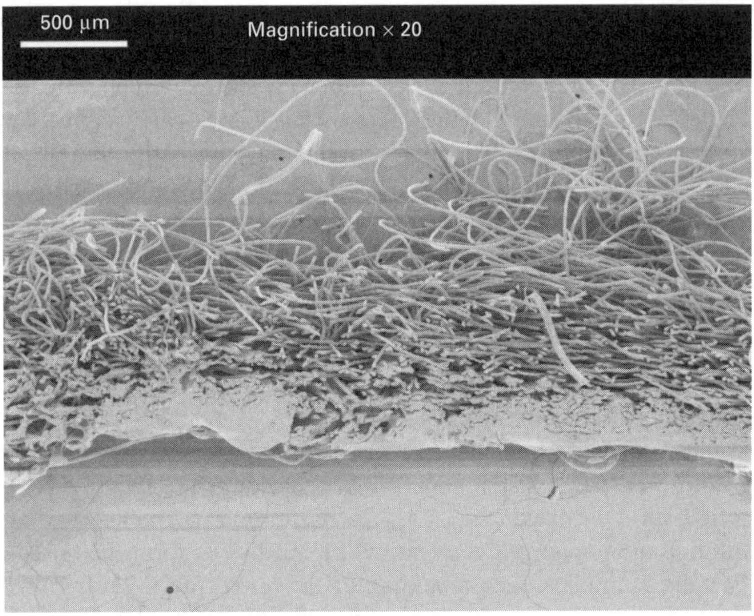

(a) Flat needlepunched

(b) Needlepunched random velour

4.3 Structure of latex-backed PET needlepunched automotive floorcovering facing fabrics.

4.4 Colback® mouldable primary backing fabrics (courtesy of Colbond bv, The Netherlands).

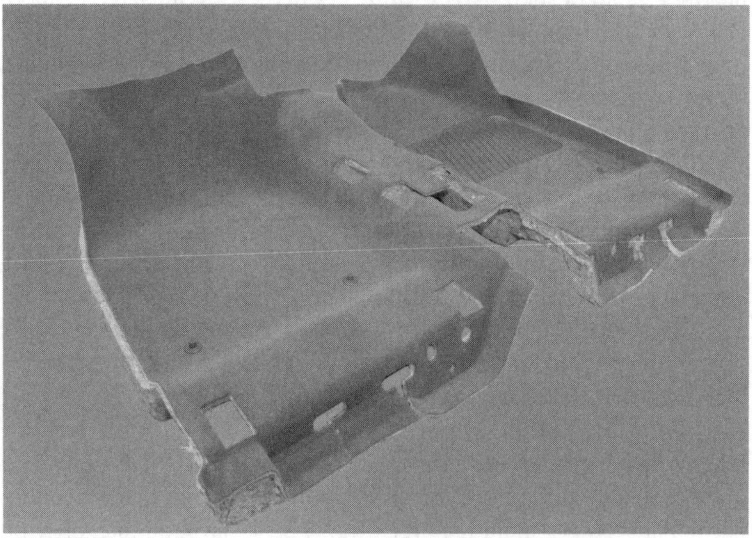

4.5 Moulded main floor module (courtesy of Ideal Automotive).

material to the transmitted sound energy (dB). In practice, the transmission loss in the bulkhead between the engine compartment and the cabin is important in controlling cabin noise and in other areas of the vehicle interior, sound absorption is of major importance. The sound absorption coefficient α is the ratio of sound energy absorbed by a surface to the sound energy incident upon it. The value depends on the angle of incidence of the sound wave, and since in a typical vehicle cabin, there are many different angles of incidence and different surface materials these have to be averaged. In a vehicle cabin that contains various surface materials and angles of incidence, the average absorption coefficient α may be determined for different, n surface materials as follows (18):

$$\alpha = \frac{\sum_{i=1}^{n} S_i \alpha_i}{\sum_{i=1}^{n} S_i}$$

where S = surface area.

The precise sound field inside a vehicle cabin cannot be easily determined because of the complexity of the interior shapes that are contained within it. The reduction of noise is a particularly important requirement even in low-cost vehicles and many nonwoven vehicle components contribute to its suppression including the bonnet liner, under dashboard insulators, wheel arch covers, boot liners, body-shell linings, floorcoverings, floor mats, headliner, parcel shelf, A/B/C pillars (where A = windscreen support pillars, B = vertical pillars at the side of the vehicle and C = rear window pillars), door trims and seals. In addition to reducing noise in the vehicle cabin, the incorporation of nonwoven fabrics in place of higher density materials has enabled substantial reductions in weight.

Needlepunched fabrics composed of either PET or reclaimed textile fibres, spunbond and meltblown microfibre fabrics are all used in acoustic insulation. Low density nonwoven insulation fabrics laminated to thin scrims are also important. Normally, to produce a highly efficient acoustic barrier a high density material is needed, which does not favour nonwovens. Porous nonwoven fabrics are particularly well established in sound absorption and, historically, such fabrics have been produced from fibres recycled from mechanically pulled waste clothing known as shoddy. This is still a cost-effective source of raw material for the production of relatively thick, dry-laid acoustic insulation fabrics. Chemically bonded jute nonwoven fabrics and other low grade fibre materials are also encountered depending on the particular vehicle manufacturer. Fig. 4.2c is an example of a production layout for producing nonwoven automotive sound insulation fabrics based on air-laying and needlepunching. Recycled fibres are incorporated into mouldable nonwovens for sound absorption in vehicle body shells (under the floorcovering) and boot liners. A typical moulded nonwoven fabric composed of reclaimed fibres is illustrated in Fig. 4.6. Nonwoven acoustic pads composed of recycled fibres must incorporate thermoplastic fibres to facilitate thermal moulding of the entire laminated component. Sometimes, microporous films are incorporated with the acoustic pad and in so doing

4.6 Airlaid nonwoven insulation fabric composed of recycled fibres for installation in to the inner dashboard (courtesy of Rieter Automotive).

there is potential to reduce the overall weight of the acoustic absorber, which is a major advantage. Unfortunately, in nonwoven acoustic fabrics, the sound absorption is frequency-dependent over the audible range and low frequency sound absorption is usually worse than at high frequency.

To improve the low frequency sound absorption and transmission loss performance, sophisticated multi-layer nonwoven fabrics have been developed containing, for example, high-density fibre materials such as glass in at least part of the construction. Increasingly, interior vehicle modules consist of laminated fabrics consisting of at least two distinct layers. The first layer is normally a nonwoven fabric that provides high sound absorption and is laminated to a high density layer providing high sound reflection. In relation to nonwoven fabric dimensions and structure, sound absorption varies with airflow resistance, which is influenced by fibre diameter, fibre cross-sectional area, fibre alignment relative to the incident sound wave and fabric porosity. The restriction of air movement in porous materials due to air resistance produced by fibre surfaces normally has a beneficial effect on sound absorption. Improved sound absorption at low frequency has been observed in nonwoven fabrics produced by sandwiching a low density layer between two higher-density surface layers (19). Acoustic insulation fabrics containing hydrophobic meltblown microfibres such as Thinsulate™ are intended for incorporation in vehicle door panels, headliners, wheel wells, pillars, and instrument panels. In reverberation room studies, the density of fabrics containing microfibres has been found to have a larger effect on sound absorption than fabric thickness or weight. In one study, the noise reduction coefficient increased to its highest value at a fabric density of about 0.14 g/cm^3 and then decreased (20). Compared to flat fabrics, hydroentangled nonwovens containing longitudinal voids within the cross-section (Hydrospace) are claimed to increase sound absorption providing potential use in, for example, automotive headliners.

In addition to passive sound absorption, work on adaptive technology based on sound cancelling or active noise control has been developing. The automotive cabin is a three-dimensional sound field enclosure. While active noise control in such environments is difficult, progress can be made in reducing the reverberation in such enclosures by the incorporation of active absorbers along the walls, which has been useful in the cancellation of engine noise (21). The incident sound is received by microphones which connect to the loudspeakers whose acoustic input impedance is matched to the sound field. Four-stroke internal combustion engines have an inherent imbalance at twice the rotational speed known as the 'second engine order', which often coincides with the frequency of the fundamental cabin resonance of vehicles, leading to a booming noise (21). Noise synchronisation facilitates active sound cancellation controlled by computer and appropriate software. While this approach can deal quite effectively with engine noise, there are other sources of noise and vibration within and outside the cabin that are difficult to control. Therefore, cabin noise persists, particularly in small vehicles.

4.9 Automotive nonwoven filters

Nonwoven filter media are important in fuel, lubrication oil and air filtration. The latter relates both to cabin air filtration and to vehicle air intakes. Cabin air filters protect vehicle occupants from air-borne pollutants and allergens by removing suspended particles drawn in to the cabin from outside; they play a major role in improving cabin air quality. Cabin filters are produced from either pleated synthetic fibre nonwovens, including laminated, multi-layer fabric constructions or paper fabrics. Efficient removal of small particle contaminants can be accomplished by increasing the fibre specific surface area, usually by decreasing fibre diameter. The microfibre nonwoven fabric layer is sandwiched between a pre-filter fabric and a carrier fabric. An example of a synthetic fibre HVAC cabin air filter is micronAir®. The concentration of gases such as ozone, NO_2, SO_2, benzene and unwanted odours entering the cabin are reduced by the incorporation of an activated carbon layer within or laminated to the nonwoven media. Such media are referred to as combination or combi-filters (22). Electret filter media are capable of high air filtration efficiency with a low pressure drop (Fig. 4.7). Electrostatically charged air filter media include Filtrete™ meltblown PP treated by a corona discharge and dry-laid electrostatically charged media such as Technostat®. The latter is produced by triboelectrical charging of finish-free fibres during

4.7 Filter panels containing pleated Technostat fabric for installation in vehicle passenger cabin ventilation systems (courtesy of Hollingsworth & Vose Filtration).

carding. Two fibre types occupying different positions in the triboelectric series such as polypropylene and modacrylic or acrylic are intimately blended and then carded: this produces multiple separations of the constituent polymers leading to electrostatic charging at the fibre surfaces.

The efficiency and increasing longevity of vehicle engines is partly due to improved air and oil filter media technology. Air intake filters help to protect engine components from wear and their size, porosity and air permeability vary according to the type of vehicle. Traditionally, these have been produced from wet-laid, resin impregnated cellulose papers although synthetic fibre nonwoven fabrics provide an effective alternative to paper in air intakes. In contrast to paper, such nonwovens are more durable and by controlling the fabric thickness, porosity and fibre diameter, either depth filtration or surface filtration conditions can be created. Improved burst strength is also possible by increasing the synthetic fibre content in the fabric. In one example of a tri-laminate filter medium containing synthetic fibres (23), a top layer of trilobal PET spunlaid fabric is bonded to a middle layer of 100% PET homopolymer and co-polymer PET filaments (either 4.5 dpf (diesel particulate filter) trilobal or 2.2 dpf filaments with a circular cross-section) and a bottom layer of bonded high-loft PET crimped fibre fabric with an isotropic fibre arrangement. The middle layer acts as a 100% PET scrim facilitating thermal lamination to the upper and lower surface layers. The tri-laminate fabric is mechanically integrated by needlepunching and also by thermal-bonding. A density gradient through the filter cross-section facilitates depth filtration by selective removal of particles of different size. QualiFlo® is an example of a polyester fibre intake filter that relies on depth filtration to remove dust particles and a fire retardant version has also been introduced. Other advances in air filtration include cellulose fibre fabrics surfaced with nanofibre webs to improve the particle holding capacity and minimise the rate of blinding (24, 25) by facilitating surface rather than depth filtration.

In some high temperature-resistant engine oil filter media, traditional phenol formaldehyde resin-impregnated paper fabrics have been replaced with multi-layer laminated synthetic fibre nonwoven fabrics to increase temperature stability, burst strength and durability (26). High performance engine development is expected to further increase opportunities for nonwoven filter media. Common rail-diesel injection technology subjects the fuel to very high pressure necessitating its thorough filtration prior to entering the engine's injectors. Wet-laid cellulose and meltblown composite media have been developed to address these demands (27). Continuing concern about the environmental impact of automobile emissions has already led to legislation providing a stimulus for development of high efficiency nonwoven filter media. Important regulatory developments include the European Euro V vehicle exhaust emission limits relating to gas (NOX, HC and carbon monoxide) and particulate matter emissions for both petrol and diesel vehicles. Diesel particulate filters (DPF) that are designed to be regenerated have been developed based on ceramic wall-flow filters but low-cost, replaceable filter media based on nonwoven fabrics composed of high temperature resistant fibres could provide an alternative solution (27). Nonwoven oil coalescing filter media are intended to contribute to lower environmental emissions by removing oil droplets from air flows. One example is in crankcase ventilation (CCV) where oil is removed from the mist and coalesced on to the fibre surfaces to prevent its emission to the environment (27). Nonwoven fuel filtration media include those designed for diesel and petrol engines in which there are multiple filter units. Fuel filter media are based on either 100% wet-laid cellulose fibre fabrics, blends of cellulose fibre with polyester or microfibre glass and multi-layer constructions containing cellulose fibre and meltblown glass fabrics (27). Glass paper fabrics for hydraulic and lubrication oil filtration in large vehicles may be stabilised by latex binder impregnation or by lamination to a polyester spunbond fabric, which also facilitates pleating.

4.10 Other components

High performance vehicle engines operate at high temperature, which affects the design and selection of thermally stable materials for incorporation in the engine compartment, the firewall barrier and other areas such as the exhaust system where thermal shielding is required. Temperature-resistant nonwoven materials are also required in electrical wire insulation, hoses and housings to protect sensitive parts of the vehicle from thermal damage. While glass fibre nonwovens are a low-cost solution for thermal shielding, they complicate recycling of the module and potential alternatives include nonwoven fabrics composed of aramid and pre-oxidised fibres. Nonwoven fabric seals fitted between adjacent surfaces, for example within plastic-plastic, plastic-metal and plastic-glass interfaces, provide a simple means of reducing rattling noises, particularly in the cabin.

4.11 Future trends

Growth in Eastern Europe, Russia, China and India is expected to increase nonwoven fabric consumption in the automotive sector (1). The rising cost of raw materials, particularly synthetic polymers and the continuation of OEM price-down agreements with suppliers is fuelling the desire for more cost-effective materials, vehicle components and assembly processes. As a result of the globalisation of automotive manufacturing, geographical differences in the choice of fabrics for vehicle interiors are expected to diminish. Advances in the construction of headliners, floorcoverings (including sound absorption), boot liners and parcel shelves are expected in the areas of fabric composition, construction and three-dimensional module design. There are already developments in the field of non-structural composites reinforced with natural fibres in headliners and floor modules that are likely to increase the importance of nonwoven fabrics in the future. More durable, lighter-weight nonwoven fabrics and laminates that provide potential for further significant weight reductions in the vehicle are important. The integration of electronic components within fabrics prior to module installation, electro-magnetic shielding of the interior (1) and the introduction of functional coatings developed elsewhere in the nonwoven industry to improve aspects of performance such as stain resistance and odour control are progressing.

In the European Community, the ELV Directive is intended to reduce the waste from vehicles when they are scrapped. It tightens the environmental standards for vehicle treatment sites, requires that the last owners have to dispose of their vehicles free of charge from 2007 (and requires producers to pay all or a significant part of the free take-back from this date), sets higher targets for reuse, recycling and recovery and restricts the utilisation of hazardous substances in both new vehicles and replacement vehicle parts (28). Specifically, the 2000/53 EC Directive has required the automotive industry to increase re-use and recovery to 85% with re-use and recycling being increased to a minimum of 80%. By 1 January 2015, the re-use and recovery target increases to 95% with re-use and recycling increasing to a minimum of 85%. The directive expects future vehicles to be designed and manufactured for efficient dismantling, re-use and recycling at the end of life. Also addressed is the requirement for continuous improvement in the handling and recycling of plastic materials. While the directive is improving the environmental impact of the automotive industry, partly as a result of the ELV, there is an increasing amount of short fibre textile waste being recovered from vehicles at the end of life and opportunities for converting this into useful products including nonwovens are being explored. Improving the ease of recyclability at the end of life, which is associated with the ELV encourages the adoption of a homopolymeric construction by eliminating blended fabrics and multi-laminate materials composed of different polymers and adhesives. To promote improved recyclability there is continued interest in polypropylene nonwoven fabrics, particularly in view of developments in elastomeric polyolefin foam technology, which provides scope for moulding modules from a single type of polymer.

4.12 Acknowledgements

The authors gratefully acknowledge the valuable assistance of Rieter Automotive, Colbond bv, Hollingsworth & Vose, Oerlikon Neumag, Ideal Automotive, Hemcore, Fleissner GmbH, Freudenberg Nonwovens and Adrian Wilson (Editor, *Nonwovens Report International*) in the preparation of this chapter.

4.13 References

1. A. Wilson, *Automotive Textiles: The Global Supply Chain to OEMs*, Textile Media Services, 2007, pp. 13–18.
2. W.C. Smith, Nonwovens in Automotives – More than Just a Pretty Face, *International Nonwovens Journal*, Fall, 2004, p. 60.
3. D.G. Dawson Jr., S.B. Winter, US Patent 6,737,114, 2004.
4. D. Schechner, *We are Living in a Material World*, Auto Interiors, 2001.
5. E. Haque, US Patent 7,279,059, 2007.
6. G.B. Byma, T. Greiner, US Patent 6,832,810, 2004.
7. S. Yan, L. Sun, O. Chiparus, I. Negulescu, V. Yachmenev, M. Warnock, Kenaf/Ramie Composite for Automotive Headliner, *Journal of Polymers and the Environment*, Vol. 13, No. 2, April 2005, pp. 107–114.
8. C. Kalogeridis, Stage is set: a new range of advanced nonwoven fabric technologies is gearing up to challenge the market dominance of conventional textiles in the automotive interior supply chain, *Automotive Industries*, Jan, 2005.
9. Y. Hangland, L.E. Florberger, S. Andersson, S. Valkenburg, J.I. Svensson, US Patent 5,788,270, 1998.
10. J. Peng, T. Subbian, K. Ewing, B.B. Xu, US Patent 7,152,873, 2006.
11. K.M. Anderson, M.P. Kadzban, M.A. Kalis, S.T. Williams, A.G. Greene, L. Norland, US Patent 7,182,383, 2007.
12. R. Thompson, US Patent 7,258,172, 2007.
13. www.rieter.com/main/.
14. www.fosterneedleusa.com/.
15. www.orvmanufacturing.com/prodotti_eng.htm.
16. www.alcantara.com/en/experience/applicazioni/automotive/automotive.shtml.

17. C.G. Bevan, European Patent 1763602, 2007.
18. J.H. Smith, *An Introduction to Modern Vehicle Design*, Butterworth-Heinemann Ltd, ISBN-10: 0-7506-5044-3, 2000.
19. D.H. Müller, J. Müssig, *Automotive Interiors using Hybrid Fleeces*, INTC Conference Proceedings, INDA, USA, 2004.
20. Y.J. Na, J. Lancaster, J. Casali, G. Cho, Sound Absorption Coefficients of Microfiber Fabrics by Reverberation Room Method, *Textile Research Journal*, Vol. 77, No. 5, 2007, pp. 330–335.
21. D. Guicking, *Active Control of Sound and Vibration From a Laboratory Curiosity to Commercial Products*, University of Gottingen, Germany, 2003.
22. O. Kievit, V. Neveling, J. Valkenburg, J. Waqué, Development of a realistic loading aerosol for cabin air filter testing, *Journal of Aerosol Science*, Vol. 27, Supplement 1, 1996, pp. 633–634.
23. J.F. Baker, No binder required to create composite media, *Filtration & Separation*, Vol. 38, No. 3, April 2001, pp. 26–28.
24. P. Trautmann, M. Durst, A. Pelz, N. Moser, *High Performance Nanofiber Coated Filter Media for Engine Intake Air Filtration*, AFSS 18th Annual Conference, April, 2005.
25. www.donaldson.com/en/engine/support/datalibrary/000165.pdf.
26. I. Hutton, *Handbook of Nonwoven Filter Media*, Elsevier, ISBN 978 1856 17441 1 2007, pp. 370–374.
27. J. Barrillon, Automotive filtration: a new horizon for nonwovens, *Filtration & Separation*, Vol. 42, No. 9, November 2005, pp. 28–30.
28. Department for Business, Enterprise and Regulatory Reform, UK, www.dti.gov.uk/innovation/sustainability/elv/page30591.html.

5

Recycling of automotive textiles

X NORMAND, Institut Français du Textile et de l'Habillement, France

Abstract: Sustainable production is increasingly important in the European automotive sector, including the need to produce components that are 100% recyclable. This chapter discusses the development of nonwoven automotive textiles made from thermoplastics which are both functional and easy to recycle. These new nonwovens can be used in engine air filters and as automotive insulation materials.

Key words: sustainability, automotive nonwovens, thermoplastics.

5.1 Introduction

Nowadays, environmental requirements are increasingly important in all industrial fields and notably in the European automotive sector, owing to the European Directive on end-of-life vehicles. This necessitates working more upstream to develop some new products that are 100% recyclable. At present, automotive products are composed of many raw materials which are difficult to separate and only partially recyclable.

The principal objective of the studies presented in this chapter consisted of developing textile structures (nonwovens) composed of thermoplastic (recycled or not) mono material having the same characteristics (acoustic, comfort, mechanical behaviour, filtration efficiency, dust holding capacity, pressure drop, etc.), or even better than existing products in the automotive market. This new concept will enable the reduction of end-of-life waste production in the automotive sector and also the definition of a new method of valorisation in the production cycle of an automotive product.

The aims of the studies were:

- To replace the classical engine air filter [1], manufactured using several raw materials which are difficult to separate and only partially recyclable, with a filter made from a thermoplastic (recyclable or not) mono material. The recycling of this mono material was studied for automotive applications or for other fields.
- To replace classical acoustic insulation products [2], manufactured using several raw materials which are difficult to separate and only partially recyclable, by textile structures (nonwovens) made from a thermoplastic

(recyclable or not) mono material. The recycling of this mono material was studied for automotive applications or for other fields.

5.2 Legislative considerations with regard to the disposability of used vehicles

A European Directive [3] lays down minimum thresholds for the reuse, recycling and recovery of the component parts and materials of new vehicles with the aim of facilitating the reuse, recycling and recovery of parts in order to fulfil the planned 2015 objectives for recycling and recovering end-of-life vehicles. From 15 December 2008, vehicles which do not comply with the requirements of this Directive may not be granted EC type approval or national type approval. From 15 July 2010, moreover, the marketing of new vehicles which do not comply with the requirements of this Directive will be prohibited.

By the Resolution of 14 November 1996, the European Parliament called on the Commission to legislate on waste streams, in particular end-of-life vehicles, on the basis of product liability. The Commission took the view that a specific directive was necessary, given the importance of this type of waste.

The Directive defines an end-of-life vehicle as any type of vehicle which is waste within the meaning of Directive 75/442/EEC. The scope of the Directive therefore covers:

- any end-of-life vehicle designated as category M1 or N1;
- two- or three-wheel motor vehicles and their components.

The provisions of this new Directive apply to cars, station wagons and people carriers (category M1 vehicles) and to light-duty trucks (category N1 vehicles), new models and models already in production in accordance with a timetable set out in the Directive.

The new Directive does not apply, however, to special-purpose vehicles (armoured vehicles, ambulances, etc.), to multi-stage built light-duty vehicles (provided that the base vehicle complies with this Directive), or to vehicles produced in small series (fewer than 500 vehicles a year in each Member State).

Waste prevention is the priority objective of the Directive. To this end, it stipulates that vehicle manufacturers and material and equipment manufacturers must:

- endeavour to reduce the use of hazardous substances when designing vehicles;
- design and produce vehicles which facilitate the dismantling, reuse, recovery and recycling of end-of-life vehicles;

- increase the use of recycled materials in vehicle manufacture;
- ensure that components of vehicles placed on the market after 1 July 2003 do not contain mercury, hexavalent chromium, cadmium or lead, except in the applications listed in Annex II of the Directive. The Council or the Commission may amend the Annex where scientific and technical progress allow these substances to be avoided.

The Directive also introduces provisions on the collection of all end-of-life vehicles (Article 5). Member States must set up collection systems for end-of-life vehicles and for waste used parts. They must also ensure that all vehicles are transferred to authorised treatment facilities, and must set up a system of deregistration upon presentation of a certificate of destruction. Such certificates are to be issued when the vehicle is transferred, free of charge, to a treatment facility.

The storage and treatment of end-of-life vehicles is also subject to strict control, in accordance with the requirements of this Directive and those of Annex I to the Directive. Establishments or undertakings carrying out treatment operations must strip end-of-life vehicles before treatment and recover all environmentally hazardous components. Priority must be given to the reuse and recycling of vehicle components (batteries, tyres, oil).

At the moment, 75% of end-of-life vehicles are recycled (metal content). The aim of this Directive is to increase the rate of reuse and recovery to 85% by average weight per vehicle and per year by 2006, and to 95% by 2015, and to increase the rate of reuse and recycling over the same period to at least 80% and 85% respectively by average weight per vehicle and year. Less stringent objectives may be set for vehicles produced before 1980.

Member States must ensure that producers use material coding standards, which enable identification of the various materials during dismantling. The Commission must establish European standards on material coding and identification.

Economic operators must provide prospective purchasers of vehicles with information on the recovery and recycling of vehicle components, the treatment of end-of-life vehicles and progress with regard to reuse, recycling and recovery.

5.3 Shortcomings in the textile materials used today

The demand for enhanced comfort and improved safety in the automotive sector will result in a significant increase in the use of textile materials in a typical mid-size car over the next five years – from less than 21 kg in 2005 to 26 kg in 2010. Furthermore, by 2020 textile usage is predicted to reach 35 kg. These are the findings of a study by the Saxony Textile Research Institute in Germany ([4]).

5.3.1 The raw materials used in passenger cars

In the manufacturing of passenger cars (Fig. 5.1) in 2000, the most important raw material used was steel (65%); textiles represented only 2% of the raw materials used.

5.3.2 The textiles used in passenger cars

The application of textile materials in motor vehicles is becoming widespread, owing to the fact that in regions like the USA, Japan and Western Europe customers expect more comfort and better safety. There is still high growth potential in the business sector of equipping motor vehicles.

Textiles have long been used in cars. Looking at both the expected future number of cars and the developments in the field of quality, textiles stand a good chance of being more and more widely used (Table 5.1). It is up to the textile industry to convincingly outline the advantages that textiles offer with regard to manufacturing, raw materials, design, combinability, functionality and recycling.

Table 5.2 gives an overview of the proportions of textiles used in different applications in motor vehicles, adding up to an overall total of about 20 kg.

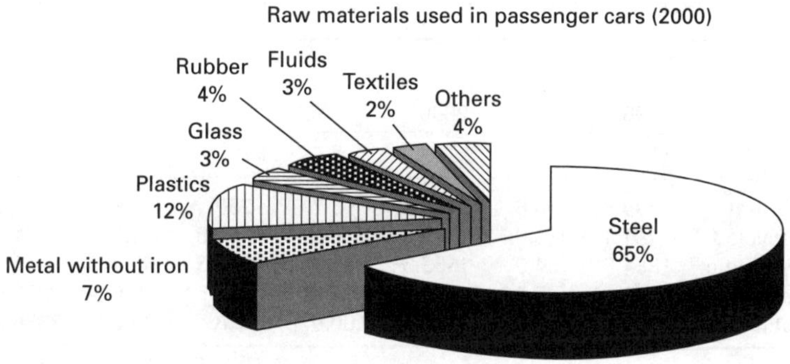

5.1 Raw materials used in passenger cars.

Table 5.1 Textiles used in passenger cars

Amount in kg in a typical car	Year		
	2000	2010	2020
Total	20	26	35
Product and application like 2000	20	14	10
New products for known applications	–	9	15
New products for new applications	–	3	10

Table 5.2 Automotive textiles according to usage

Application	kg
Carpets, interiors	4.5
Upholstery	3.5
Belts, tubes, tapes	1.6
Tyre cord	1.5
Safety belts	1.3
Airbags	1.2
Composites	4.5
Others, e.g. filters	2.2
Total	20.3

Table 5.3 Worldwide production of automotive textiles: type of textile [5]

	Production in kt				
	1985	1990	1995	2000	2005
Fabrics, knits	953	1125	1170	1221	1214
Nonwovens	95	109	107	111	114
Composites	284	443	521	736	976
Others	76	97	119	152	179

Table 5.4 Worldwide production of automotive textiles: type of fibre [6]

Type of fibre	Years							
	1995		2000		2005		2010	
	kT	%	kT	%	kT	%	kT	%
Natural	139.8	6.6	166.6	6.7	250	8.8	399.8	12.0
Recycled	81.1	3.8	75	3.0	64.8	2.3	71.2	2.1
Synthetic	1354.3	64.0	1553.5	62.7	1722.5	60.9	1941.6	58.2
Inorganic	542.3	25.6	683.8	27.6	790.6	28.0	925.5	27.7
Total	2117.5	100.0	2478.9	100.0	2 828	100.0	3338.2	100.0

From a design-related aspect and looking at quantities, a number of classical knits and wovens rank first. Textile composite materials are expected to have the greatest growth potential as they can be tailored to serve the applications in question (Table 5.3).

5.3.3 The fibres used in passenger cars

More and more natural fibres are being used in passenger cars. The use of recycled fibres has remained at the same level since 1995. The use of other fibres (synthetics and inorganics) has also increased (Table 5.4).

5.3.4 The engine air filters used in passenger cars

In the European market, engine air filters are made using classical cellulosic filters made with wood pulp. These filters are manufactured using the wet-laid nonwoven process. The main manufacturers of these filters are:

- Sogefi/Fram
- Mann & Hummel
- Mécaplast
- Knecht/Mahle
- Tecafiltre

In the international market, engine air filters are made from classical cellulosic alone by manufacturers making synthetic filters for the trademarks Toyota, Suzuki, Hyundai and Mitsubishi. Synthetic filters are manufactured using the dry-laid nonwoven process.

5.3.5 The acoustic insulation structures used in passenger cars

The acoustic insulation structures used in passenger cars are made from foams or nonwovens. Often the foams are manufactured using polyurethane. Nonwovens are manufactured from recycled or virgin fibres. These filters are manufactured using the dry-laid nonwoven process.

5.4 Conception of recyclable engine air filters for the automotive industry

5.4.1 Market analysis

Almost all the filter elements for engine air admission are manufactured from cellulosic media, whether for the original equipment or its replacements.

At the end of 2001, the only engine air filtering elements made from a synthetic medium that were sold for original equipment were manufactured in Japan and the USA by the Japanese equipment supplier Denso, mainly for the Toyota, Suzuki, Hyundai and Mitsubishi ranges. Various designs are on the market which have been analysed and tested, the most advanced being a mono material and thermo-formed filter element with an integrated seal.

In the replacement and spare parts market sectors, filter elements in a synthetic medium have also appeared, whether mono material or not, manufactured mostly by Asian companies. In Europe, manufacturers are taking a close look at these filtering elements owing to their filtration performances, and to their potential for integration into the under-the-bonnet architecture. Many projects, some very advanced, are under development and the first marketing of these by European manufacturers will probably be seen in the near future.

5.4.2 Bibliographical analysis

The various mechanisms for trapping particles in a fibrous network were listed to understand the operation of a filtering medium, and to define the textile structure and fibre composition giving the best filtration results. A comparative analysis using laboratory and airborne dust types was made. Finally, an analysis of the incidence of the ingestion of particles into an internal combustion engine has made it possible to classify them, and treat any risks on a hierarchical basis.

Pollutants

As an example, those particles influencing the wear on an engine have a diameter of between 1 and 40 microns. They are responsible for the wear of the segments, the sleeves and the bearings. Most harmful are those whose size is close to 5 to 10 microns. As well as the size of the particles, their hardness and their type are the other parameters that have a great influence on engine wear. The most harmful particles are dusts of a mineral type, while the carbonaceous types present no danger to the engine.

The dust types used in the laboratory and approved by the car manufacturers are the purely mineral dust types (SAE fine and coarse). Their representivity compared to airborne dusts is relatively low, which is confirmed by clogging tests, both in bench and vehicle testing. *Ashrae* dust is additionally interesting as it also contains carbonaceous soot types.

Air in the atmosphere is composed of SO_2, NO_2, NO, O_3, CO; hydrocarbons (benzene, xylene); heavy metals and dust types in suspension. Their proportions and their concentrations differ according to the weather conditions and location. Road dusts sucked in by the engine come primarily from the vehicles themselves (exhaust fumes, brake linings, tyre wear, etc.), or from the roadway. Their concentrations vary from 0.1 mg/m^3 (clean roads) up to 50 mg/m^3 (tracks), and even up to 1000 mg/m^3 in dusty countries.

Factors influencing filter working life

The principal requirement of an air filter is to maintain its filtration performance throughout its working life (Fig. 5.2). This can be estimated in the laboratory, or calculated, but with variable precision as it depends enormously on the running conditions encountered.

Patents: a bibliography was created for the patents on the media and on the filter manufacturing processes.

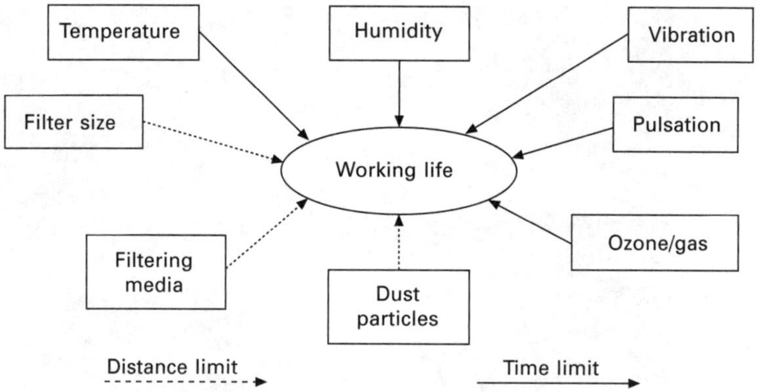

5.2 Factors influencing filter working life.

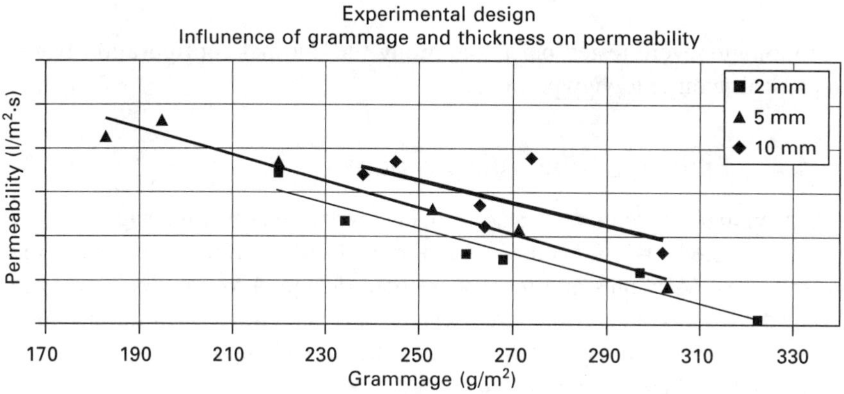

5.3 Influence of grammage and thickness on permeability.

5.4.3 Design of the filtering medium

An analysis of 204 synthetic medium filter types on the market was carried out, and the most interesting were subject to close analysis. This study also led to research on the choice of multi-layer filtering media, and on the use of fibres of different grades and section types. An experimental design was then made up and samples were tested on a laboratory scale by IFTH-Lyon. These media were tested and ranked in line with their filtration performances, while the parameters influencing these performances could then be targeted, quantified and optimised (Fig. 5.3).

Production of several media on an industrial scale was then carried out by Mécaplast (Fig. 5.4) and using these industrial media, the influences of the parameters affecting the various stages of their manufacturing methods could be refined. Lastly, at each stage in the manufacturing of the media, several

5.4 Mécaplast media manufacturing line.

technologies were tested before adopting the optimal configuration from a technical-economic viewpoint.

5.4.4 Filter manufacturing process

Thirteen major patents were analysed, and in parallel several design concepts were invented and protected, with three of them offering more obvious advantages. Machines and tools were then designed to produce prototypes using the technologies of the three design concepts. The technologies involved are very diverse but they call upon techniques already mastered by Mécaplast, or similar to those already used.

The first stage of the process consists of folding the cellulosic medium to increase the filtering surface, but supplementary modules had to be added to the usual pleating machines used for cellulosic media (Fig. 5.5). The following stage consisted of stabilising the pleating geometrically, while the final stage involved completing the filtering element while ensuring, if this was not already achieved, that the sealing function for the tank and lid of the filter was working.

5.4.5 Checking the filters

Blocking with various dust types

The filtration tests quantified the primary function of the filtering element. They were carried out on aeraulic benches with laboratory dust types, on an endurance bench with external air, and on vehicles. The various test conditions selected made it possible to explore the behaviour of this medium type. The development of specific media proved necessary for the different vehicle

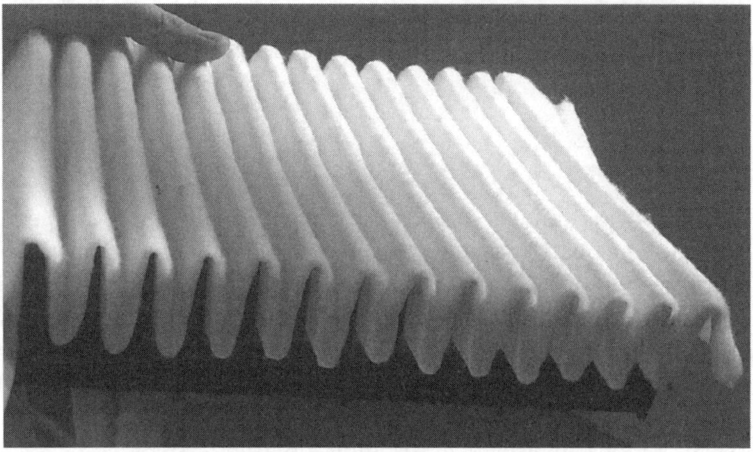

5.5 Folding synthetic media filter manufactured by Mécaplast.

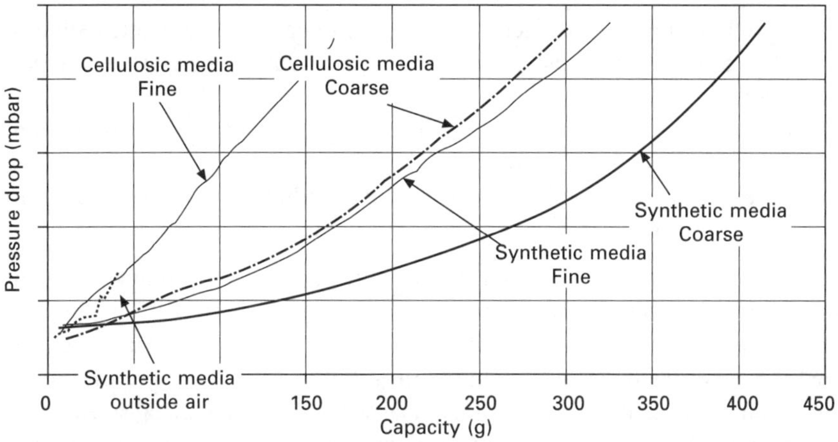

5.6 Blocking with various dust types.

applications (room available, environment, etc.) and manufacturers' specifications. A comparison could, in particular, be made between the filtering elements in a cellulosic medium and those made of a synthetic medium. Figure 5.6 illustrates this, showing the curves of clogging achieved with various pollutant types.

Water resistance

On the basis of the car manufacturers' specifications, water ingestion tests were carried out on a specific Mécaplast bench and these demonstrated a high fold stability using the synthetic media, even when impregnated, with regard to depression in the admission line. The risk of seeing chevrons and stripes appear as with cellulosic medium filtering elements, definitively deforming the folds, is nil.

Fire behaviour

At present, a limited number of manufacturers apply a specific self-extinguishing treatment to their cellulosic medium filtering elements. However, following various accidents and catastrophes occurring in Europe which implicated the car, fire safety regulations will be reinforced, both for the inflammability of materials and for their toxicity. Two examples of accidents already observed show the risk of a vehicle catching fire caused by the filtering element: the suck-in of a still-burning cigarette end, and to a lesser extent, a flame light-back in the admission line.

These situations were reproduced on a calorimetric cone and comparative tests of synthetic/cellulosic media were carried out, with the tests showing:

- a fire flare-up much earlier with a paper rather than a synthetic medium;
- a less energetic combustion with a paper medium compared to a synthetic medium;
- smoke emission with the combustion of a paper medium less significant than with a synthetic medium; and
- toxicity from the combustion of a paper medium higher than that from a synthetic medium.

The synthetic fibres involved did not have any specific flame-retardant treatment. However, these treatments do exist, either via a fire-proofing agent (chemical product) fixed on the fibres, or via an *in-situ* treatment of the fibres during their manufacture. In addition to their price, the major disadvantage remains the environmental aspects of such chemical substances. Research programmes by Mécaplast, partnered by chemical manufacturers, aim to optimise these fire treatments from both economic and ecological standpoints.

5.4.6 Environmental aspects

Working life

Cycles of cleaning/clogging synthetic medium filters showed interesting performances: a gain in initial efficiency after cleaning, and an increase of 10% to 30% in the dust loading capacity according to the cleaning parameters. The total working life of a filter can thus be prolonged, which corresponds to an appreciable lengthening of replacement intervals (between 30 000 and 80 000 km according to the car manufacturers). Over the life of a vehicle, the number of filters to be recycled can thus be reduced, but the application of this technique on an industrial scale does require some new thinking for the car manufacturers, notably in terms of logistics and organisation.

Recyclability

Shredding tests (Fig. 5.7) and crushing tests (Fig. 5.8) on shear crushers were carried out, and they demonstrated the industrial feasibility of these recycling techniques. The transformation of the residues into thermoplastic pellets by fusion/compounding/extrusion is a proven technique and applied on an industrial scale, in particular, for wall-to-wall carpeting. As for the recovery aspects, techniques have already been patented by certain recyclers who have had several material grades approved for cars. Other grades with lower performance properties are already used in less technical applications such as horticulture, etc.

A profitability study based on simulations of waste volumes has been undertaken, and it demonstrated the viability of such recycling operations provided that a regular minimum volume could be treated.

Lastly, the need to set up a sizable logistics organisation to collect all the waste, and then treat it in the optimal manner, remains a major issue for study by the industrialists concerned.

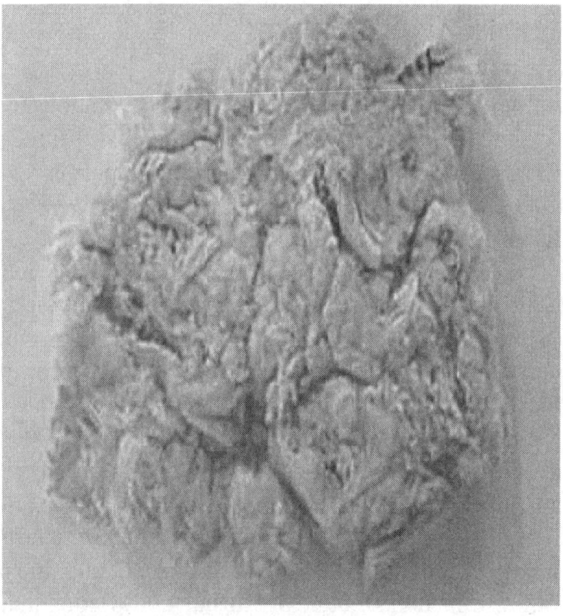

5.7 Shredding material of blocked filter.

5.8 Crushing material of blocked filter.

5.5 Development of recyclable textile acoustic insulation structures for the automotive market

5.5.1 Market analysis and state of the art

In order to orientate this study, it is necessary to know the existing position of products in the automotive market. Mechanical characterisations (resilience, compression, etc.) and acoustic characterisations (absorption, intrinsic parameters) were carried out on a material set of products, porous and fibrous, present in the automotive market. At the same time, an analysis of the state of the art was carried out.

Some mechanical characterisations (carried out by IFTH) and some acoustic characterisations (carried out by Mécaplast) were carried out on a panel of 12 representative samples of the automotive market. The mechanical characterisations consisted of determining the energies (compression, indentation, resilience, etc.) and the deformation combined with expressing the compression behaviour of damping thick materials. The acoustic performances were obtained by some measurements of absorption coefficients (measured with an impedance pipe) and intrinsic parameters (air passage resistivity, tortuosity, and porosity, thermal and viscous characteristic lengths) which enabled the materials to be assessed and the acoustic behaviour (absorption, impedance, etc.) of fibrous and porous materials to be predicted.

Simulation software provided fast, low-cost results for complexes that matched the car manufacturers' requirements.

The acoustic and mechanical characterisations demonstrated that:

- the more a fibrous material is compacted,
  - the more the driving in, the porosity, characteristic lengths and the absorption performances decrease,
  - the more the resilience, the air passage resistivity and the tortuosity increase;
- the more alveolar material (such as a foam) is dense, the more the driving in decreases and the resilience increases.

The analysis of the state of the art, based on consulting several databases on patents and articles and using specific key words, highlights that:

- The 100% recyclable acoustic textile materials that are used in the automotive industry are mainly polyester, used in headliner applications.
- Generally, textile structures are nonwovens with thermobonding fibres used for conversion of the web in automotive products, their mechanical characteristics and their resilience. There is also interest in double-faced textiles (wovens, knitted fabrics and also nonwovens (3D web linker such as NAPCO technology)) for the products being analysed. Within this framework, several kinds of fibres are used: microfibres for their acoustic properties, fibres with a good reputation for their cost and their mechanical properties, and highly crimped fibres for their resilience.
- Two or more layers of different densities are present in some materials; the more dense for their mechanical characteristics and soundproofing, the less dense for their acoustic absorption properties.
- The parameters having an influence on nonwoven resilience are mainly the grammage, the presence or not of several layers, the orientation of fibres, the fibrous mixing used, the bonding technology (needling and/or thermal bonding and/or chemical bonding), the density of each layer and the fibres' characteristics (length, title, crimp, shape, surface state, spin finish or treatment, etc.).

5.5.2 Determination of virgin fibre complexes by an acoustic study

Following on from the benchmarking and determination of the state of art, a design experiment was carried out, taking into consideration the specifications of car manufacturers (the idea of products taking the environment into account, passenger and pedestrian safety and responding to the specifications for lower price and lighter weight). This experiment's plan allowed for evaluating the fibres' influence, the fibrous mixing and web structures, on acoustic performance and other characteristics (mechanical, climatic ageing, etc.).

Twenty or so fibrous structures were made. These webs were nonwovens of the same grammage (500 g/m^2) and the same thickness (25 mm), needled or not, bonded in an oven or not. The first parameters studied were the raw fibres' materials (influence of the fineness, the form, and the chemical composition) and the web structures (superposition of webs or intimate mixing).

Several fibres were used; fibres of different densities (0.9 dtex microfibres, 7 dtex medium fibres and 5 dtex large fibres), of different shapes (multilobed, cylindrical and empty fibres) and different raw materials (PP or PES). In the same way, several intimate blends were studied; 100% (to evaluate the real influence of fibres) and 50%/50% (to evaluate the influence of intimate blends). Other blends were studied by adding 20% of thermobonding fibres to the original compositions for the thermal bonding.

The process used to carry out laying, needling, superposition of webs and needling for the bi-layer nonwovens used an oven passage (climatic chamber) for the products with thermobonding fibres.

Acoustic measurements and prediction calculation

The different absorption measurements carried out by Mécaplast quantified several tendencies, illustrated by the figures below. The presence of microfibres in a fibrous structure significantly improves the acoustic performance (Fig. 5.9). Compared with mechanical bonding, heat bonding improves the acoustic performance (Fig. 5.10). Finally, a structure composed of complex shape

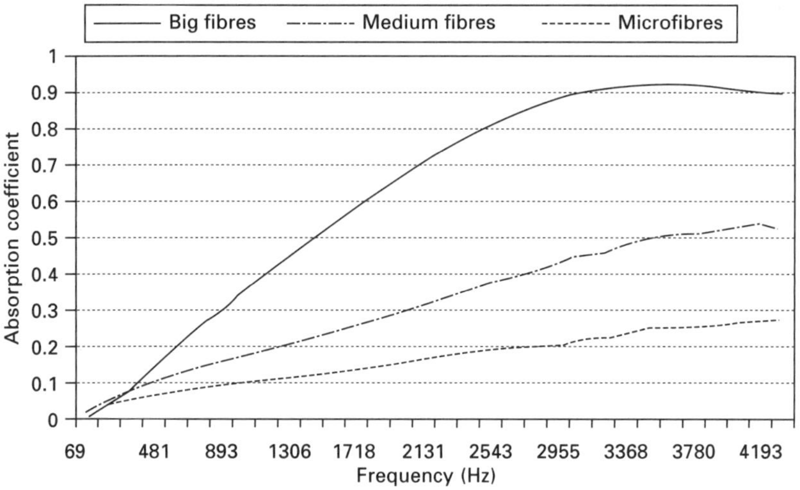

5.9 Fibre fineness.

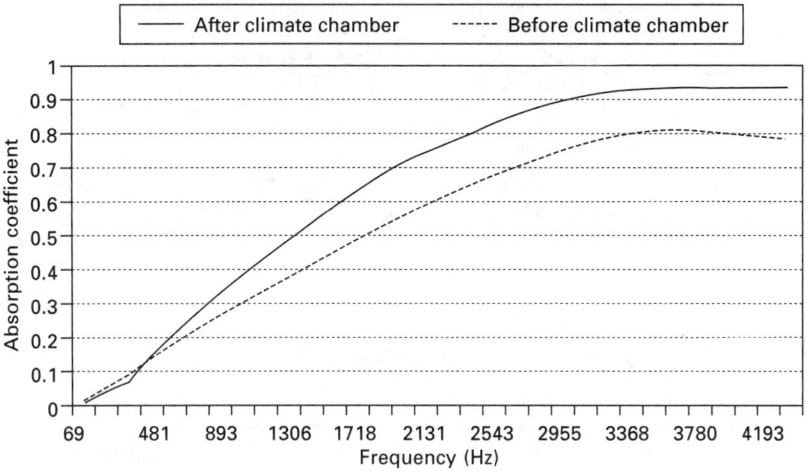

5.10 Mechanical/heat bonding.

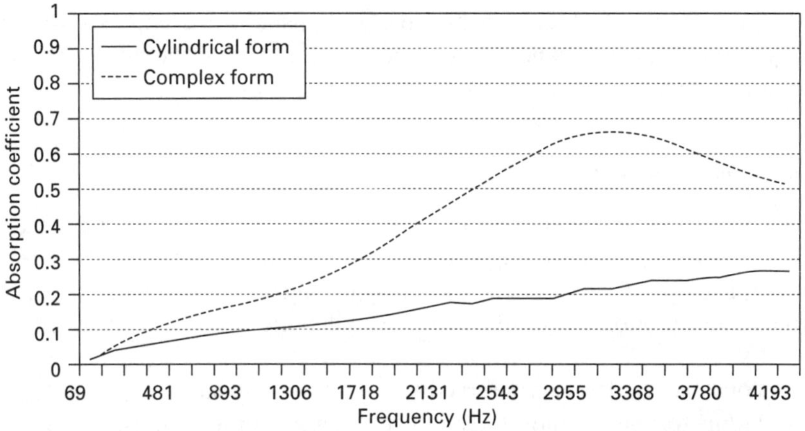

5.11 Fibre shape.

fibres absorbs more than a structure composed of cylindrical fibres (Fig. 5.11).

In the same way, measurements of intrinsic parameters and a prediction calculation were made to evaluate the correlation between measurements and calculation (Fig. 5.12). The five intrinsic parameters (air passage resistivity, tortuosity, porosity, thermal and viscous characteristic lengths) were measured on precision equipment. A correlation failure for a measurement of absorption coefficient is necessary to improve parameter values (in particular the thermal characteristic length that is obtained with respect to a multiple of the viscous characteristic length).

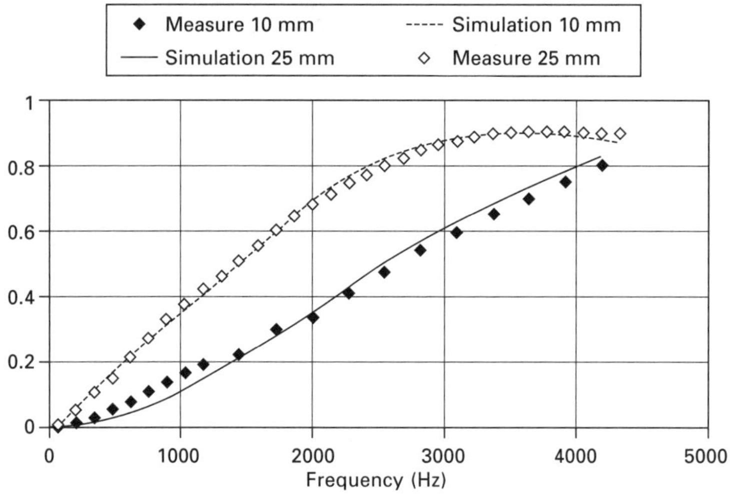

5.12 Measurement correlation.

The prediction calculation enables the grammage of a complex bi-layer (web + aspect) for a thickness to be optimised, avoiding the necessity for making prototypes and reducing the development time.

Acoustic measurements

Following this acoustic study, the acoustic absorption measurement was carried out by Mécaplast for the headliner application. A comparison between the complex optimised by this study and the technology usually used, a thermoset felt (cotton waste felt with 30% phenolic resin), is illustrated in Fig. 5.13.

No acoustic degradation is observed for a grammage reduction by 2 (passage of 1450 g/m^2 for the thermoset version at 750 g/m^2 for Mécaplast solution).

Mechanical, thermal and chemical characteristics

Laboratory validations were carried out for the headliner and floor carpet applications according to the car manufacturers' specifications.

Bi-layer and tri-layer complexes were designed for the headliner application. The most important characteristics for this application are engine fluid behaviour (water, antifreeze, diesel fuel, cooling liquid, oil, essence, etc.), the combustibility test and the peeling test of the fibrous structure. The engine fluid behaviour is confirmed by its appearance. This appearance is treated oilphobic and hydrophobic. More polyester fibres, principal components of the fibrous structure, have a reduced absorption. A fireproof product is

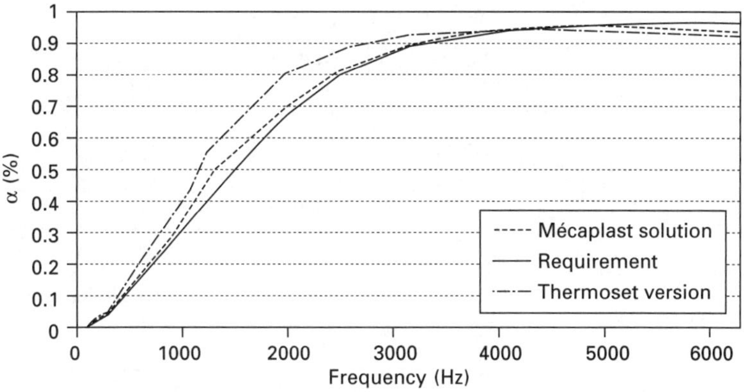

Mécaplast solution: nonwoven aspect 60 g/m^2 + fibrous web 600 g/m^2 + nonwoven back 90 g/m^2
Thermoset version: nonwoven aspect 60 g/m^2 + thermoset felt 1300 g/m^2 + nonwoven back 90 g/m^2
Requirement: car manufacturer specifications

5.13 Acoustic comparison between thermoset technology and Mécaplast solution.

obtained when a fibrous structure composed of polyester fibres contains Panox fibres in its make-up. The heat bonding of the fibrous structure is an important parameter because this bonding enables a good cohesion to be obtained and addresses the requirements of the peeling test.

For the floor carpet application, composed of aspect, septum and spring, a specific study on the spring (injected foam or felt) has been carried out. The spring deformation, when a pressure is applied, is a very important factor. In particular, the floor carpet must not be crushed when passengers get into the vehicle. For that two tests exist: the lift test and the residual compression deformation test.

The lift test consists of applying a pressure to the spring to compress it at 25% and 50% compared with the initial thickness. The samples are compressed between two plates (Fig. 5.14) fixed on traction equipment. To avoid crushing of the floor carpet, it is necessary that the value of the lift test increases. The springs that have a high lift test are the fibrous structures developed in this programme. The higher the grammage of the fibrous structure, the better is the lift test (Fig. 5.15).

The residual compression deformation enables the springs to be characterised in static compression. This test consists of placing a weight of 5 kg on the samples for 22 h at 23°C or 70°C (Fig. 5.16). The weight is removed and the residual compression deformation (thickness loss percentage) of samples is measured after 15 min of relaxation and after 7 days of relaxation.

To avoid having crushed floor carpet, the residual compression deformation must be as small as possible. The fibrous structures developed in this programme

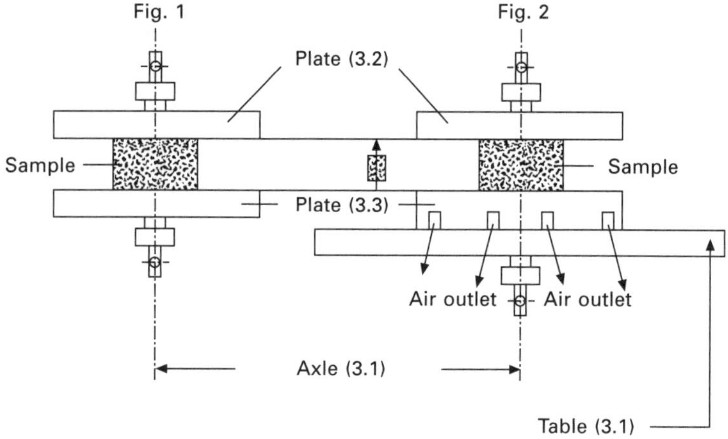

5.14 Measurement method lift test at 25% and 50%.

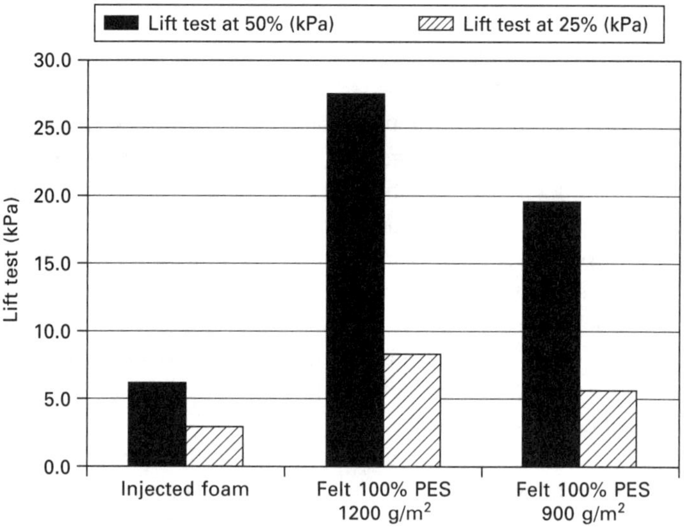

5.15 Results of lift test at 25% and 50%.

(grammage of 900 at 1200 g/m^2) have a residual compression deformation better than that of injected foam but worse than that of a thermoset felt of grammage 1300 g/m^2 (Fig. 5.17). The higher the fibre fineness the worse the deformation.

Passenger vehicle comfort is very important for the car manufacturers, particularly with regard to odour. Indeed, a norm exists. Samples are put in containers with water, then they are placed in a climatic chamber at 80°C for 2 h. When the containers are opened, the odour emission is classified (numbered

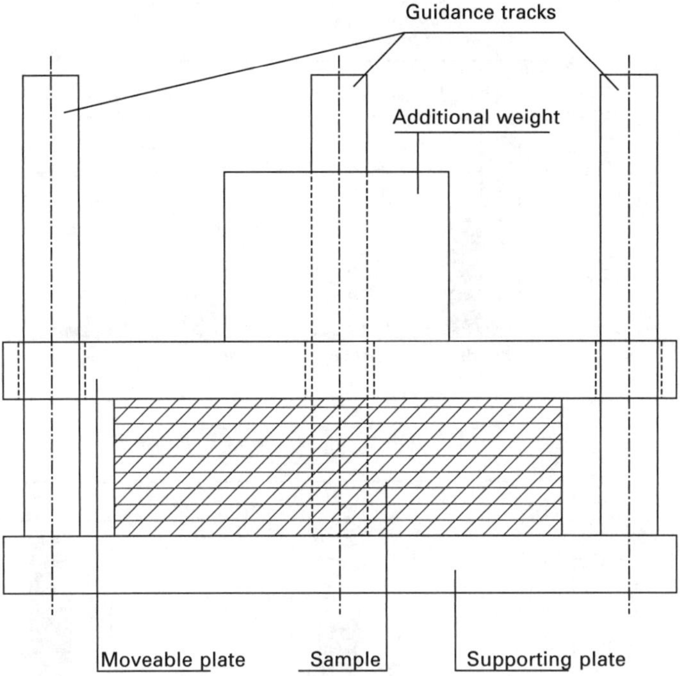

5.16 Measurement method.

from 1 to 7). The higher the number, the more unpleasant or even intolerable the odour emitted (1 for odourless, 7 for untenable).

The technologies in current use for the floor carpet application are injected foam PUR (polyurethane resin) and thermoset felt (cotton waste with a percentage of phenolic resin). The thermoset felt products emit a very disagreeable odour, caused by the presence of phenolic resin. The PUR foam products emit an odour that is strong but bearable. The products developed in this study (felt 100% PES) are almost odourless.

5.5.3 Headliner: recycling and use of recycled and shredded fibres

The acoustic study presented above was carried out solely using virgin polyester fibres. For the headliner application, the use of virgin fibres increases the cost of the part, making our offering uncompetitive. So to reduce costs, a search for suppliers of cheaper fibres was performed. This search allowed us to take delivery of a cheaper fibre set in order to create some prototypes and to test them according to the automotive manufacturers' specifications. The fibres tested were mainly recycled and shredded fibres.

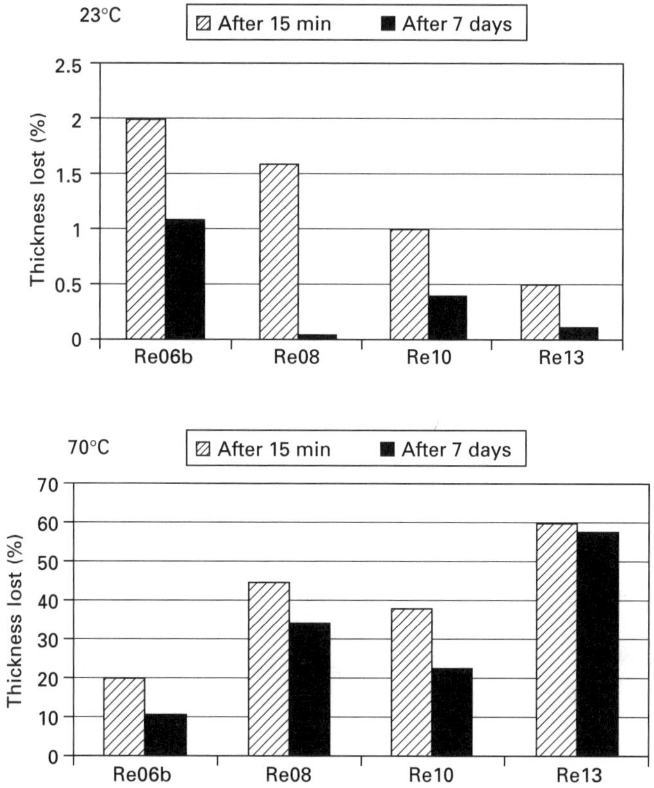

Re06b: thermoset felt 1300 g/m^2
Re08: felt 100% PES 1200 g/m^2
Re10: felt 100% PES 900 g/m^2
Re13: injected foam 50 kg/m^2

5.17 Residual compression deformation at 23°C and 70°C.

Determination of fibre supply sources

A search of polyester fibre suppliers was performed for various fibres:

- Virgin: all densities are available, whatever the quantities. These fibres are mainly used for technical products and for exteriors.
- Recycled: these fibres are created mainly from PET bottles. The density varies between 6.7 dtex and 5 dtex, whatever the quantities. These fibres are mainly used for undercoats.
- Of transition: these fibres are created from coloured transition bales (fibres manufactured between two stages of production of two colours). The density varies between 6.7 dtex and 17 dtex, whatever the quantities. The downside is that sales of these are 'spot' sales, according to the stocks available. These fibres are mainly used for undercoats.
- Substandard: these fibres are fibre blends of different densities and lengths.

The downside is that sales of these are 'spot' sales, according to the stocks available. These fibres are mainly used for undercoats.

- Shredded: these fibres are created from the shredding of polyester wastes. The density is around 6.7 dtex and the length varies. These fibres are mainly used for undercoats.

Acoustic performance

Some measurements of absorption coefficients have been carried out by Mécaplast (Fig. 5.18). In the majority, the presence of high quantities of shredded fibres in the structure contributes towards an improved acoustic performance of the product.

Recycling of automotive products at the end of vehicle life

Some recycling trials for headliner applications have been carried out in-house on some 100% PES fibrous webs composed of thermobonding fibres and virgin fibres. The treatment envisaged for this complex is divided into four stages: washing (automotive products in the end of vehicle life cleaning, presence of chemical agents in the under-bonnet application, for example), crushing, densification (micronisation, manufacture of compounds, etc.) and drying.

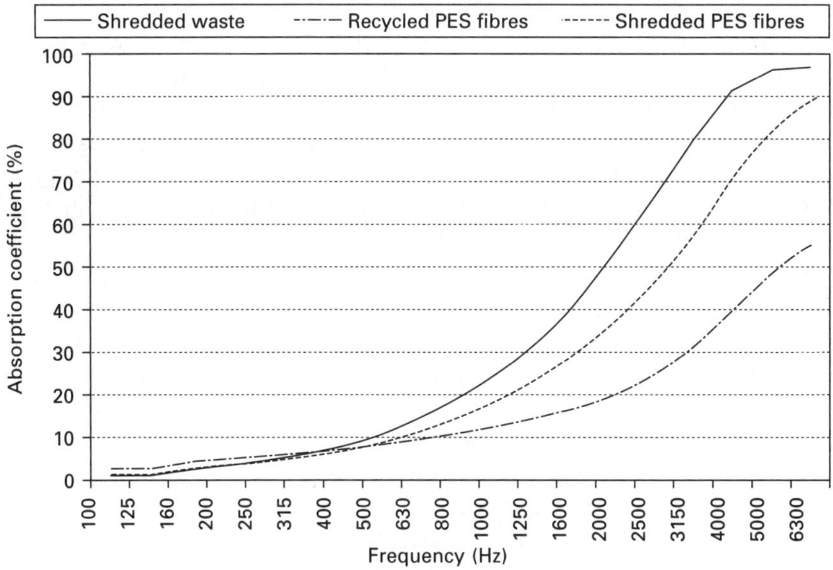

5.18 Acoustic measurements.

To show whether the under-bonnet products developed in this study can be recycled by shredding, some trials have been carried out with different prototype products dirtied with different liquids to discover the behaviour of chemical agents present under the bonnet. Some products have been dirtied so as to be similar to what their state would be at the end of vehicle life. These shredding trials have been conclusive; the odour of engine liquids persists.

The apparatus performing the shredding and cutting of the headliner operates as follows. The raw material is fed in by a conveyor, then pinched at the entry between the feed cylinder rotating at low speed and the static part (metallic static part taking on the exact shape of the feed cylinder) and shredded by an opener cylinder rotating at high speed. On exit from this machine, the raw material enters a cyclone to separate the fibrous material and the heavy pollutants such as wood or metallic materials.

5.6 Future trends

The two studies with Mécaplast were carried out as a result of the European Directive which increased the rate of reuse and recovery to 85% by average weight per vehicle and year by 2006, and to 95% by 2015, and increased the rate of reuse and recycling over the same period to at least 80% and 85% respectively by average weight per vehicle and year. Since the textile materials used in a typical mid-size car represent only 2% of the raw materials used to manufacture the car (from less than 21 kg in 2005), more and more car manufacturers are asking their suppliers, such as Mécaplast, to supply them with easily removable and easily recyclable textile products. They are also asking suppliers to take back their textile products and to recycle them.

Nowadays, automotive textile products are made from several raw materials, making them partially recyclable. The main objective of these studies was to produce some 100% recyclable automotive products having the same characteristics as the existing products currently on the automotive market.

Different methods (measurement, calculation and bibliography) have demonstrated and evaluated the influence of raw material parameters (fineness, shape, rugosity of fibres, etc.) on their acoustic and filtration performances.

For the acoustic structures, the lower the fibre density, and the more the shape and the fibre surface are complex, the better is the acoustic performance. Some headliners based on recycled or shredded fibres have been designed and validated according to the automotive manufacturers' requirements. The presence of a lot of shredded fibres in the structure notably improves the acoustic performance of the product.

The key advantages which manufacturers agree unquestionably that these synthetic media have are their increased capacity and their recyclability. However, they will also be studying the new requirements such as their fire behaviour, resistance to humidity, behaviour with soot types, etc. In these areas the synthetic media offer certain advantages (integration potential into the under-the-bonnet architectural constraints, cost gains in the air filtration function and so on).

As for recycling, which is the major advantage of these synthetic media and acoustic structures, the crushing and shredding tests carried out on industrial units allow us to conclude that they are industrially recyclable with such technologies. The recycling methods for products at the end of their life, and for production waste, are numerous, while the interest of the recycling professionals (recyclers or users of recycled materials) has been demonstrated.

Following the good results of the two studies, Mécaplast must manufacture synthetic engine air filters and recyclable acoustic insulation structures. This society must put forward these recyclable textile products to car manufacturers.

Owing to the European Directive on used vehicles, the results of these studies are duplicated for other products (textiles or not) used to manufacture automotives. This eco-friendly idea is being used more and more in the manufacture of cars.

5.7 References

1. Study with IFTH and Mécaplast financed by ADEME: To replace the classical engine air filter manufactured from several raw materials, which are difficult to separate and partially recyclable, with a filter composed of a thermoplastic (recyclable or not) mono material. The recycling of this mono material was studied for automotive applications or for other fields.
2. Study with IFTH and Mécaplast financed by ADEME: To replace classical acoustic insulation products manufactured from several raw materials, which are difficult to separate and partially recyclable, by textiles structures (nonwovens) composed of a thermoplastic (recyclable or not) mono material. The recycling of this mono material was studied for automotive applications or for other fields.
3. European Parliament Directive 75/442/EEC.
4. Study by the Saxony Textile Research Institute in Germany: 'Verband der Automobilindustrie (VDA) Jahresbericht 2005'.
5. Market study: 'Marktstudie Technische Textilien 2003, DRA'.
6. Market study of David Rigby: The technical textiles 2005.

Part II

Automotive interiors

6

Design of automotive interior textiles

N B POWELL, North Carolina State University, USA

Abstract: The challenges for designers across the automotive supply chain for textile products include the development of textile characteristics within high performance and low cost parameters. Consumer lifestyles and fashion trends also influence the development of distinctive brand offerings.

This chapter describes the textile design process in automotive and identifies the fabric characteristics for interiors application and the changes in the market factors influencing the need for innovative products. The range of textile characteristics developed within performance and cost parameters considered are fabric constructions, pattern, color, hand, luster, and scale. The demand for rapid prototyping, cost effectiveness and streamlining of supply systems is also identified.

Key words: design management, interior textiles, automotive trim selection process, color and textile design.

6.1 Introduction

Transportation textiles are considered high performance technical fabrics but must also meet the esthetic demands of changing markets. Designers are often the first to envision innovation and initiate new concepts in vehicle exteriors and interiors. Textile designers and the design services they provide to OEM stylists and designers are a critical first step in the introduction of new products for vehicle interiors, and potentially for composite parts. Textiles will be utilized to solve existing and future challenges in building innovative means of transportation. In this chapter, the challenges for designers across the automotive supply chain for textile products, where art meets engineering through the design process, will be considered. The range of basic textile characteristics developed within high performance and low cost parameters to be considered are fiber characteristics, fabric constructions, pattern, color, hand, luster, and recyclability. Furthermore, the influences of consumer lifestyles and fashion trends continue to grow in the differentiation of brands in the wide choice of vehicle offerings.

The Journal of the Textile Institute has kindly granted the author permission to use excerpts from the article Powell, N.B. 'Design management for performance and style in automotive interior textiles', *Journal of Textile Institute*, Vol. 97, No. 1, pp. 25–37 in this chapter.

6.2 Current automotive environment

By 2007 there were an estimated 806 million cars and light trucks on the world's roads with 244 million of those operating in the US. The comfort and safety of drivers and passengers have been greatly enhanced by functional textile products. As the automotive industry developed and vehicle exterior styling created various body shapes and sizes, highly engineered fabrics for performance, safety, and appearance were required. Personal transportation has been influential in the development and integration of products and services including entertainment, fashion, communications, and food services.

6.2.1 Global consolidation

In the changing automotive environment, there has been a drastic consolidation of global manufacturers. Fewer large OEMs are controlling the major volumes of automotive production, and the major players continue their expansion into existing and new markets. Nieuwenhuis and Wells (2003) estimate market growth by 2020 to 25 million in North America, 13 million in Western Europe, and 27 million in Asia Pacific region with a world total forecast of 91 million (p. 35). As new drivers come onto the world's highways, more consumers will experience the emotion of purchasing, owning, and changing their car. As one global car style will not suffice, the multicultural markets will demand vehicles targeted toward their own regional preferences. Designers of these products require knowledge of local markets and their cultures even if the product development and manufacturing takes place elsewhere.

6.2.2 Cost pressures

More and more component engineering, development, and testing is being outsourced to suppliers. During economic crisis, the pressures to reduce costs fall on the suppliers at the same time as the demand for new differentiating products increases. Innovative products introduced at a lower volume level add to the complexity of the manufacturing mix. Internet sourcing also opens up opportunities for new suppliers to enter the global markets. These OEMs publish their component requirements on the Internet and suppliers across the world bid for the business placement (Trott, 2002). This competitive action further increases the pressure on the global suppliers to reduce costs and increases the consideration of textiles as commodity products (WAW's Supplier Survey, 2003). Although the target price is set by the OEM, frequently the business contract is awarded based on the lowest quoted price. Price negotiations are based on current costs and business conditions including the perceived value of the product. Over the development and product launch periods, reengineering or design adjustments may prompt heavy investments in the refinement of the product (Powell, 2006). Once committed to the product, the OEM avoids the difficulties and costs involved in replacing an adopted textile, particularly in bodycloth (upholstery and door panel fabrics).

6.2.3 Speed to market

A company's ability to design and prototype rapidly provides a competitive edge in responsiveness to changes in the market demands. The evolution of digital systems such as Computer Aided Design and Computer Aided Engineering, virtual simulations, electronic supply chain and customer systems, and electronic procurement will be instrumental in expediting this flexible response (Nieuwenhuis and Wells, Table 15.1 by Wells, 2003, p. 199). Reducing time to market without compromising quality can be accomplished through the development of technology 'platforms' which can be common across a collection of products with shared components (Bruce and Cooper, 2000). However, this does not mean that the life cycle of a product can be extended indefinitely if it does not remain viable.

6.2.4 Quality

Consumers demand vehicles and the textile products used in them to be durable and attractive over the period of ownership and to maintain resale value. New products should not be submitted for a production vehicle unless they meet the OEM's standards. Meeting those qualifications on every sample and every shipment is a given if a supplier expects to continue as a meaningful source. The industry requires QS 9000 (Anon., 1996) or TS 16949 certification including Design Control guidelines and documentation of procedures to consistently meet industry standards (Powell, 2006). The New Technical Specification ISO/TS 16949:2002 was initiated by an International Automotive Task Force in conjunction with the International Organization for Standardization (ISO) to coordinate the various OEM requirements into a universal set of management standards. This specification aligns existing American (QS 9000), German (VDA6.1), French (EAQF), Japanese (ISO 9001/2) and Italian (AVSQ) automotive quality systems standards within the global automotive industry. It does not replace existing quality system requirements, but has been accepted as an equivalent, along with customer specific requirements. It has become a mandatory set of requirements for most major OEMs (www.qmi.com/information_center/standards/iso16949).

6.2.5 Changes in lifestyles

As the global availability of affordable personal vehicles grows for various types of travel and functions, so will the expectation for interiors. In the current market, consumer preferences reflect the changes in demographics and lifestyles as diverse as individual personalities. No longer can marketers forecast the majority's choice and build according to generalizations (McKenna, 1988).

Beyond simple commuting, the modern mobile society attempts to socialize, eat, drink, conduct business, and entertain passengers in their cars. A contemporary lifestyle is an attitude, beyond the socioeconomic, demographic statistics including age, income, and family size. Although similar products may be sought across traditional segments of consumers, a single product may not suit all customers or even a large group of customers any more. Products must be differentiated to appeal to specific consumer desires.

Increasingly the consumer has become more discriminating through access to information. More and more new car buyers use the Internet at some stage of the purchase process (Flynn *et al.*, 2002). OEM internet sites will allow you to create your preferred vehicle by selecting standard models and reviewing available colors, options and accessories. This may be considered a first step towards customization within a limited array of factory options. After purchasing, individuals have been influencing the marketing and sales of vehicles by word of mouth, quickly and broadly through internet blogs and independent webpage communications. Understanding how to reach these consumers as they make decisions based on the emotional aspects of purchasing and customizing will be a growth opportunity in targeting niche markets (Cagan and Vogel, 2002).

OEM designers and marketers have responded to these consumer behaviors by shaping the message and the products to appeal to their interests. Toyota recognized the importance of personalization to a new group of independent thinking buyers by offering value and unique features through a marketing strategy. By addressing directly young customers through internet videos and sponsored musical events, the 'mono spec' Scion offers a basic vehicle at an affordable price with the potential to customize with an offering of 40 factory accessories. The standard vehicle becomes very attractive for dealers selling accessories worth on average \$1000–\$3000 per vehicle (www.scion.com, *Toyota Scion Review*, 2003). The value of factory delivered options has been carried over to other vehicles giving the consumer the choice to adapt their vehicle with authentic parts. The 2007 Toyota Tundra was introduced with over 60 factory accessories.

As our diverse societies become more global, manufacturers must respond to the challenges that the desire for individualization and customization bring. A strategy based on volume will not support a continuing competitive advantage.

6.2.6 Brand

An understanding of the current position and any potential repositioning of the brand in the market is critical in developing the contributing components.

Designers are being challenged by the expectation that every aspect of the vehicle should reflect brand character or personality.

6.2.7 Sustainable products

As consumer ecological awareness grows, global consumers will have conflicting perspectives on the transportation industry and its role in the environmental issues from vehicle design and manufacture through use and the end of life cycle. Government legislation, environmentally conscious consumers and their advocates may bring the industry to task in the future on corporate responsibilities in green issues.

Although environmentally friendly products have been explored in the North American market, it has not been seen as a value for which OEMs or consumers have been willing to pay a premium (Anon., 1997). However, there are indications that the OEMs are recognizing there is potential in considering environmental stewardship as a market strategy, particularly as the price of petroleum rises. The use of textile composites, especially from natural fibers, will have a major role in developing sustainable products for all types of vehicles in the future (Powell, 2006).

As automobiles are a major contributor to greenhouse gases and pollution, there is an essential role that OEMs and their suppliers have to play in affecting important change in the global environment. Textiles can provide safety and comfort with renewable materials and composites which can lighten the weight of the vehicle, thus improving fuel efficiency and reducing the carbon footprint.

6.3 Automotive textiles market

Four main types of technical textiles dominate the automotive industry: airbags, preimpregnated fibrous materials for composite molding, tire cord, and interior trim. According to the *Journal of the Textile Committee*, about two-thirds of automotive textiles are categorized as interior trim, including headliners, door panel fabrics, carpet and seat covers and approximately 45 square meters of textiles are used in each vehicle (Anon., 2004a). In the interior of the vehicle alone, textiles are used in the headliner, instrument panel, floor, pillars, package shelf, and seats (Rodgers and Powell, 2006). The development of airbag and seatbelt textiles which protect all occupants of the vehicle cabin continues to provide added consumer confidence.

6.3.1 Design's role

Design contributes to the business success in designing new products, in improving existing products both in their manufacturability and cost effectiveness, in marketing the value, and in advancing research for future innovations in products and processes. Design is not only a differentiator of products and services, but has become a part of a company's intellectual property, a value protected through copyright and in some cases, patents. Establishing and maintaining design standards reflected in products, services, communications, and other activities of the organization is part of design's role in developing the corporate image and brand personality. Strong design management can build the organization's design leadership position in the market.

6.3.2 Design strategy

A corporate strategy for innovation must include design responsibilities aligned with priorities in manufacturing capabilities, product portfolios, marketing, customers, and management goals. As active members of cross-functional teams within the organization and interacting with the market and the supply chain, design is an initiating force for quality in style and function. Working with marketing, the design team can play a key role in developing the right products for the targeted customers to achieve the objectives of the company including attention to the life cycle of products. Without design participation in the early stages and throughout the development process, expensive errors in decision making can affect responses to the customer and productivity of the company. A single poor choice by design can create complexity in the processing of textiles, quality issues, cost issues, and lower productivity. Although design may be the first evaluation point with the customer, the stringent standards for quality and cost parameters are rarely compromised for the sake of design.

6.3.3 Understanding the customer

Although the textile supplier may be separated from the OEM by the first, second and sometimes third tier supplier, fabrics may be developed and purchased directly by the OEM (Powell, 2003). Textile suppliers may be required to ship their goods to laminators, cut and sew companies, seat manufacturers or directly to the OEM. The textile manufacturer remains responsible for any fabric issues and technical services to the customer (Rodgers and Powell, 2006). OEMs prefer to maintain control over selection, pricing and delivery of textile products to support the vision of the branded vehicle. Although some programs have shifted responsibility to the Tier One supplier as an integrator, this has not been adopted as an industry practice (Powell, 2006).

6.3.4 Approaches to different customers

The OEM's color and trim design team are responsible for developing the colors and trim materials such as bodycloth, headliner, carpets, vinyls, leathers, paints, plastics, woods and finishes for vehicles. Color and trim teams communicate to suppliers new material objectives to maintain brand image and to match the quality and design integrity of the vehicle to the needs of the market. The criteria for the vehicle development recognize three distinct stakeholders in the product: design, purchasing and engineering.

In the launch of new vehicle development or freshening of an existing model, the OEM color and trim team announces the opening some two to four years in advance of the model year. During the ongoing relationship with the customer, the textile supplier will prepare for build out or changeover of production vehicles whether the business is their own material or supplied by a competitor. The design team may anticipate new directions through continuous work with the production or concept vehicle teams and with the OEM's advanced studios. Global OEMs have established advanced studios in areas such as northern Italy and California, some distance from their manufacturing plants. These centers for advanced development serve not only in the development of concept cars but as a leader for trends and forecast for consumer preferences for production vehicles.

A preferred technology or fabric construction may be specified in requests for material submissions. The engineering standards and price parameters are established as critical targets of the total specification. Specifications on lamination or additional requirements may be communicated (Sirvio, 2003). Engineering and purchasing have priorities such as performance based on testing standards specific to the OEM and price targets which will be part of the total design brief for suppliers and eventually the sales contract.

Color standards, type of samples, and the format for submissions within a strict timeline are set by the OEM. In order to be awarded the business, the supplier must provide an exceptional performance characteristic or other added value in the product offering. It is common for several suppliers competing for the placement of their fabrics in the vehicle to develop hundreds of samples for submission for each opening. It is also common in the US to have as many as 30 to 40 different openings per year with at least three trim levels (Crabtree, 2005). European vehicles will have as many as seven trim levels in an interior.

The OEM color and trim team communicates information about the new opening through 'reverse presentations' to potential suppliers. As potential suppliers of interior components and seats, Tier Ones are included in the new vehicle launch presentations to trim suppliers. The seat design and seat manufacturer may or may not be chosen or disclosed to the fabric supplier at this point. The seat development may include various manufacturing processes which will dictate specific fabric characteristics, like increased stretch or abrasion qualities for fabric application on certain seat designs. Fabric development may require further engineering based on whether a seat is to be trimmed with a cut and sewn cover or molded. The seat manufacturer will know the expected levels of trim – base, mid level or luxury, but it may be very late in the seat development process when the fabric is finalized. Any innovative textile product may affect the seat design or change established trim processes which may not be aligned with the quality and cost objectives of the Tier One.

6.3.5 Effect of customer priorities on design strategy

Costs

Limitations on cost may affect the choices made by designers on technology, yarn type, fabric construction, color, and finish. The textile designers' challenge is to achieve the original design intent of the OEM's objective within the cost and performance parameters.

Performance and quality in technical textiles

Durability and maintaining an attractive appearance are primary in a consumer's perception of the quality of an interior. In order to become most familiar with the standards and processes of their customer, the textile design and development team may be assigned exclusively to a few or even one OEM and/or Tier One. It is unusual for engineering to compromise on the level of a testing standard to meet design's color or patterning preferences. A color may be developed that is a fashionable color, but the color will not be considered if it does not meet lightfastness standards.

Performance and quality are integrally linked. An expectation of quality is a given, and if it is not met, the customer will quickly have many other options and a choice of competitive suppliers. Consideration of quality standards while under time and cost pressures requires careful attention. Quality is the priority and must not be compromised when adjusting fabric content for other criteria.

Brand esthetics

Craftsmanship in the trimming of the interior and the combination of color, texture, hand, and luster of the materials create the atmosphere of the vehicle cabin and contribute to the overall experience of the brand. The textile designer must be aware of the vehicle brand image to create fabrics that provide strong visual and visceral cues supporting the character of the brand.

Studying the previous models of the particular brand and related accessories and merchandise, advertising and promotional materials can begin to establish a profile of the customer's aspirations. Observing the consumers' use of the vehicle and discussing the perception with the consumer in surveys and focus groups can provide a more comprehensive view of the brand's impact in the market. Brands like the VW Beetle and the BMW Mini have revitalized their brands in a retro/modern way to attract new consumers.

6.4 Establishing the target consumer

6.4.1 Lifestyle

If a textile design and development team does not clearly understand who the targeted consumer is and *will be* by the time the model arrives on the market, the efforts can be misguided. Consumers today are active, enjoy traveling, are concerned about their health, and wealthier than their parents may have been. Beyond traditional socio-economic gender, age, income, education, location and marital status, the designer must consider the psychological and emotional factors influencing buyers' decisions. The values, interests and psychological needs of consumers do not always match up to traditional segmentations.

As global populations acquire wealth or address changes in their lifestyles, product preferences may alter. Designers must stay attuned to the preferences of consumers who can have an impact on the market. Recognition of this fast-paced flow of lead users' preferences and the potential following of larger numbers of consumers can provide valuable input to automotive designers. Attention should also be paid to influential consumer products that are representative of trends in consumer lifestyles. The growth of personal consumer electronics has prepared the consumer to expect their automobiles to support and deliver similar services.

6.4.2 Fabric and fashion

Fashion continues to influence the fabric appeal in interiors of cars and trucks. Automotive companies leverage the recognition of fashion products to influence the style and promote the image of a vehicle. Fabric designs in US automotive markets have traditionally been more akin to menswear than home furnishings patterns in both scale and color. Even though the car seat and interior may be considered as personal an environment as the home, the smaller enclosed space kept interior fabrics plain, simple stripes or checks, and small scale geometrics in neutral colors. In Europe and Asia, even though the vehicles are generally smaller and diverse model production volumes lower, the opportunities for special editions, multiple trim levels and unique multicolored patterns are captured by textile suppliers (Powell, 2006).

6.4.3 Consumer experience

Beyond basic functions of transportation, consumers desire vehicles that provide entertainment and pleasurable experiences. If the aspirations of the consumer are to acquire a higher standard of living, to reflect personality or status to the public, or to escape the stress of everyday life, the automobile can be seen as a means to achieving these goals, even if only in the mind of the consumer during the experience. The interior of the vehicle and the materials can create an ambiance that is essential in this perception.

6.5 Trim selection process

6.5.1 Target vehicles

If supplier corporate priorities are communicated to product development and design, resources can be focused on vehicle openings appropriate to achieving business objectives. Within the established strategy for business growth; the sales, marketing, and development teams can create a concise plan for pursuing the best opportunities for winning the vehicle placements. A valuable component of this plan is a thorough review of all customer input on expectations for the targeted vehicles.

6.5.2 Market research

In benchmarking of competitive vehicles, only current production vehicle information and products may be available, but this will provide an important background for future direction of the vehicles. A study of competitive concept vehicles and the materials employed may lead to insights on future directions in the market place.

6.5.3 Fabric history

Another important source of information is the ongoing tracking of the actual fabrics used in previous models of the targeted vehicle. The evolution of a vehicle and its customer preferences for certain colors and types of fabrics can assist in shaping fabric submissions supportive of the brand image and performance parameters. Collaboration with OEM designers and market research may provide input on a change in direction for future vehicles. The fabric designer can also bring new influence from leading edge markets such as fashion, sports, and other consumer products. Trade events such as the SEMA (Special Equipment Marketing Association) Show and sources like Material ConneXion® can stimulate new ideas and innovative concepts which are very valuable to designers.

Product development teams may compile various research and detailed information about products and competitors. The cross-functional nature of the team demands an orderly system of reviewing information as a means of making effective decisions and gaining consensus. These systematic tools can be utilized to reflect the voice of the customer or customer requirements organized on a matrix, evaluating the perception of the customer in comparing competitive products. Charts and matrices, conceptual maps, conjoint analyses and other tools can be used effectively to chart the development path forward. The design and development team may create helpful visual charts to map the competitive vehicles and the fabrics, for example a conceptual map comparing characteristics of sporty perception to luxury relative to a cross axis of cost.

6.5.4 Design brief

Once this input has been evaluated and a strategy has been developed, the designer will create a design brief including priorities that must not be compromised (performance and cost) and esthetic variables based on customer input. Color, patterning, scale of pattern, luster, and hand should be indicated along with testing requirements and targeted cost parameters. Based on this brief, the development team will generate samples and finalize a collection to be submitted for the targeted vehicle opening.

6.6 Impact of seat design requirements

Interiors and specifically seats are considered highly complex components and have a significant impact on the comfort and safety of the users (Cagan and Vogel, 2002). Seat systems typically represent approximately 40% of the cost of the total automotive interior (Ward's, 2003) and the trim cover represents about 20% of the seat cost (Rodgers and Powell, 2006). As the complexity of the design of seats and interior compartments has grown, these components are outsourced by the OEMs to Tier One suppliers such as Faurecia and Johnson Control who have specific targets for design, performance, costs, and cycle time.

Within the interior, seat coverings have the highest potential for failure in material durability. The seat system comprises three basic components: a metal frame or support structure, a foam cushion and a covering material. Many manufacturers find that a three-layer composite fabric is most successful in performance, costing, and in the process of upholstering or trimming the seat. The three layers are composed of a top layer which is polyester; a middle layer of polyurethane foam; and the bottom layer, a polyamide warp knit scrim (Ishtalque *et al.*, 2000; Fung and Hardcastle, 2001). Many seat manufacturers have eliminated the textile scrim, substituting cost-effective polyethylene film which minimizes fabric stresses during production cutting and sewing of the cover. Lamination techniques are used to form various composite types and thicknesses depending on the substrates and the required component part's performance criteria. The stiffness or drape of the fabric can affect the ability to properly upholster the seat. This trimmability or ease of application of the fabric to the seat can be a major factor in the acceptance by others in the supply chain. Not only the appearance of the final seat but also the cost of the seat set can be affected by a fabric that needs special requirements in handling in the seat trimming process (Fung and Hardcastle, 2001).

Although many OEMs and suppliers use CAD systems to map and drape a simulation of a fabric on a virtual seat, the final decision is usually based on actual seat designs and trim applications. Tier One manufacturers are given the opportunity to review the candidates for typical manufacturing properties such as cutting ability, sewing ability, ease of assembly, and final appearance before releasing full approval for production.

Within an estimated 42-month vehicle development period, the evaluation, selection, and validation of a seat fabric (bodycloth) occurs early in the process, and the production intent seat system evaluation happens near the end of the development cycle. The selection and approval of automotive bodycloth can require from 10 to 18 months while complete seat system testing on a production seat may require 30 months. The final seat design and sewing pattern and associated fabric requirements may not be finalized or revealed to the textile designer during the development of the samples. The OEMs have full responsibility for selecting the interior materials and typically select numerous fabrics and combinations for seat evaluation to prove functional performance (Dickey, 2007; Rodgers and Powell, 2006).

6.7 New product development process for automotive trim

Each supplier develops a range of designs for each vehicle trim opening or 'carry over' production refinements either as computer aided design simulations or fabric samples in usually one or two customer standard colors. Samples are reviewed by the color and trim team, and a group of fabric styles is selected to go forward to the next review or decision point by the OEM (Powell, 2006).

The initiation of samples is based on ideation of a theme or design brief based on the customer original product launch input. To create desired fabric features to interpret these concepts a designer considers hand, pattern, scale, luster, texture, color, performance, and price of the fabric (Powell, 2003). A proactive strategy is to include advanced design concepts and inspirational samples as a design service to the customer. The textile engineer responds to a more defined brief interpreting the 'qualitative' characteristics into technical, quantifiable requirements (Vincenti, 1990). Delivering numerous design concepts, performance test data, theoretical costs and engineering yardage can be very costly to the fabric manufacturer, and it is important to rely on the cross-functional perspectives of the entire team.

The textile designer initiates pattern development on the computer aided design (CAD) system and selects colors to be dyed in yarn or fabric to match the customer's plastic standard. These simulations and sample blankets are designed as the engineer may be facilitating the yarn and sample production requests, running finishing trials through the plant, and expediting the samples' progress through the mill and testing labs. Samples are evaluated to meet specific customer standards, and the results are archived as a reference for future designs. The textile engineer plays a critical role in the selection of yarns, constructions, back coatings, finishes or lamination required to meet specific customer certification and manufacturing processes. The engineer must prepare the required construction test packet and certification information supporting the final fabric submission to the OEM's engineering department. According to the Chemical Fabrics and Film Standards (2005), a production part approval process (PPAP) packet on a new product would include a process flow, a process FMEA (failure modes effects analysis), a control plan, complete performance and dimensional test results, and for esthetic parts, an appearance approval report. Any change in raw materials or process would prompt a new submission of documentation and may also be requested as part of an annual inspection (www.chemicalfabricsandfilm.com/pdfs_researchSection/standards/ppap, 2005). The engineer may also prepare a theoretical cost for the sample constructions as the designer creates a merchandizing package or appropriate presentation of the sample fabrics to deliver the total offering to the OEM (Powell, 2003). OEMs prefer and often demand that samples be confined and developed exclusively for their projects. Maintaining the integrity of the design vision through this confidentiality is the foundation of outsourcing of components. Selected designs progress through several refinements as the time schedule will allow and are reviewed by the customer decision makers (Powell, 2004).

6.8 Sample development

6.8.1 Yarn type

Although polyester is the predominant fiber in automotive for interiors, other parts of the vehicle will use nylon, polypropylene, and even carbon fibers (Fung and Hardcastle, 2001). Wool, polypropylene and acrylics may be used in other modes of transportation such as aircraft, rail, and coach; but polyester remains the leader in world automotive markets because of its performance in durability, UV resistance, cleanability, and its relative low cost and availability. Continuous filament air textured polyester yarns' high abrasion resistance provides a consistent strength for weight advantage over nylon and polypropylene fibers (Fung and Hardcastle, 2001). Designers seek a variety of fiber lusters, textures, dyeability, and novelty yarn structures to differentiate fabrics and deliver added value in performance or look (Powell, 2006).

The fiber and fabric properties that yield the desirable esthetic characteristics are often conflicting with performance and cost requirements. Many applications require soft hand that requires lower DPF, denier/filament. For good abrasion and resiliency higher DPF is required. For better coverage you need larger yarns or more density, which means more weight and cost (Manley and Powell, 2004). Although other non-petroleum-based fibers may be of interest to the designer, none have met the critical performance standards at a competitive price.

6.8.2 Textile technology

The automotive industry focuses on just in time manufacturing techniques and expects its suppliers to respond quickly to any demand changes in volume, style, and color. Flexibility in manufacturing is key to responding to the fluctuations in the assembly line, and can be very costly to all in the supply chain if the choice of textile technology is not capable. Fabric suppliers must produce and deliver the exact yardage required in the specified technology and color within very short delivery times. As performance standards evolved, every major textile technology and many types of fibers have been engineered to meet testing requirements and consumer expectations (Powell, 2006).

From flat woven and velours to circular knits and microdenier nonwoven materials, automotive designers have explored the gamut of textile technologies. By 2002, flat woven fabrics held 47% of the European market versus 14% in the US and 12% in Asia. The US market preferred woven pile fabrics at 30%. Double needle bar knit pile fabrics also captured 23% of the US bodycloth market, and flat circular knitted fabrics held 21% of the European market. Tricot knits dominated the Asian market at 44% during this period. Flat wovens continue their high market share in Europe for interior fabrics with growing strength in US markets (Anand, 2003). By 2003 flat wovens continued to grow at approximately 55% of the world market versus warp knits at 25% and circular knits at 20% (Rigby, 2003, p. 10.8). Leather continues to be perceived as a luxury cover in the US but remains at approximately 20% of the market as delivered from the assembly plant (Anand, 2003). The introduction of sueded or sanded nonwoven products from Asian and US suppliers at various price points and weights complemented leather trims and responded to the desire for a non-pile textile product with a smooth, skin-like hand.

The analysis of the use of major textile technologies by product type is shown in Table 6.1.

6.8.3 Knit

Both warp and weft knit technologies are used to create various types of structures in automotive fabrics. Tricot, double needle bar, and circular knits from both flat and circular machines are employed for flat and pile fabrics. The stretch requirements for better trimmability combined with mounting cost pressures encourage the application of circular flat knits provided that they meet all performance characteristics. Knits are perceived by the automotive industry to deliver the desirable characteristics of low cost, low weight, and high performance (Fung and Hardcastle, 2001). The European automotive industry views weft knits favorably due to their design flexibility, stretch characteristics, appearance, and comfort. This design flexibility may provide the type of short run color and pattern complexity with minimal changeover time required by European markets (Anon., 2004b). Knitted fabrics' stretch characteristics provide ease of application especially in shaped parts like head rests and molded parts like headliners, and door panels (Rodgers and Powell, 2006).

6.8.4 Woven

Designing for flat and pile woven fabrics begins with yarn and color development for warp planning which will be the foundation for the subsequent pattern and filling yarn trials. Standard loom set ups and constructions, whether dobby or jacquard, are developed to meet the cost, color, and performance standards of the customer. Editing from computer aided design

Table 6.1 End use consumption analysis by product and fabric (vol., '000 tonnes)

End use product	Woven fabric	Knitted fabric	Nonwoven fabric	Total*
World totals, 2000				
Woven & knit trim	5.6	22.2		27.8
Nonwoven trim			109.7	111.6
Upholstery	53.2	53.8		106.9
World total forecast, 2010				
Woven & knit trim	4.2	16.9		21.2
Nonwoven trim			150.5	152.5
Upholstery	74.3	53.6		127.9

*includes other fabrics
Source: David Rigby Associates, Mobiltech: Detailed Forecast Tables, 2003.

simulations and sectional warping of sample blankets can expedite the efficient development of woven samples. While velours have been more popular in the North American market, a move to flat fabrics will depend on the acceptance of the hand and cleanability of the fabric by consumers and the amount of stretch provided for trimming seats.

6.8.5 Dyeing

Piece dyed fabrics have dominated the market with the convenience of quick color development from small sample dyeings, and the economies of large volume production. When more patterning was desired, piece dyeable fabrics could be woven or knit with a combination of yarns of different luster levels or yarns which dyed differently or resisted dyestuffs, such as deep dyeable polyester, bright nylon or accents of solution dyed polypropylene. Piece dyed pile fabrics used for inserts (the more decorative panel used down the center of the seat) can be patterned with cut out, omits or areas where there is no pile and the backing construction shows. Small scale repeating motifs, usually geometric or stripes controlled by the fabric constructions can be formed as dobby or jacquard structures. It may be better for color matching to other interior materials and more cost efficient to take the bolster or plain fabric and emboss, print, or etch the pattern through mechanical or chemical methods (Powell, 2006).

The interest in multicolored products has been influenced by fashion directions, jacquard home furnishing fabrics, and female consumers' preferences. The use of computer aided design systems to simulate possibilities accelerated the development of larger scale or random repeat patterns in kaleidoscope colors. Jacquard technologies provide pattern capabilities in geometric, abstract and organic designs, and novelty motifs. As most true pile fabrics are produced by weaving or knitting a double layer sandwich, the designer must balance the design to accommodate top and bottom fabric pieces with mirror image differences. This issue and other yarn wastes and total weight of these fabrics push the limits of most targeted cost brackets (Powell, 2006).

6.8.6 Color development

Organizations such as The Color Marketing Group® and the UK Color Group® bring together industry professionals whose main responsibilities are developing and marketing color trends in products as diverse as building materials, home textiles, and cosmetics. The automotive group within Color Marketing brings together OEM, Tier One, textile, and raw materials suppliers who also tap into other consumer product color forecasts as a resource.

Customer color standard

Color development is carefully controlled by the OEM and color standards, small plastic rectangular plates or chips with a smooth and a textured surface are provided to suppliers. Fabric master standards with a limited range of variation are set to which the sample submissions and final production carefully adhere.

All materials will be matched under specific lighting and reviewed in an area specified by the OEM such as a closed room with a light box and neutral grey walls and controlled lighting. A further visual check may be done by draping the fabric over a seat in daylight. Matching of multiple materials to one color standard and then coordinating finished components in the interior or on the individual seat may present problems, such as metamerism, where materials may match under certain lightning conditions but not under other lighting situations. Many color and trim studios will ask for fabric samples to be developed in a medium shade of their standard neutral color, and in the darkest value in order to evaluate the pattern under specific lighting conditions.

As dyestuff technology has moved forward to meet increasing performance standards, colors have moved from dark shades of blues and browns to lighter colors. To enhance the cockpit closeness of the smaller interior with increasing number of electronics, gages, and switches facing the seats, neutral colors are matched on plastics, fabrics, carpets, and other trims materials. A grey or beige interior could be standard for any exterior paint colors. Fewer stock keeping units or component parts streamlines the production process for assembly.

Dyed yarns

In the early 1990s, the market saw a drastic move to yarn dyed products. This opened up a palette of expressive patterns and colors for the color and trim designers. Yarn dyed products brought matching, performance, and development time issues to the suppliers. A few fabric suppliers invested in their own dye houses while others relied on sample equipment and the expertise of their suppliers to meet complexity and quick response challenges. Companies such as Unifi, Trevira (Neckelmann) and Autofil responded to the volume and performance expectations of the automotive industry with accompanying risks and rewards.

The approval of yarn colors from a yarn pom, skein or small knitted 'sock' of dyed yarn is a time-consuming process relying on the communications between the dye house, the designer, and the customer in interpreting adjustments in the color range. The art and science in refining a shade to match the customer's input is complicated by fabrics that contain multiple colors and yarn types.

Solution dyed polyester products have been considered, but realistically are most viable in large volume colors such as black. Specific color development on lower volume shades is still an issue as minimum quantities would be prohibitive as most yarn dyeing companies require 50 ton lots or merges. The partial solution to this problem is to develop the colors as package dyed yarns, and then transfer the high volume shades (usually the predominant background color) to solution dyed products before the final approval and product launch. Small makes may be available at a premium price, but package dyed is still the viable option for inserts and most bolster fabrics (Pfeiffer, 2007).

Accent colors

Textile designers may use the customer's standard neutral color – beige, gray, or black – as a background shade and use various colors to create patterns, highlight motifs, or enrich the complexity of the fabric color. These accent colors may be dictated by the exterior paint colors or by the overall theme of the interior color story. Care must be taken not to cast the overall shade away from the standard by the amount or intensity of the accent colors.

Space dyed yarn is a fancy effect yarn which was introduced in apparel and home furnishings. Intermittent lengths of as many as five colors in a single yarn brought a random multicolor effect to flat and pile fabrics. A designer creates a color combination in a single yarn which could include a range of accent colors such as the vehicle's exterior paint colors. This simplified the necessary yarn s.k.u's for the style construction and to be carried in yarn inventories. However, the inherent patterning in the package dyeing process created a repetitive banding or flaming across the fabric face. Also the amount of yarn waste to minimize banding causes an increase in the real cost of the yarn. This led to cost controlled limits in the use of the space dyed yarn and also the plying of this type of yarn with other solid filament or spun yarns (Powell, 2004).

Printing

Although the printing process is an effective mechanism to deliver color to fabric, the technology has had a limited success in automotive fabrics. Print bases can be knit, woven, or nonwoven constructions, but the piece dyed plain fabrics are preferred for their cost effectiveness. In the late 1980s printed automotive fabrics were reported to be about 30% in the Japanese market and companies such as Milliken and Guilford lead the development of print technology for the North American market (Milliken, 1997). Ford was one of the first US OEMs to evaluate print systems' effectiveness using

Guilford's print replication of a yarn dyed circular jacquard knit pile fabric but this was replaced with a conservative patterned piece dyed knit. Ford also placed a Milliken rotary screen printed fabric, a traditional geometric in its design and color, which ran in production. More prints were adopted by the market but never met the expectation of cutting development lead times and possible mass customization potential (Powell, 2006).

The development of pigments and other chemicals which can be screen printed allow the creation of patterns, textures and three-dimensional characteristics in the fabric surface but must maintain a pleasing hand and pattern integrity during actual wear. The latest development in printing has been the development of dyestuffs suitable for digital or inkjet printing of automotive fabrics requiring high standards for lightfastness. A Japanese manufacturer, Seiren, established Viscotec® as the first production level supplier of digitally printed automotive fabrics to the global markets (Powell, 2006).

Other surface processes related to printing would be etching or removing the fiber or yarn from the fabric in a pattern to give three-dimensional effects. Embossing is also a patterned pressing of the fabric used to create high and low areas in the fabric. Although these special effects are more visible on pile fabrics, they are used on different types of knitted, woven, and nonwoven fabrics.

6.8.7 Finish

Fabric finishing can enhance a plain or patterned fabric with improved surface effects, and is critical in those areas showing a wide expanse of fabric to the consumer at close range, such as headliner and bench seat upholstery. Specialized equipment and technical expertise in finishing different constructions of fabrics are a major investment for a textile company. Most manufacturers maintain this process within the assets of the company, but others may choose to outsource this process to other expertise, often aligned with a lamination line.

As the economic pressures and the dictates of fashion move away from heavier pile fabrics, more flat wovens and flat knits are seen in the market. OEM designers want clearly defined textures and patterns to complement the technical appearance of highly engineered vehicles. The visual complexity of a fabric must be balanced with the feel of the material. It is critical to avoid a hard and raspy hand in the fabric in achieving a crisp design appearance. Achieving an acceptable hand on a flat fabric relies primarily on finishing techniques such as napping or needling to soften the surface feel but at the risk of deteriorating the integrity of the design or the structure. There remains a high expectation of comfort and the durability which automotive fabrics must meet, especially in any tactile or visible wear or loss of color or pattern.

Leather interiors have been perceived in the market place as a luxury level option. Achieving the feel or hand of sueded natural leather is a technological challenge to meet the stringent automotive standards for lightfastness and crocking of fashion colors within narrowing price parameters. Toray's Ultrasuede™ brand is well established in the apparel market and a version of this nonwoven sueded product is trimmed across many vehicle platforms in Europe as ALCANTARA™ with various product weights for application in different parts of the interior (Hansen, 2005). Milliken's Preferred Suede™ and other competitive suede-like polyester products are available to be trimmed out in combination with leather or fabric in the interior.

Textiles are coordinated with leather, plastic, wood, and metal finishes, and are expected to complement the tones, luster, and visual cues of these non-fibrous elements. Within the trim materials, a combination of different types of fabric constructions may be used in vehicle interiors. Bolsters are lower cost plain fabrics usually used on side and back panels of seats and in door panels. When used in combination with more expensive materials, a more interesting seat may be created and the total cost of the seat covering may be lowered. For example, a yarn dyed jacquard woven may be used on the insert or central front panel of the seat and a piece dyed tricot knit used as a cost-effective bolster (Powell, 2004).

6.8.8 Fabric testing

Textiles are evaluated to assure quality, analyze any potential failure of the material, and lower the risk of product failure in the market place. Evaluation of materials can also be seen as a way to identify opportunities for product development and improvement (Rodgers and Powell, 2006). An interior is expected to maintain its appearance over the lifetime of the vehicle's use and to increase resale value. The median age of automobiles in operation in the US in 2000 was 8.3 years (Ward's, 2002) and vehicles like SUVs and other light trucks are expected to travel global roads for more years and many miles. The level of expectation for a highly functional and durable interior during the ownership of a vehicle demands extreme performance of the seat and its cover (Powell, 2006).

The textile producer conducts the initial basic physical performance testing of the automotive textile in accordance to the OEM engineering specification. The OEM materials engineer reviews an evaluation of the tested samples and data and may conduct repetitive or additional testing based on the initial results. Most fabrics are screened by width, thickness, weight, and construction (Bhagwat, 2005) and the physical tests are conducted in three directions (warp/fill/bias). Each OEM has a preference on physical performance tests (Rodgers and Powell, 2006) but most are from the primary list of over 30 individual tests certified by British (BS), American (ASTM) and German

(DIN) standard methods including evaluation for characteristics such as flammability, abrasion/durability, lightfastness, seam strength and dimensional stability (Fung and Hardcastle, 2001).

Beyond the standard tests, there may also be additional tests for specific technologies and the OEM may determine the method or format of fabric submission. Fabrics can be tested face only rather than as a laminate, or as a laminate. Physical tests for abrasion, pilling, and snagging can be the most critical for flat fabrics. Additionally, pile fabrics are particularly reviewed for pile crush, lint retention, pile whitening, and resulting seat application issues caused by ingress/egress of the seat occupant (Fahy, 2005).

6.8.9 Application testing

Beyond the basic material evaluation, each fabric application has a unique set of requirements that affect the overall performance of the final component. In the case of bodycloth, it is the final seat product that is evaluated by the consumer. Tier One seat engineers have developed further methods for testing the materials in its final application. Two particular tests evaluate the overall performance of the fabric within a seat system: jounce and squirm and ingress/egress.

Jounce and squirm testing is widely used to evaluate the foam cushion integrity and trim cover attachments. This test is designed to simulate an occupant sitting on a vehicle seat during use. During 'jouncing', the seat part (loaded by a body form representing an occupant's back (for seat back evaluations) or buttocks and thighs (for seat cushion) is attached to a table that can be moved up and down a specific distance at a set rate. With the body form loading the seat, it is rotated left and right around a pivot point ('squirm'). Numeric measurements of the seat contour and foam integrity are taken, typically at 100 000 cycles. If the fabric is visibly worn and/or the shape or integrity of the foam is reduced beyond a specified standard, further engineering of the components and testing is required (www.seattesting.com).

The ingress/egress test, also known as sliding entry, involves simulating an occupant's movements while entering (ingress) and exiting (egress) a vehicle. A metal or plastic buttocks form is attached to a multi-axis robot, which is programmed to simulate human movements based on pressure mapping of a statistical sampling of moving occupants in a designated vehicle (Lentz, 2005). The robot duplicates this movement for testing of each vehicle's unique path. Entry and exiting of a vehicle is dependent on the door opening, vehicle height, and seat design position (Rodgers and Powell, 2006). Although this test can be adjusted to a specific vehicle for load, cycle count or additional squirm features, this method is essentially the same for most OEM manufacturers. A team of engineers evaluates the automotive seating materials and the seats' performance throughout the required cycles. Successful completion of this critical evaluation releases the material into a production ready status. The common failures are pile distortion, pile loss, hole formation, severe creasing, and seam separation (Rodgers and Powell, 2006).

6.9 Communicating value and design services

Design service provides value to the customer in the interpretation of trends, consumer behavior, new technologies, new materials, and other influential factors on automotive materials. An OEM evaluates the supplier's design capabilities based on design expertise, sample quality, technology proficiency, market research and professional merchandizing of the products. The supplier's design group may also provide updates on exhibitions, fashion, media, and other events in a relevant design package to automotive designers.

6.9.1 Presentation

Critical opportunities to deliver the value story of the textile range may occur in a small or large venue with days' or months' notice. The design team is prepared to formally present to the customers including OEM color and trim designers, marketing managers, engineering and purchasing, and possibly the Tier One representatives. A well crafted script should introduce the product, describe the response to color and trim's launch direction, and explain refinements on subsequent preliminary submissions. The presentation should focus on design solutions to the initial launch goals, and provide optional perspectives on the vehicle trim. If an interpreter is to be used, it is important to rehearse timing, clarify any misunderstandings, and set the pace of the presentation. Immediate feedback from the audience may be a general indication but the editing and refinement of samples will take place working specifically with the design team.

The collection's concept must be fresh and stimulating while delivering function, performance, and style. The developed theme should be comprehensible at an obvious visual level and in the subtle development of color and fabric design complexity. The presentation should include an understanding of the brand personality, its current status in the market relative to its competition, and any new direction for the brand. Competitive fabrics and colors currently used in leading production vehicles should be benchmarked. The textile design team may have collaborated on advanced vehicle projects or concept show vehicles, and the final presentation is an opportunity to build on the success of these projects.

A range of textile designs and samples from the basic bolsters and coordinating plain or textured fabrics to the advanced 'reach' conceptual fabrics can be showcased. Large hanging samples of approximately 18 × 36 inches (46 cm × 91 cm) are submitted to the customer along with some type of archival swatch on A4 (8.25 × 11.75 inches) sample cards, or other swatch documentation notebooks. Reference samples must be clearly and accurately identified to facilitate development and future trim orders. CAD simulations of the fabric designs applied to a generic seat may be utilized. If time, logistics, or budget do not allow full seats to be trimmed, simple shaped forms or other framing may be mocked up to show the fabrics in a three-dimensional application. These samples, presentation boards, promotional materials, and color stories may be used as the color and trim's recommendations to internal management or may set the standard in interpreting objectives for other suppliers.

6.10 Future trends: innovative materials

In a market where cell phones are named after flavors of chocolate and sportswear is merchandized as having intelligent properties to enhance performance, the consumer has high expectations of automobiles delivering value on a significant investment of personal income. The smart consumer wants ease of use, affordable price, and products that reflect status and individual personality. The textile designer may consider influential consumer products aligned with the targeted consumer in the fabric development, enhancing the vehicle appeal to consumers.

OEM designers responsible for seat and interior components are seeking innovative textile components to match their advancements in multiple materials performance, ergonomic safety and comfort, improved acoustics, and the increasing electronic interface with the vehicle (Rodgers and Powell, 2006). The development of technical fabric incorporating antistatic, antimicrobial, or even self-cleaning attributes into interiors can provide further product convenience and satisfaction to driver and passenger. Interactive features will provide interfaces between drivers and their vehicles, other vehicles, and other ports of communication. New opportunities exist for e-textiles or smart materials to provide and react to valuable information; for example, sensors may alert the seat to adjust to occupant's body size, temperature or cognitive status and driving alertness. The development of fabrics with thermo physiological properties is also a focus of continued research.

Global economic development will face the conflicting goals of profitability and quality of life, and may balance these objectives through advanced technologies (Powell, 2006). The development of green products may be based on not only corporate responsibility statements and global regulations, but also on the real business opportunities coming from a market desiring sustainable products and processes. The design of fibrous components will be a major factor in developing greener products. Renewable raw materials are being sought that meet performance standards, but at a competitive price. Natural fibers such as flax, hemp, coconut fibers, Ingeo™, and kenaf continue to be evaluated in exterior and interior panels as composites or utilitarian materials to reduce weight resulting in lower emissions through fuel efficiencies (Anon., 2007, p. 8). Also the reduction of back coating and foam lamination used in seating products is a concern for recycling. The further application of spacer or sandwich knitted fabrics to replace foam is growing in practicality (Ishtalque *et al.*, 2000; Fung *et al.*, 2001). Open structured elastomeric fabric which can be applied to a seat or head restraint and minimize the need for foam and springs, thus reducing weight and increasing space in the interior cabin have been introduced by companies like Milliken's M-Flex™ and Quantum Group's Tech-Style™ fabrics (Fig. 6.1; www.Quantum, 2007; Milliken, 2000).

The reconfiguration of the industry and the demands of customers at every point in the complex supply chain are leading to increased importance of automotive advanced materials and the development of innovative products. Future developments in this area will be centered on textiles as the appropriate sustainable solution to provide safe, comfortable, lightweight, cost effective, and high performance materials.

Design's role will continue to expand in understanding the spoken and unspoken needs of the customer, interpreting these needs into viable and attractive products, and communicating that value to the market place. Sophisticated consumers will demand seamless technology which provides a new interface between humans and personal transportation systems. This

6.1 Quantum fabrics interior shown on Mazda Kabura concept car shown at the 2005 NAIAS. Source: Quantum Group, 2007.

interaction is facilitated by the interior flexibility, comfort, and safety provided by well designed textiles.

6.11 Sources of further information and advice

6.11.1 Books

Fung, W. and Hardcastle, M., 2001. *Textiles in Automotive Engineering*, Woodhead Publishing Ltd., Cambridge, England.
Wilson, A., 2007. *Automotive Textiles: The global supply chain to OEMs*, Textile Media Services, Keighley, UK.

6.11.2 Publications

Automotive Engineering International www.sae.org/automag/technewsletter
Car Design News www.cardesignnews.com
Automotive News www.autonews.com
Automotive Design Line www.automotivedesignline.com
MobileTex www.mobile-tex.com
Interior Motives www.interiormotivesmagazine.com
View on Color Magazine, United Publishers, S.A.

6.11.3 Major trade/professional bodies, research and interest groups, web sites

American Association Textile Chemists and Colorists www.aatcc.org
Wards Auto Interiors Show www.autointeriors.com
Society of Automotive Engineers www.sae.org
Color Marketing Group www.colormarketing.org
The Color Group www.city.ac.uk/colourgroup
CAUS www.colorassociation.com
Society of Dyers and Colourists www.sdc.org.uk
Industrial Fabrics Association www.ifai.com
Automotive Aftermarket Suppliers Association www.aftermarketsuppliers.org
Division of MEMA www.mema.org
Material ConneXion www.materialconnexion.com

6.11.4 Education

Art Center College of Design, Pasadena, California www.artcenter.edu
Center for Creative Studies, Michigan www.ccscad.edu
NCSU College of Textiles, North Carolina www.tx.ncsu.edu
Royal College of Art – Vehicle Design, London www.rca.ac.uk

Philadelphia University – Philadelphia, Pennsylvania www.philau.edu/textiledesign/ms.html
Rhode Island School of Design, Providence, Rhode Island
Reutlingen University, Reutlingen, Germany
University of Huddersfield, Queensgate, Huddersfield, England
Central Saint Martins College of Art and Design, London, England
Loughborough University, England www.lboro.ac.uk

6.11.5 Events

International Engineered Fabrics Conference and Expo www.inda.org/IDEA08/
TechTextil http://techtextil.messefrankfurt.com
Specialty Equipment Market Association www.sema.org

6.12 References

Anand, S., 2003. Recent Advances in Knitting Technology and Knitted Structures for Technical Textiles Applications. *International Isparta Textile Congress (Istek) Conference Proceedings*, May, 2003.

Anon., 1996. *Quality System Requirements* – QS 9000, International Organization for Standardization.

Anon., 1997. 'Car Seating Fabrics – Suppliers & Trends', Automotive Seating Review, *Inside Automotives International*, pp. 28–37.

Anon., 2004a. Technical Textiles: Emerging Opportunities, December 9, 2004, *Journal of the Textile Committee*. Downloaded from www.bharattextile.com/library/003035.php

Anon., 2004b. Weft Knits Drive Forward, *Knitting International*, July, Vol. 111, Issue 1316, pp. 28–29.

Anon., 2007. Nothing New about Natural Fibres in Cars, *MobileTex*, Vol. 2, no. 1, January, p. 8.

Bhagwat, V., 2005. Testing of Automotive Textiles, *Asian Textile Journal*, Vol. 14, No. 1/2, p. 79.

Bruce, M. and Cooper, R., 2000. *Vantage Case Study*, Creative Product Design, pp. 106–132, Chap. 6. John Wiley & Sons, Chichester, England.

Cagan, J. and Vogel, C., 2002. *Creating Breakthrough Products*, Chapter 9, Automotive Design, Prentice Hall, Upper Saddle River, NJ.

Crabtree, N., Correspondence with Crabtree who retired from Chrysler Color and Trim, 2005.

Dickey, C., 2007. 'Developing New Products: The Competitive Advantage' Textile Apparel Manufacturers and Merchants Annual Conference, Savannah, Georgia, May 15.

Fahy, L., Personal Interview, Automotive Design Director, Melba Industries, May, 2005.

Flynn, M., Belzowski, B. and Haas, S. 2002, 'E-CRM and the Automotive Industry: Focusing on Customers', Office for the Study of Automotive Transportation, Ann Arbor, Michigan.

Fung, W. and Hardcastle, M., 2001. *Textiles in Automotive Engineering*, Woodhead Publishing Ltd., Cambridge, England.

Hansen, O., 2005. Telephone Interview, Automotive Business Manager, Toray Industries.

Ishtalque, S., Yadav, P. and Sharma, N., 2000. A New Approach to Improve the Performance of Spacer Fabrics for Automobiles. *Man–Made Textiles in India*, March, pp. 67–74.

Lentz, J., 2005. KUKA Robot Handles Ingress/Egress Seat Testing, http://www.roboticsonline.com/public/articles/index.cfm?cat=104, Retrieved April 24, 2006.

Manley, R. and Powell, N., 2004. The Development of Woven Velours for the Transportation Market, www.tx.ncsu.edu/jtatm/volume3issue4/Articles/Powell/manley_full.

McKenna, R., 1988. '*Marketing in an Age of Diversity*', Markets of One, Harvard Business Review Book, Boston, MA.

Milliken Prints, 1997. *Promotional print booklet*, Milliken & Company, Spartanburg, SC.

M-Flex™, Milliken, 2000. Promotional Materials, Milliken & Company.

Nieuwenhuis, P. and Wells, P., 2003. *The automotive industry and the environment: A technical, business and social future*, CRC Press, Woodhead Publishing Limited, Cambridge, England.

Pfeiffer, M., 2007, Guilford Europe Automotive, Telephone interview, June, 2007.

Powell, N. B., 2003. Mass Customization in Transportation Textiles through Shaped Three Dimensional Knitting, *Proceedings from International Textile Design and Engineering Conference*, Edinburgh, Scotland, Heriot-Watt University, Netherdale, Galashiels TD1 3HF, UK, September, 2003.

Powell, N. B., 2004. Design Driven: The Development of New Materials in Automotive Interiors, *Journal of Textiles Apparel and Technology Management*, *Transportation Textiles Issue*, Vol. 3, Issue 4, www.tx.ncsu.edu/jtatm.

Powell, N.B., 2006. Design management for performance and style in automotive interior textiles, *Journal of the Textile Institute*, Vol. 97, No. 1, pp. 25–37.

Rigby, D., 2003. Mobiltech Chapter, *Technical Textiles Report*, 2003.

Rodgers, S. and Powell, N., 2006. 'Automotive Textile Durability 2D Testing in a 3D Application', Autex Conference Proceedings, June 13, 2006, College of Textiles, Raleigh, N.C.

Sirvio, K., Personal Interview. General Motors, June 5, 2003.

Trott, P., 2002. *Innovation Management and New Product Development*, 2nd Edition, Prentice Hall, Pearson Education Limited, Harlow, England.

Vincenti, W. G., 1990, *What Engineers Know and How They Know It*, The Johns Hopkins University Press, Baltimore, MD.

Ward's Automotive, 2003. WAW 24th Annual Supplier Survey. Southfield, Michigan.

Ward's Automotive, 2001 & 2002. Motor Vehicle Facts & Figures 2002, Southfield, Michigan.

www.chemicalfabricsandfilm.com/pdfs_researchSection/standards/ppap.pdf, downloaded July 15, 2005.

www.quantum5280.com/index.html, retrieved July, 2007

www.qmi.com/information_center/standards/iso16949, 2002 retrieved April 24, 2006.

www.Scion.com, 2003. *Toyota Scion Review*, downloaded August, 2003.

7

Three-dimensional textiles and nonwovens for polyurethane foam substitution in car seats

H ERTH and B GULICH, STFI, Germany

Abstract: Seat cover materials are present as multi-material composite systems of textile and foam layers with permanent fixed connection. Three-dimensional textile structures offer novel possibilities in construction of seat cover materials. These structures can be manufactured by different technologies. Nonwovens and warp-knitted spacer fabrics with strong three-dimensional properties will be described within this chapter. The achieved elastic properties of such materials are comparable with actual used material. When PUR foam is replaced by a textile structure, the seating climate comfort will be improved. Furthermore, the system consists of 100% textile materials. This fact permits the recycling and re-use in textile structures. It is even possible to manufacture the material system as a single-material composite system.

Key words: upholstery, PUR foam substitution, composite materials, recyclability, stitch-bonded nonwovens, warp-knitted spacer fabrics, lamination process, performance requirements for seat-cover fabrics, environment.

7.1 Introduction

Textiles have been used in cars for a long time. Looking at both the expected future number of cars and the developments in the field of quality, textiles stand a good chance of being more and more widely used (see Table 7.1). It is up to the textile industry to convincingly outline their advantages since textiles are selected with regard to their manufacturing processes, raw materials, design, combinability, functionality, recycling and cost. In areas like the

Table 7.1 Trends of textile use in passenger cars

	Year		
Amount of textiles in a typical mid-range car (kg)	2000	2010	2020
Total	20	26	35
Products and applications like those in 2000	20	14	10
New product for known application (e.g., cushioning by textiles)	–	9	15
New product for new application (e.g., shielding by functional cushioning)	–	3	10

USA, Japan and Western Europe, customers expect more comfort and improved safety, which could be achieved by the use of textiles. A particularly high growth potential is expected for the business sector that equips higher-class motor vehicles. One trend is the substitution of synthetic and other materials in common applications with well-known functions by textiles, e.g. the substitution of polyurethane (PUR) foam in upholstery components by polyester nonwovens [1].

The structure of composite materials used for upholstery in vehicle interiors, in particular for seat-covers, has remained unchanged for more than 30 years. Three different components are connected by means of adhesive layers. Various woven, warp-knitted and knitted fabrics are used in order to produce textile top-layer materials, polyester (PES) being the predominant fibre (about 90 per cent of these particular materials) [2]. A 2 to 8 mm thick (average: 3 to 5 mm) PUR foam layer is added beneath the top layer, followed by a meshed material made from PES or polyamide (PA). The adhesive agent for the two adhesive layers of the laminated composite is PUR foam.

Why is the automotive industry questioning the use of a composite seat-cover material that has been optimised and well-proven for decades? The two reasons are that the car driver is asking for greater comfort and that there are now stricter laws on the protection of the environment. The advantages of PUR foam upholstery, such as excellent elasticity and a well-proven manufacturing technology, are now being called into question owing to a number of drawbacks that have come to light in recent years, in particular the fact that as a result of the flame lamination process used, toxic fumes are produced from combustion during manufacturing. Their disposal into the air is subject to stricter regulation and they may cause emissions in the completed vehicle, resulting in fogging and unpleasant odours. Furthermore, the low air and water-vapour permeability of composite materials can make passengers less comfortable [3].

Another reason for finding a substitute for the current seat-cover composite material is its non-recyclability. In particular, it is impossible to re-use the production waste from the manufacturing cycle since the composites are made of several materials which cannot be separated. This is also important in the context of achieving eco-friendly disposal [4].

7.2 Properties and performance requirements for car seats

Properties and performance requirements are set and evaluated for both the complete composite of the seat-cover fabric and the cushioning layer. All these requirements are influenced by the different targets of customers' comfort, manufacturers' guidelines, laws on the environment and total cost that should be fulfilled by each product. The main elements of these requirements are listed below.

Composite

- number of layers
- emissions
- electrostatic behaviour
- recyclability
- microclimate of seats
- manufacturing cost.

Cushioning layer

- thickness
- mass per unit space density
- elasticity to compression
- compressive strength
- permanent deformation
- elastic elongation
- air permeability of cushioning layer
- moisture transport
- resistance to ageing
- combustion behaviour
- necessity of fire retardant finishing

Every car manufacturer sets its own requirements for seat-cover composite properties in its delivery conditions. Some selected and summarised numbers and data for the requirements listed above are shown in Table 7.2 [5–7].

7.3 Types of materials used as cushions

The substitution of PUR foam is, in theory, achievable by a wide range of textile materials including:

- thermal bonded/heat-set nonwovens
- needle-punched nonwovens or nonwoven composites (Fig. 7.1)
- velourised needle-punched nonwovens
- vertically lapped and thermal-bonded nonwovens, e.g. Struto or Wavemaker
- stitch-bonded nonwovens, e.g. Malivlies
- stitch-bonded nonwovens, e.g. Kunit, Multiknit, Caliweb®
- spacer woven fabrics
- spacer circular-knitted fabrics
- spacer warp-knitted fabrics.

There are some specific technical and financial restrictions on the range of materials that can be used for PUR foam substitution. With regard to their

Table 7.2 Selected and summarised requirements for seat cover composite and cushioning layers

Property of cushioning layer	General range possible	Recommended range
Mass per unit area (g/m^2)	170–600	230–330
Thickness (mm)	2–8	3.0–4.5
Density (kg/m^3)	50–100	70–75
Remaining deformation, dry (%)	15–25	17–20
Remaining deformation, dry (%)	10–25	20–23
Compressive strength (kPa)	2–20	5–7
Elasticity to compression (a30-5H) (%)	30–50	35–40
Elasticity to compression (a30-E3) (%)	90–98	92–95
Air permeability (l/m^2s)	> 200	500–1200
Short-term water vapour absorption (g/m^2)	> 2.5	> 4.0
Combustion behaviour according to SMVSS 302 longitudinal/crosswise	A/A	A/A

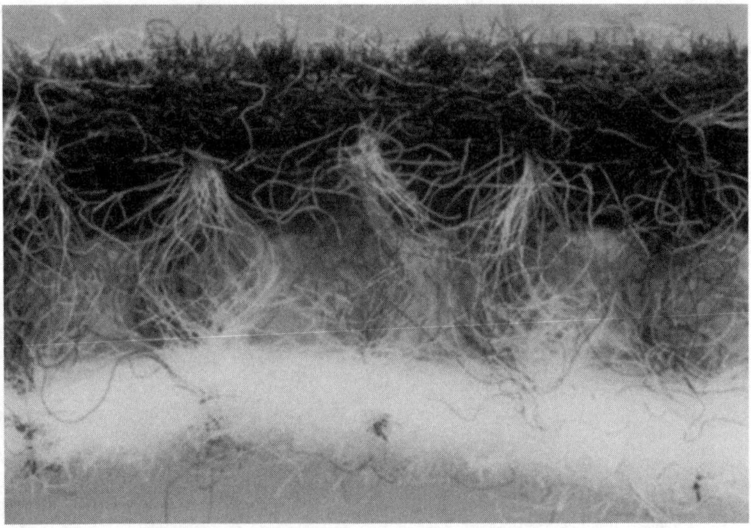

7.1 Needle-punched spacer nonwovens composite type NAPCO®.

price and their flexible properties, nonwovens are reasonably suitable for use as textile structures for upholstery material, being directly produced from fibres or filaments, without any production of yarn. In order to provide the upholstery characteristics required, however, only a few voluminous kinds of nonwovens are available. A good rebound to pressure is only shown by products which contain fibres arranged perpendicularly to the length of the material.

In addition, some other three-dimensional knitted and woven textiles, especially warp-knitted spacer fabrics, show good elasticity properties.

7.4 Key technologies for replacing polyurethane foams in car seats

Several flat nonwovens, such as stitch-bonded or needle-punched products, are already in use in car seats. Mercedes-Benz has used the Malivlies type of material as a supporting layer, glued directly onto the seat-cover fabric, since 1992. However, such materials do not function as a cushioning layer in the same way as PUR foam.

Textile materials with thicknesses from 2 to 3 mm up to 7 or 8 mm are required for upholstery use. Until now, such textiles have only been produced for end-use application from stitch-bonded and thermal-bonded nonwovens and warp-knitted spacer fabrics.

Investigations into the use of textiles as substitutes for PUR foam in car seats have been carried out by several European car manufacturers. Today, the three-dimensional, stitch-bonded nonwovens, Multiknit/Caliweb®, are used in several types of car from Audi, BMW and Mercedes. The nonwovens are manufactured by TWE Vliesstoffwerke GmbH & Co. KG, Emsdetten/D and TECHTEX GmbH Vliesstoffe, Mittweida/D (Figs 7.2 and 7.3).

J.H. Ziegler GmbH & Co. KG, Aachern/D, produces a three-dimensional thermal-bonded nonwoven, called Haco®, as an upholstery material for Audi car seats. 3mesh®, a special type of warp-knitted spacer fabric made by Müller Textil GmbH, Wiehl/D, is current used where perfect comfort in

7.2 Three-dimensional stitch-bonded nonwoven type CALIWEB® laminated with cover fabric.

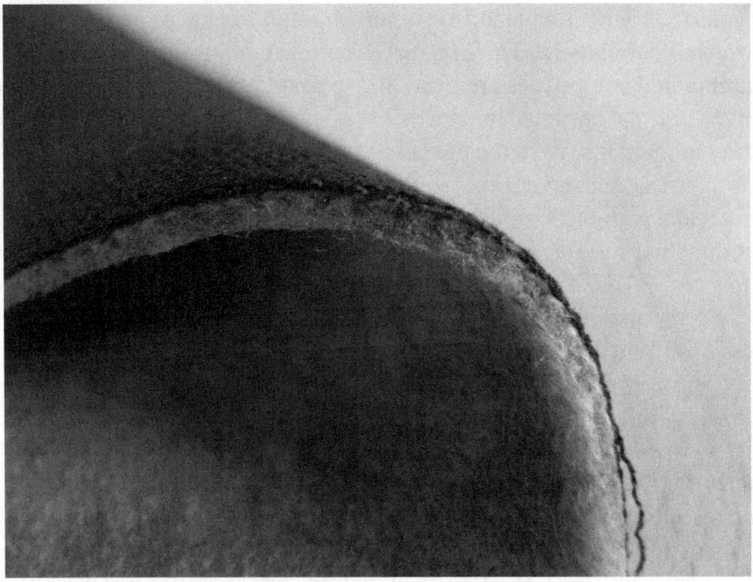

7.3 Three-dimensional stitch-bonded nonwoven type CALIWEB® laminated with cover fabric.

climatic and ventilated seats is requested. This material fulfils all the requirements such as the highest possible distribution of air, soft cushioning effects and excellent rebound after pressure, but, because of its higher price, it is currently only used in higher-class cars, like the Mercedes S-Class and Audi A8.

7.5 Environmental advantages of using substitutes

Between seven and nine million used cars are disposed of annually within the European Union. The quantity of recyclable materials currently only amounts to about 75% of the total vehicle mass. The remaining 25% are hazardous waste products from light shredded material, which consists of plastics, textiles, glass and metal waste. The disposal procedure is regulated by a number of national and international directives, such as the European Directive on the Transfer and Eco-friendly Disposal of Used Cars (Directive on Used Cars – AltautoV) of 1997 and the Self-commitment concerning the Eco-friendly Recycling of Used Cars (limousines) in the framework of the Law on the Economic Cycle of 1996 (both Directives being effective from 1 April 1998) as well as the EU Directive on Used Cars of 2000. They require both material- and design-related alterations to the components currently used, with regard to effective material recycling.

That is why recyclability is an impetus to the search for a substitute, which is currently hard to find, for the current composite seat-covering material. Laminated composites are available to seat producers as lengths of material. Even optimal cutting cannot avoid waste which, depending on the seat shape, may amount to five to ten per cent. As the materials used all display different properties, and separation from the polyurethane foam is not practical, it is not really possible to re-use this waste in the production cycle. Great importance is attached to this when considering the eco-friendly disposal of used seat covers.

Furthermore, emissions may result from the flame lamination process used for PUR foam layers. Their disposal into the air is being regulated more and more and they may also cause problems in a new vehicle, resulting in fogging and unpleasant odours.

As mentioned previously, PES is applied as a fibre material for the manufacture of nonwovens in a similar way to the material used for the upper fabric. The use of equal polymers guarantees that most of the seat-cover composites can be re-used. This has led to an investigation into the use of reclaimed fibres. Reclaimed PES fibre waste can be purchased for use in motor vehicles and it may be used as a substitute for up to 50% of primary fibres in the manufacture of nonwovens, depending on the quality of the reclaimed fibre. The physical properties will remain well inside the requirements for seat-cover composite materials and the cost of manufacturing will also be reduced. Almost all the processes available for the application of hot-melt adhesives may be used to bond the upper material and the nonwoven upholstery. The advantages of wholly textile composites include eco-friendly laminating, low emissions, and good air and water-vapour permeability [8].

7.6 Future trends

The use of textiles as upholstery material beneath the seat-cover fabric is strongly influenced by the price. Despite any advantages that might be gained from new constructions for seat-cover composites, they must not cost more than 10–15% more in comparison with today's commonly used foam-cushioned materials.

Because of the large quantities used, costs are of special interest. Nonwoven-based composite materials, in particular, will be fully competitive compared to the PUR foam composites that have been used until now. This is because the warp-knitted backing side of the foam layer is no longer necessary, and neither is the adhesive layer that was needed for it. The resulting process optimisation and improved recyclability will bring about a further reduction in cost.

Technical trends are influenced by the integration of functional materials to improve comfort or, alternatively, to install more features in the car. The use of anti-static fibres in the cover or the cushioning layer will improve anti-static behaviour. The integration of activated carbon in the cushioning layer could create a good microclimate in the seat. The increasing numbers of requirements for driving comfort, safety and infotainment lead steadily to a higher proportion and a rising packing density of electronic car components and control devices in cars. Thus, more requirements for ensuring electromagnetic compatibility (EMC) for the entire car system as well as for car components have to be fulfilled. Safe, economic and light-weight shielding for electronic components is needed and metallised fibres or areas in the cushioning layer will assist this shielding.

Warp-knitted spacer fabrics, which today are only used in high-priced cars in thicknesses of 8 to 12 mm (Fig. 7.4), will be used in single-layer materials containing the seat-cover fabric and cushioning layer all in one product (Fig. 7.5). Furthermore, warp-knitted spacer fabrics will be installed more and more as substitutes for the foam core in the seat. This will lead to the replacement of foam by nonwovens for the cushioning layer of seat-cover fabrics and by warp-knitted spacer fabrics from 30 to 60 mm thick for the foam core squab (Figs 7.6 and 7.7). Within ten years, the amount of foam used in most car seats will be reduced to nil [9–11].

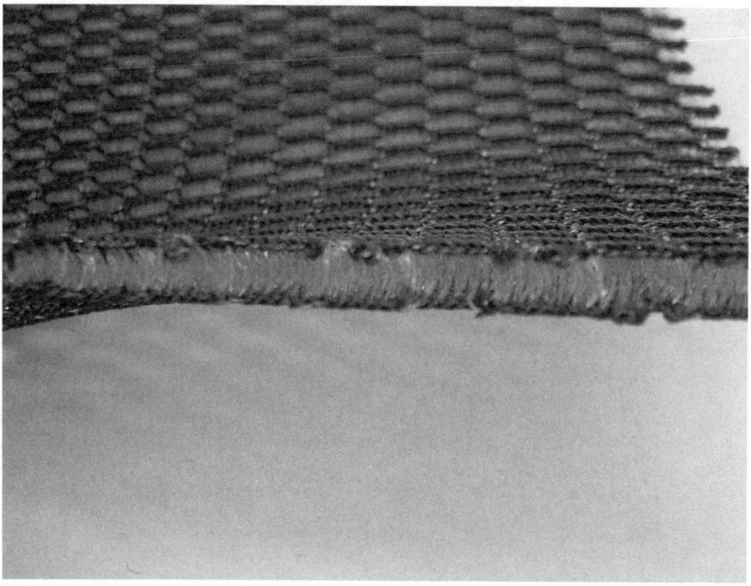

7.4 Warp-knitted spacer fabric.

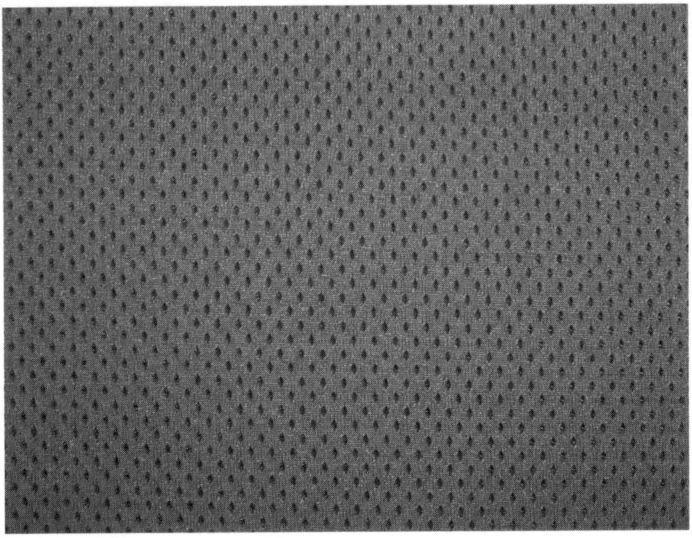

7.5 Warp-knitted spacer fabric with closed surface containing seat cover fabric and cushioning layer in one product.

7.6 High distance warp-knitted spacer fabrics for replacement of the foam core squab – cross section.

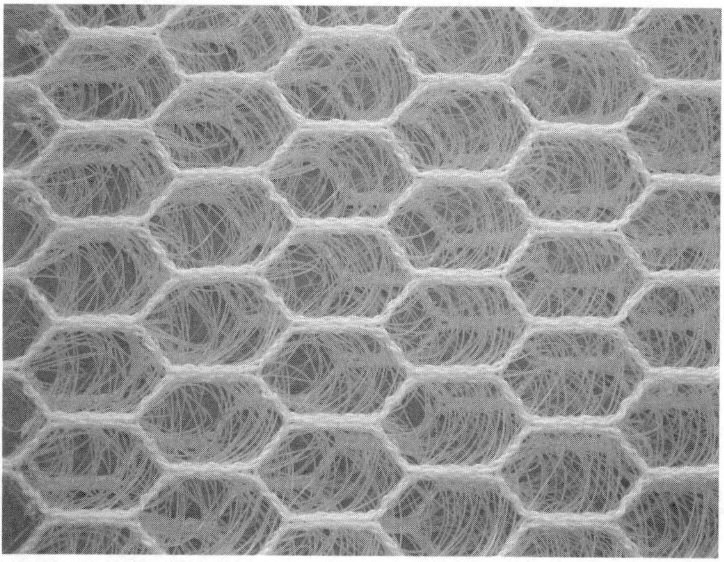

7.7 High distance warp-knitted spacer fabric – top view.

7.7 References

1. Schmidt G F (1992), 'Kaschierung von Autopolsterstoffen mit Faservliesen', *Melliand Textilberichte* 6, 479–486.
2. Nagl F (1998), 'PES-Filamentgarne für textiles Autointerieur', *Technische Textilien* 11, 184–185.
3. Coll-Tortosa L, Schilde W, Erth H (2000), 'Development of Recyclable Upholstery Textiles for the Automotive Industry and Public Transport', EU Research Report EU-BRPR-CT 96 0292.
4. Erth H (1999), 'Recycelbare Polstervliesstoffe für die Automobilindustrie', *STFI-Kolloquium 'Reißfaser '99'*, Chemnitz/D 1999.
5. Schmidt G F (1999), ' "CALIWEB" substituiert Schaumstoff', *Kettenwirk-praxis* 4, 31–34.
6. Erth H, Fuchs H, Wegner A (2000), 'Recyclable three-dimensional nonwovens for upholstery in car seats', *Textiles for Mobility – Hightex from Germany*, Atlanta/USA 2000.
7. Lutz S (2007), 'Technische Anforderungen an Textilien im Automobilbau', *Textilveredlung* 5/6, 26.
8. Erth H, Schilde W, Schmalz E (1998), 'Einsatz von Wirkvliesstoffen mit Polstruktur als Unterpolstermaterial für Fahrzeugsitzbezüge unter besonderer Berücksichtigung deren Recyclefähigkeit', *13th. Hofer Vliesstoffseminar*, Hof/D (1998).
9. Erth H, Gulich B, Fuchs H (2005) 'Mobile Textiles – Future Trends to 2010', *44th International Man-Made Fibres Congress*, Dornbirn/A 2005.
10. Erth H, Gulich B, Fuchs H (2006), 'Development of Nonwovens for Transportation', *8th International Dresden Textile Conference*, Dresden/D 2006.
11. Stegmaier T, Erth H (2007) 'Market Data and New Developments in Textiles for Mobile Applications', *Techtextil Symposium*, Frankfurt/D 2007.

8

Physiologically optimised car seats

V T BARTELS, Hohenstein Institutes, Germany

Abstract: A particularly important aspect of comfort in cars or other vehicles is the seats. The performance of a driver over long distances significantly decreases if car seats do not support posture and heat balance. This chapter reviews how seats can support the thermoregulation of the body via heat and moisture transport, ways of evaluating the thermophysiological aspects of vehicle seats at the developmental stage, and the use of textile materials to optimise the comfort of car seats.

Key words: car seats, comfort, thermoregulation.

8.1 Introduction

Today, comfort has become a major quality criterion of cars. Comfort in a car is a complex phenomenon and comprises such different aspects as, for example, noise, driving behaviour, or ease of handling (ATA, 1997). One of the most important factors influencing passenger convenience is thermal comfort. Therefore, car manufacturers are paying a lot of attention to this aspect, as can be seen by an increased application of air conditioning in the car, for example.

A particularly important aspect of vehicle comfort is the seats. Seats do not only have to have an attractive design or meet specific design criteria for safety reasons, they must also have optimum comfort properties. But seat comfort is much more than just passenger convenience. Scientific findings show that the performance of a driver over long distances significantly decreases if the car seats do not support posture and heat balance as required (Umbach, 1999, 2000a, see Fig. 8.1). This leads to exhaustion and loss of concentration, which, in extreme cases, could result in serious accidents.

In addition to ergonomic considerations, again the climatic or thermophysiological comfort of the seat is of particular importance. This indicates whether the seat is able to support the thermoregulation of the body via heat and moisture transport, as described in Section 8.2.

Today, the thermophysiological comfort of car seats can be obtained by a set of laboratory test apparatus, which are introduced in Section 8.3. Using measurements taken with the skin model and seat comfort tester (cushion testing device), it is now possible to evaluate and optimise the thermophysiological aspects of vehicle seats at the developmental stage.

Using a number of examples, it is demonstrated in detail in Section 8.4, how the physiological seat comfort of vehicle seats can be improved in real

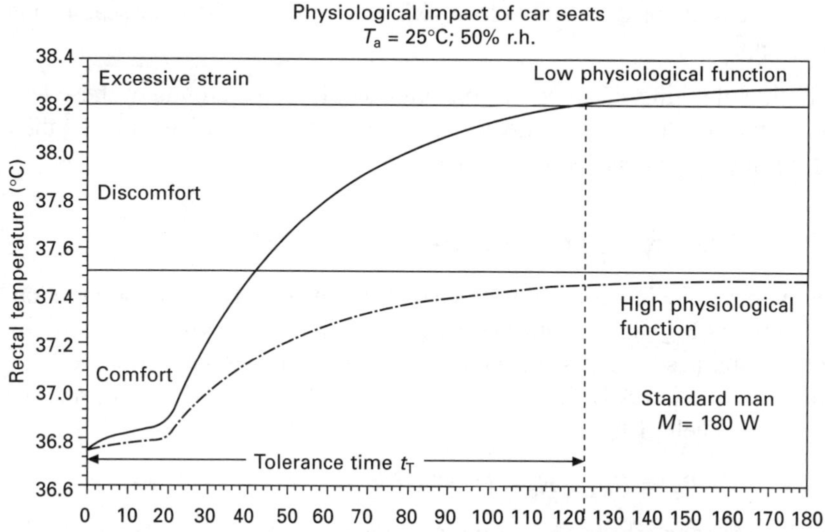

8.1 Rectal temperature of a car driver as a function of time while sitting on a seat with high physiological function (lower line) or a seat with low physiological function (upper line) (Umbach, 1999, 2000a).

terms by using modern technical textiles. Only if the material and design of a seat are optimally coordinated will the driver enjoy optimum seat comfort. Otherwise, heat and moisture can accumulate, which feels unpleasant and can have a negative physical effect on the driver, limiting performance and concentration.

8.2 Thermophysiological comfort of car seats

From the physiological point of view, seat comfort comprises the following four parameters:

1. The initial heat flow following the first contact with the seat. In other words, the sensation of warmth or cold in the first few minutes or even seconds after entering the car.
2. The dry heat flow on long journeys, i.e. the amount of body heat transferred by the seat.
3. The ability, known as 'breathability', to transport any perspiration formed away from the body. In so-called 'normal' sitting situations, there is no perceptible perspiration, but, nevertheless, the human body constantly releases moisture (so-called 'insensible perspiration'), which has to be taken away from the body.
4. In the event of heavy perspiration (a car in the summer heat, stressful

traffic situations) the ability to absorb perspiration without the seat feeling damp.

In the following sub-sections, the physiological background of these four parameters are described. In Section 8.3 their testing, and in Section 8.4 their optimisation by advanced textile materials is discussed.

8.2.1 Energy balance of the human body

Thermophysiological comfort is based on the principle of energy conservation. As shown in Fig. 8.2, all the energy, which is produced within the body by metabolism, has to be dissipated in exactly the same amount from the body (Umbach, 1986, 1999). In terms of a mathematical formula, this principle gives (Mecheels, 1998):

$$M - P_{ex} = H_{res} + H_c + H_e + \Delta S/\Delta t \tag{8.1}$$

On the left-hand side of this formula, the energy production is given by M, the metabolic heat production. In a car, M usually ranges between 150 and 300 W. P_{ex} is the external physical work, which in a car is mainly due to steering, shifting the gears, etc., and which is much smaller than M.

On the right-hand side of equation 8.1, the heat dissipation is quantified. Here, H_{res} is the respiratory heat loss because of breathing, which is typically about 10% of the metabolic rate M. H_c is the dry heat flux comprising radiation, conduction and convection. H_c is strongly dependent on the car seat, the passenger's clothing and the cabin climate. The same holds for the evaporative heatflow H_e, which is caused by sweating.

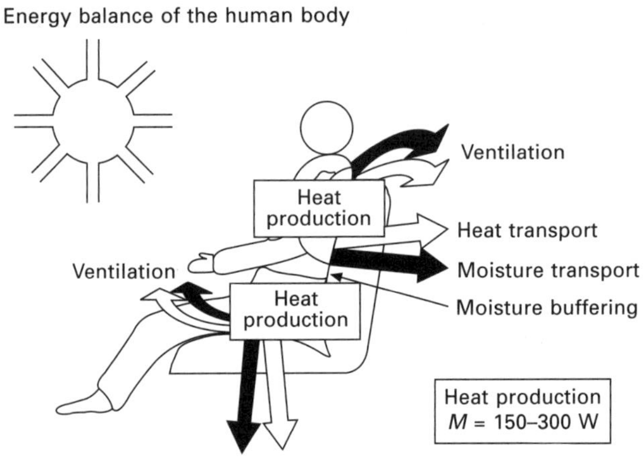

8.2 The energy balance between produced and dissipated heat as a prerequisite for a good thermal comfort (Umbach, 1986, 1999).

If more energy is produced than dissipated, the body suffers from hyperthermia. On the other hand, too high a heat loss implies hypothermia. Both lead to a change in the body's energy content ΔS with time Δt. ΔS may be either positive (leading to hyperthermia) or negative (hypothermia) and is zero for the steady state.

This steady state (i.e. $\Delta S = 0$) is desirable, and it has to be the aim of a car seat manufacturer to achieve energy balance. M, P_{ex} and H_{res} cannot be influenced by a car seat, but H_c and H_e can. Therefore, all the physiological testing apparatus described in Section 8.3 are intended to measure H_c and H_e. And all optimised constructions of car seats given in Section 8.4 are intended to improve H_c and H_e.

8.2.2 Warmth sensation

Initial perception of warmth (entering the car)

The passenger is already sensing his first thermal impression of a car seat while entering the vehicle. This initial perception of warmth after sitting depends on the heat flux between person and seat within the first few minutes or even seconds. The heat flux should be as low as possible, otherwise a car seat feels cold in the winter time or hot during summer.

Although this initial feeling may last only a few minutes, it is nevertheless very important for the user's acceptance, as it is being repeated frequently. If a car is used every day during the winter time and each morning the driver is dissatisfied when entering the car, acceptance can be significantly decreased.

For the initial perception of warmth, the dominant seat components are the cover and (if present) heating. For the covering material, the main parameter influencing the initial perception of warmth is heat capacity (Bartels and Umbach, 1997, 2005; Bartels, 2003; Umbach, 1999). For the heating, both its heating power and its position are of great importance (Hänel *et al.*, 1995, 1997).

The exact relationship between (a) the heat flux between seat and person on the one hand and (b) the subjective thermal sensation on the other hand was surveyed in Bauer and Bartels (2005) and Bauer *et al.* (2005). In Fig. 8.3 the initial thermal sensation of warmth is shown as a function of the heat flux H_c between seat and skin. These results were obtained under cool climatic conditions during the first 10 minutes after getting seated.

Figure 8.3 shows a clear correlation between the subjective and the objective data. In particular, the sensations 'pleasant' and 'slightly cool', which are most important for achieving comfort under the conditions considered here, can be attributed to heat fluxes of H_c = 86 W/m^2 and 64 W/m^2, respectively.

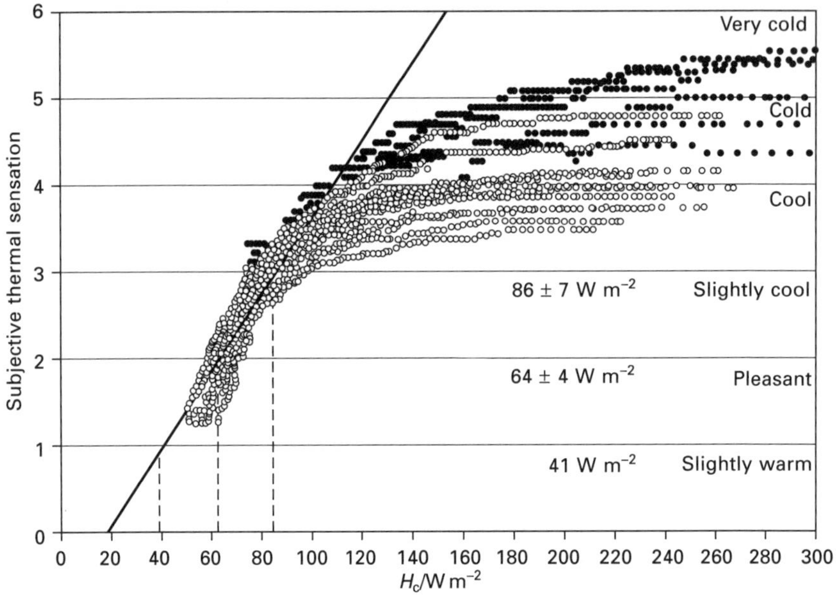

8.3 Initial subjective thermal sensation of test persons as a function of the heat flux H_c between seat and skin (Bauer and Bartels, 2005, Bauer *et al.*, 2005).

Steady state heat flux (long journey)

During long journeys it is favourable if the seat offers a *high* steady state heat flow, to minimise the tendency to sweat (Umbach, 1999; Bartels and Umbach, 2005), whereas for the initial perception a *low* heat flux is required (see page 153). Hence a conflict arises between these two scenarios.

This conflict can be overcome, because the cover, which determines the initial perception, is only of minor influence on the steady state heat flux, which is mainly determined by the thermal insulation of the seat. Owing to its greater thickness and, hence, higher thermal insulation in comparison to the cover, the cushion becomes the dominant part.

On the other hand, the heat flux is also dependent on the ventilation in the seat and between seat and passenger (see Fig. 8.2). Ventilation itself is determined by the design of the seat (side supports, surface grooves), the elasticity and air permeability of the cushion (e.g., foam or 3-D spacer fabric), and, if present, a fan to enforce ventilation.

8.2.3 Moisture sensation

Water vapour transport (normal sitting)

The moisture sensation of the passenger is very important for perceived overall seat comfort. As shown in Fig. 8.4, a seated person does recognise

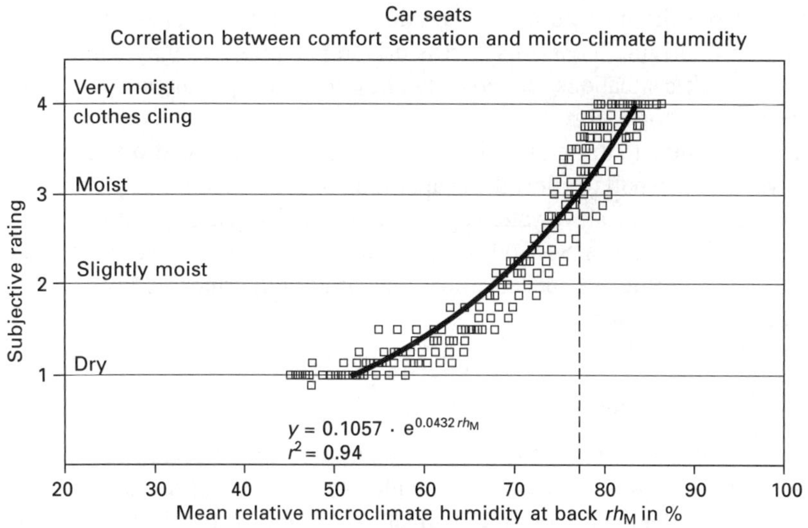

8.4 Subjective moisture sensation of human test subjects as a function of the relative humidity in the microclimate between seat and skin (Umbach, 1999).

differences in the microclimate humidity between seat and skin (Umbach, 1999). Especially a rating of 'moist' (corresponding to a mean relative humidity in the microclimate of 78%) or poorer has to be avoided. Otherwise, the acceptance of the car seat by the passenger is severely impaired.

In order to achieve a dry microclimate, the ability, known as 'breathability', of the seat to transport any perspiration formed away from the body is crucial. Not only under warm summer conditions is good water vapour transport necessary, but even when there is no perceptible perspiration. The human body constantly releases moisture, the so-called 'insensible perspiration'. As the skin is not totally water vapour tight, our body loses at least 30 grams of moisture per hour. Because a car seat covers large areas of the body, the seat has to manage a large part of the perspiration formed, and, hence, a considerable amount of moisture.

But we do not only sweat insensibly, we sensibly perspire in order to ensure the body's thermoregulation and to cool down the body core temperature by sweat evaporation (see Fig. 8.2). The amount of moisture produced can be up to one litre per hour or even more, e.g. during sports or if the ambience is extremely hot.

In the case of sensible perspiration, the sweat glands inside the skin are actively producing moisture. But only if this moisture can readily evaporate can the desired cooling effect be achieved. This is a direct demand on the seat, which has to allow this evaporation.

Apart from thermoregulation, we additionally sweat because of mental stress. While driving a car, this 'stress driven perspiration' may appear due to difficult traffic situations. But no matter how the sweat production originates, the seat has to cope with it.

Hence, for most sitting situations, it is essential that a seat offers a high water vapour transport to allow the evaporation of sweat. Technically speaking, the seat needs to have a low water vapour resistance (i. e. a high 'breathability'). As further discussed in Section 8.4, all components of the seat have to be water vapour permeable, because just one single impermeable layer would impede water vapour transport.

Sweat buffering (heavier sweating)

When entering a hot car in the summer or because of stressful driving situations, we start to sweat more heavily. This additional amount of sweat has to be 'buffered', in order to keep the microclimate as dry as possible. The dominating parts for the buffering capacity are seat cover and lining.

In general, for a good water vapour buffering capacity, both the water vapour transport and the water vapour absorbency are important. As shown in Bartels (2003), the water vapour transport is especially of interest for the seat cover. Only if the excessive water vapour can be transferred from the body and into the seat can a dry microclimate be achieved.

When looking at the whole car seat, the influence of water vapour transport on water vapour buffering becomes smaller. Just due to its thickness, a car seat is already a considerable barrier against diffusion. Therefore, for the whole car seat, water vapour absorbency determines the water vapour buffering capacity (Umbach, 1999). In order to achieve fast absorption, the covering material and the lining beneath it in particular should be hygroscopic.

8.3 Measurement of seat comfort

8.3.1 Seat trials with test persons

One way to test seat comfort is seat trials with human test subjects. They can be performed as so-called 'field trials', as they are frequently carried out during test drives. The greatest advantage of these test drive results is that they represent 'the practice'. However, these types of tests cannot be reproducible, as the weather and the traffic conditions change. If, for example, one day there are a few more clouds in the sky than another day, solar irradiation is much smaller and, hence, also the driver's tendency to sweat.

Therefore, in order to have confidence in the results of the seat trials of the investigated seat construction, it is recommended to perform the experiments in a climatic chamber, which controls ambient temperature, humidity and radiation. In a modern climatic chamber, climatic conditions can be adjusted to range from arctic to tropical.

Using a driving simulator (see Fig. 8.5), different traffic situations can be adjusted such as the type of street (small, rural roads, highways, streets in cities, etc.), the traffic density, or the behaviour of other road users. In addition, unexpected incidents can be added, like a staggering cyclist or a deer that suddenly jumps into the road.

During the trial, the test person has to steer the wheel, shift gears, accelerate and break as in real driving situations. Thus, the test person not only performs the same body movements as in a car, but also shows the respective metabolic heat production. Ventilation due to pumping effects caused by the car's movements can be simulated by a small vibration of the board on which the car seat is mounted.

8.5 Seat trial with a human test subject on a driving simulator in a climatic chamber.

A big advantage of a modern driving simulator is that it can record errors. As already mentioned in Section 8.1, if a driver feels uncomfortable on the seat or even suffers from heat stress (see Fig. 8.1), mental performance is decreased. The number and seriousness of errors is a way to quantify this decrease.

Seat trials can be easily adjusted to the problem under investigation. Not only can climatic and driving conditions be set in a climatic chamber, but also a variety of objective data can be obtained. If, for example, the 'moisture management' of a car seat is of interest, humidity sensors can be attached between skin and seat, which monitor the relative humidity in the microclimate (Frankowsky and Umbach, 1995; Bartels and Umbach, 1997; Umbach, 1999; Bartels, 2003). In addition, the test person, clothes and the seat can be weighed before and after the test to determine the sweat production of the person as well as the moisture uptake of the clothes and the seat. Rectal and skin temperatures can be measured using thermocouples. For testing the initial sensation of warmth, heat flux sensors can be adjusted to the skin (Bauer and Bartels, 2005; Bauer *et al.*, 2005). Also the metabolic rate can be determined by an analysis of the exhaled air concerning oxygen consumption and carbon dioxide production (ISO 8996). Last but not least, the heart rate is monitored via ECG electrodes. A typical set of probes attached to a test person's body is shown in Fig. 8.6.

For the evaluation of subjective comfort sensation, ISO 10551 can be employed. Via subjective judgement scales, the test persons' sensations can be quantified. For example, the humidity sensation could be assessed using the following scales: 0 = dry, 1 = back slightly moist, 2 = back moist, 3 = body moist, 4 = body moist and clothes cling, etc.

However, seat trials also have some severe disadvantages, which make them ineffective for product development or quality control. In order to obtain statistically significant results, a large number of test persons and repetitions is required. Additionally, test persons react differently in the morning than in the afternoon, hence tests have to be performed at both times. Test subjects have to be acclimatised to the climatic and activity conditions in pre-tests. Data have to be extensively statistically analysed, etc. This makes seat trials time-consuming and expensive and laboratory test methods, as they are described below, a faster and cheaper alternative for many applications.

In conclusion, seat trials with test persons are best positioned at the end of a development process, in order to prove conclusively the physiological performance of a car seat. In addition, seat trials are needed as a 'calibration' for the laboratory test procedures, which are introduced in the next sections. In particular, mathematical regression analyses between the laboratory tests on the one hand, and wearer trials data on the other hand have to be carried out, in order to ensure the practicability of the laboratory tests, and to interpret

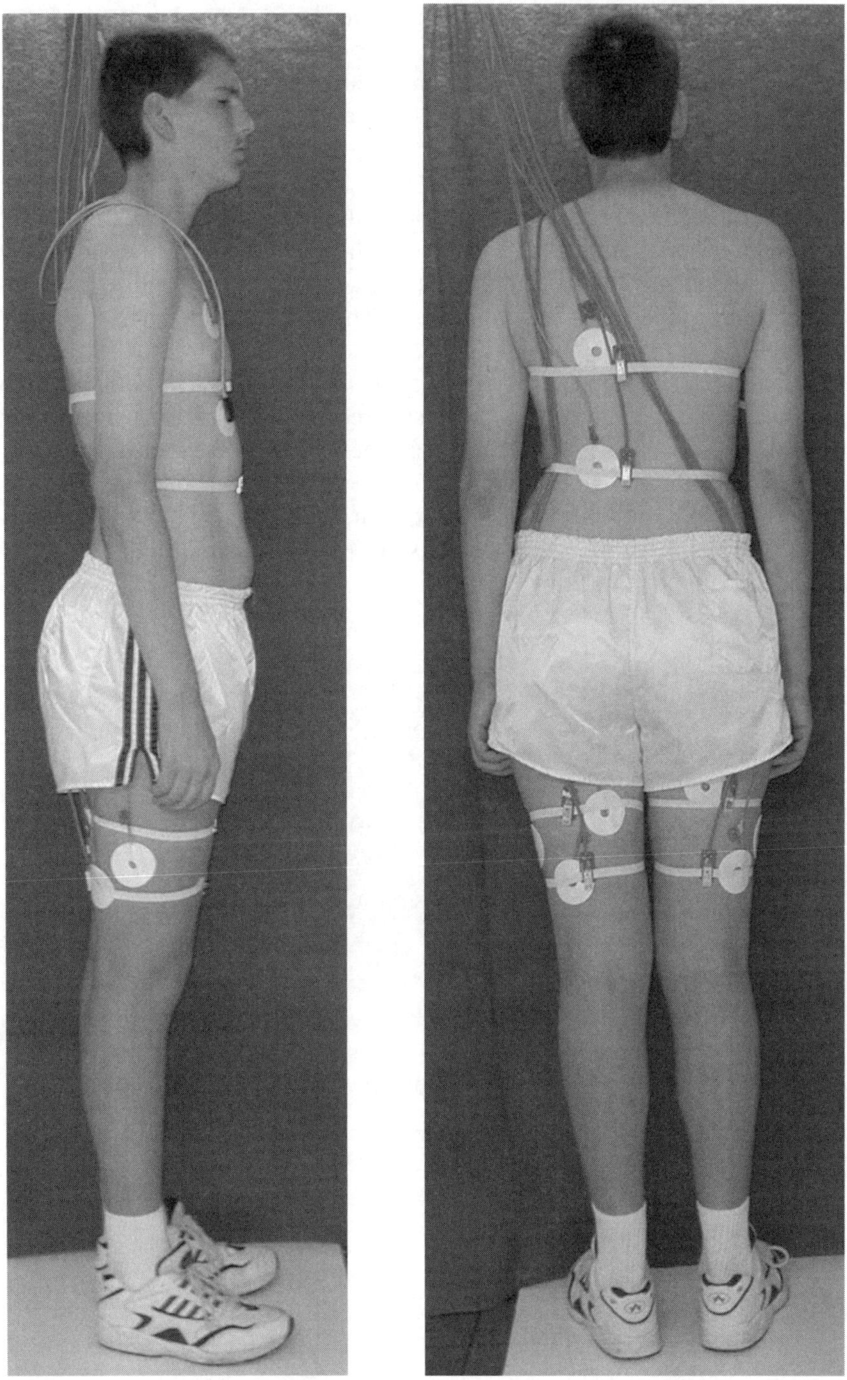

(a) (b)

8.6 Seat trials. Probes attached to the body of a human test subject.

their data (5I3M, 2003). This is a crucial criterion for laboratory techniques and, in fact, only a few test methods lead to results which correlate to real human subject data (Bartels and Umbach, 2003; Bartels, 2003).

8.3.2 Skin model

An important laboratory test method that fulfils the above mentioned criterion of correlations to seat trials data is the so-called skin model. The skin model is a thermoregulatory model of the human skin. It tests the thermophysiological seat comfort of materials and compounds. The skin model is internationally standardised in ISO 11092, EN 31092.

A photo and a schematic drawing of the skin model is given in Fig. 8.7. The measuring unit shown is made of sintered stainless steel and sized 20 × 20 cm^2. Water, which is supplied by channels beneath the measuring unit, can evaporate through the numerous pores of the plate, just like sweat out of the pores of the skin. Additionally, the measuring unit is kept at a temperature of 35°C. Thus, heat and moisture transport are comparable to those of the human skin.

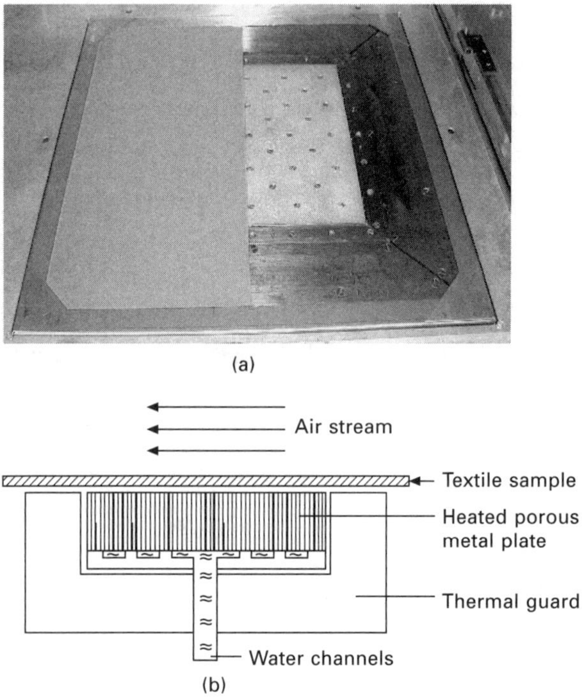

8.7 Photo and schematic drawing of the skin model for testing the thermophysiological properties of materials and compounds.

The car seat specimen is placed on the skin model with its upper surface facing the plate. To simulate the static pressure a person exerts, the sample is loaded by lead weights to a pressure of 14.7 cN/cm^2. To simulate the movements and vibrations of a car, a pneumatic pivot presses onto the specimen with a frequency of 2 Hz and a dynamic load of 5 bar.

With the skin model different seat situations are possible (Umbach, 1999; Bartels and Umbach, 1997, 2003; Bartels, 2003):

1. Normal seat situations, as they are characterised by an insensible perspiration (see page 154). For normal seat situations, especially, the thermal insulation R_{ct} and the water vapour resistance R_{et} ('breathability') according to ISO 11092, EN 31092 of the textiles are important, both should be as low as possible.

   If samples are identically constructed, the thicker one always has the higher (and thus poorer) water vapour resistance R_{et}. In order to take the different thicknesses into account, the ratio

   $$i_{mt} = 60 \text{ (Pa/K) } R_{ct}/R_{et} \qquad 8.2$$

   is defined as the water vapour permeability index, which is a measure of breathability with respect to a sample's thermal insulation. High values of water vapour permeability index are favourable.

   Last but not least, water vapour absorbency F_i is part of the picture in normal seat situations. Therefore, the test is run for exactly two hours. F_i is determined by weighing the specimen before and after the test. High values are advantageous.

   These tests can be divided into 'static' and 'dynamic' measurements (Frankowsky and Umbach, 1995; Umbach, 1999). During static tests, the pneumatic pivot is switched off, during dynamic tests the pivot is turned on. By comparing the difference between the two, the ventilation rate (in per cent) can be quantified. In the case of the water vapour resistance, it is indicated as ΔR_{et} and should be high.
2. Heavier sweating (see also page 156) is simulated with the skin model by measuring the buffering capacity against vaporous sweat F_d according to BPI 1.2 (1994). In this case, 4 cm^3 of water are suddenly evaporated, creating a 'sweat impulse'. Using a humidity sensor, which is placed between the plate and the sample, the time-dependent amount of moisture in the microclimate is monitored. The lower the microclimate humidity, the better, which is expressed by a higher value of F_d.

8.3.3 Seat comfort tester

The seat comfort tester (or upholstery tester) is a device, which was originally built by Hänel and co-workers (Hänel *et al.*, 1995, 1996, 1997). The photo in Fig. 8.8 shows a slightly modified version (Bartels and Umbach, 1997; Bartels, 2003). The seat comfort tester shows good correlation to quantities objectively measured in human test subjects, both for the initial sensation (Bauer and Bartels, 2005) and for long-term sitting (Bartels, 2003).

With the seat comfort tester, entire car seats (seat cushion as well as back rest) can be tested. Its main part is an 'indentor', an aluminium stamp with a rotational elliptical shape, which simulates the contours of the human body. The surface temperature of the indentor is kept constant at 35°C. A telescopic arm presses the indentor into the car seat with a defined force of 440 N on the cushion and 75 N on the back rest, respectively, simulating a person with a weight of 75 kg. Via four heat flux sensors distributed over the indentor's surface area, the initial heat flux H_{ci} into the car seat can be measured (which should be small), as well as the steady state heat flux H_c, which is determined after two hours of testing and should be high. From H_c

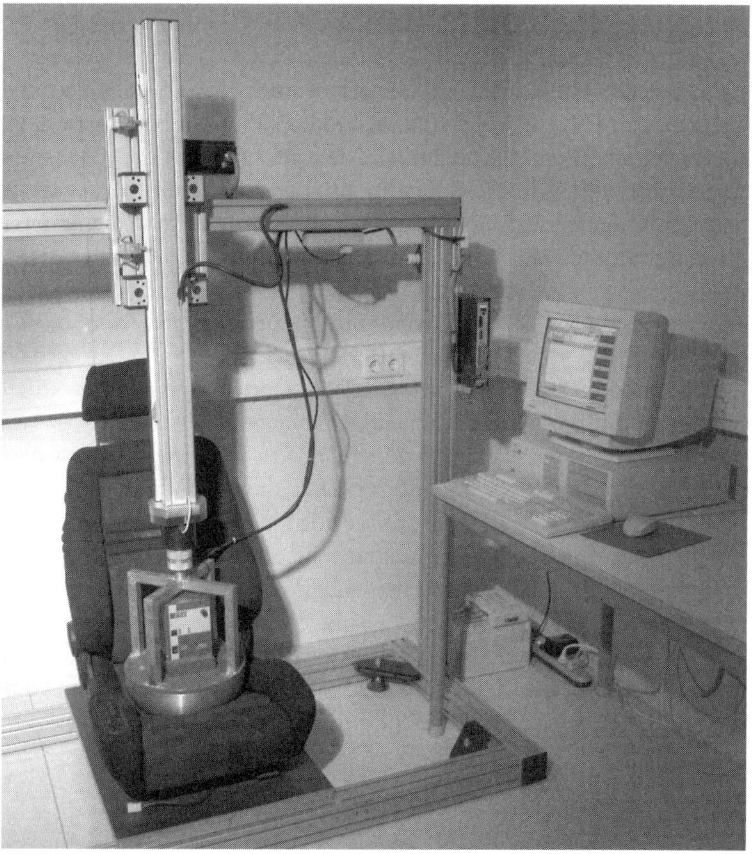

8.8 Photo of the seat comfort tester (upholstery tester) for testing the thermal properties of entire car seats.

the seat's thermal insulation R_c can be calculated (low values are favourable). The seat comfort tester is placed in a climatic chamber, which maintains the climatic conditions.

8.4 Improving comfort properties of car seats

The following shows how the described physiological test methods can be used to improve the comfort of car seats. Therefore, some recent results are given, which in particular exhibit the advantages of modern technical textiles in comparison to currently used materials.

8.4.1 Seat cover and lining

In Table 8.1 the physiological parameters as obtained with the skin model and the seat comfort tester are presented for six different covering materials (Bauer and Bartels, 2005). If one compares the textiles with the leather, it turns out that all textiles are advantageous to leather concerning nearly all physiological parameters. Because of its high thermal conductivity, leather feels significantly cooler when entering a cold car in the winter ($H_{c\,max} = 520$

Table 8.1 Thermophysiological properties of different covering materials. Abbreviations: R_{ct} = thermal insulation, R_{et} = water vapour resistance ('breathability'), i_{mt} = water vapour permeability index (see eq. 8.2), F_i = water vapour absorbency, F_d = buffering capacity against vaporous sweat impulses, $H_{c\,max}$ = maximum initial heat flux in the cold (data taken from Bauer and Bartels, 2005)

Material	R_{ct} (10^{-3} m^2 K/W)	R_{et} (m^2 Pa/W)	i_{mt} (no units)	F_i (g/m^2)	F_d (no units)	$H_{c\,max}$ (W/m^2)
Woven fabric wool/polyamide 95/5, 400 g/m^2	28	5.7	0.30	21.2	0.52	410
Flock, bulk material polyamide, flocked with polyester (65%) and cotton (35%), 270 g/m^2	25	12.5	0.12	3.8	0.32	350
Textile leather imitate, polyester with polyurethane and a cotton lining, 420 g/m^2	26	8.5	0.21	3.1	0.49	390
Velour, cotton/polyester/polyamide, 975 g/m^2	93	19.4	0.29	36.9	0.30	310
3-D spacer fabric, polyester, thickness 4 mm, 355 g/m^2	57	8.9	0.43	0.5	0.51	320
Real leather, 865 g/m^2	33	86	0.02	4.4	0.06	520

W/m^2) than the textiles. Its water vapour transport properties are poor, especially water vapour resistance of R_{et} = 86 m^2 Pa/W, which is too high. This is a consequence of the finishing process of the leather, which is usually covered with a thin film of polyurethane of which the water vapour permeability is small. Modern finishing procedures improve the water vapour resistance of leather by up to 45% (Bauer and Bartels, 2005), but textiles are still much better. An interesting alternative for those who would like to have the 'leather look', is the textile imitation, of which the physiological properties are much better than those of a real leather.

For a pleasant initial sensation of warmth, the thermal conductivity of the cover should be small. Best results are found for textiles that include a high amount of air, such as, for example, a velour or a three-dimensional spacer fabric ($H_{c\ max}$ = 310 or 320 W/m^2, respectively).

The water vapour absorbency can be improved by using hygroscopic materials like wool or cotton. The woven fabric made of 95% wool presented in Table 8.1 is able to absorb 21.2 g/m^2 of water vapour. On the other hand, the three-dimensional spacer fabric made of polyester absorbs just 0.5 g/m^2.

For the buffering capacity against vaporous sweat impulses F_d of the whole car seat, the water vapour absorbency of the cover and its lining are of great importance. In Fig. 8.9, the influence of different linings is presented (Umbach, 2004). In combination with the same cover, the three-dimensional spacer fabrics lead to significantly lower microclimate humidities than a

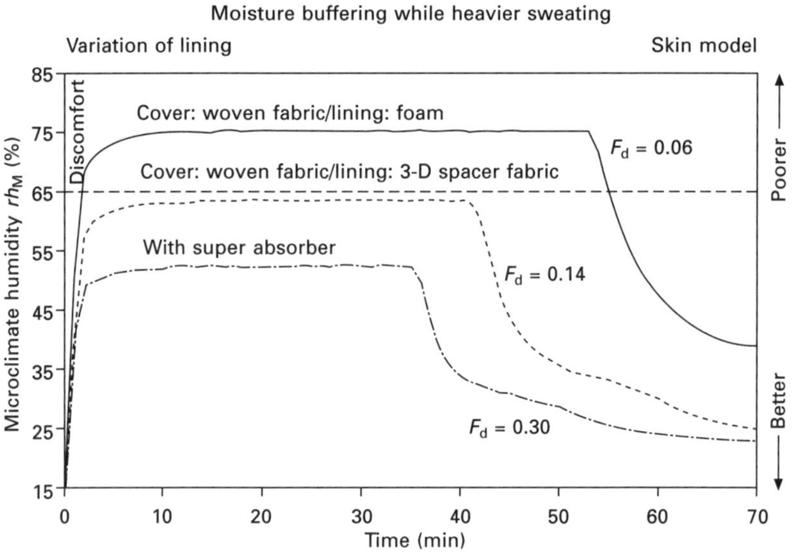

8.9 Influence of different lining materials on the buffering capacity against vaporous sweat impulses (Umbach, 2004).

foam lining, and, in addition, humidity values drop faster after the simulated sweat impulse. Consequently, the buffering capacity of the combination with the spacer fabric lining is $F_d = 0.14$, whereas the compound with the foam lining only reaches $F_d = 0.06$.

The up-to-date best lining alternative from a physiological point of view is also textile based, a nonwoven with super-absorbent-polymers (SAP) (Thill and Krause, 2005). As shown in Fig. 8.9, such a lining leads to the lowest microclimate humidity and obtains $F_d = 0.30$.

8.4.2 Cushion

Not only for the covering material but also for the cushion do textiles offer great potential in improving the physiological performance of car seats (Bartels and Umbach, 1997, 2005; Umbach, 1999, 2000b; Bartels, 2003). In Fig. 8.10 the water vapour resistance as measured with the skin model of different cushion materials is presented. These tests clearly demonstrate that moulded foam pads, which are quite popular in the automotive industry due to their comparably low costs, lead to only a poor (i.e. high) water vapour resistance of $R_{et} = 113$ m^2 Pa/W. Sometimes even higher values occur, because for the moulding process a thin film on the back of the cover may be required, which prevents foam material from penetrating through the cover, but which impedes water vapour penetration. In this case, a breathable membrane would be preferable to that film, but due to the higher costs, breathable membranes could not find their way into the actual production of moulded car seats.

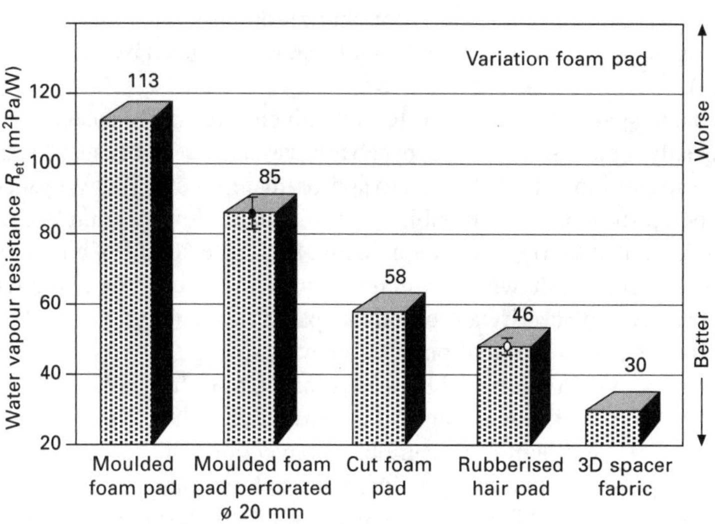

8.10 Water vapour resistance R_{et} of different cushion materials (Bartels and Umbach, 2005).

As an additional problem, many foams only have pores, which are not or are only very loosely interconnected with each other. For a water vapour molecule, this further handicaps the diffusion through the foam. A perforation improves water vapour transport. In the example shown in Fig. 8.10, the water vapour resistance can be significantly lowered to a value of $R_{et} = 85\,m^2$ Pa/W. However, each of the holes should have a diameter of at least 20 mm, otherwise they collapse when compressed by the weight of the passenger, and, thus, lose their function.

A perforation improves the water vapour transport, but a cut foam with pores, which are connected with each other, is better. It decreases the water vapour resistance to $R_{et} = 58\ m^2$ Pa/W, which is only approximately half of the non-perforated moulded foam pad (i.e., doubled 'breathability'). Hence, cut foams are frequently used in the car industry, although they are more expensive to handle than moulded foam pads.

A further improvement is achieved by a rubberised hair pad ($R_{et} = 46\ m^2$ Pa/W). Owing to its high content of hygroscopic hair, this construction can also absorb a considerable amount of water vapour. As a disadvantage, rubberised hair pads often offer only poor ergonomic comfort. Hence, they have to be combined with softer cushion or lining materials.

From the physiological point of view, today the optimum cushion alternative is a three-dimensional spacer fabric. It offers the smallest water vapour resistance of only $R_{et} = 30\,m^2$Pa/W, i.e. approximately the doubled 'breathability' in comparison to the best (cut) foam pad. This advantage is due to the very open structure of the spacer fabric that supports water vapour diffusion. Another benefit is that air can easily move through the spacer fabric, which enhances ventilation and pumping effects due to the car's or the passenger's movements and which can be enhanced by a fan (see Section 8.4.4).

Today, spacer fabrics are often combined with other cushion materials, especially foams. Just a few years ago this was necessary, because the thickness of the spacer fabrics was limited to approximately 1 cm. Today, spacer fabrics can be up to 6 cm thick (Helbig, 2006), which allows a complete or nearly complete substitution of the foam. In Fig. 8.11, an example of an experimental aeroplane seat is shown, in which the foam in the cushion and back rest is extensively replaced by a 4 cm thick spacer fabric (Bartels, 2003).

From the physiological point of view, a spring coil construction is also favourable over a foam cushion (Faust and Umbach, 1984a, 1984b; Bartels and Umbach, 1997). A spring coil construction includes a thick air layer, in which warm and humid air can obey convection. Hence, the physiological effect is similar to that of a seat including a three-dimensional spacer fabric. However, owing to their high costs and heavy weight, spring coil constructions are not found in modern cars.

8.11 Modification of an aeroplane seat by means of a spacer fabric (Bartels, 2003).

8.4.3 Surface design

In addition to the selection of the materials, the seat's design is important (Umbach, 1999, 2000b). Surface grooves enhance ventilation, which leads to a higher exchange of moist air with the drier environment. On the other hand, large side supports, as they are frequently used in sport seats, hinder ventilation. In this respect, today's car seat design trends are often in contradiction with improved physiological properties.

8.4.4 Active climatisation

During approximately the last ten years the inclusion of active elements played a significant part in car seat production. For both cold and hot climate, active elements are available, which support the physiological function of the seat and improve comfort.

Today standard in upper class vehicles – but available in nearly all car classes – is seat heating (Weiss and Ulbrich, 2003). Many of the heating systems require a carrier. For this purpose, today apart from foams various technical textiles are employed, such as nonwovens, knitted fabrics, waddings, or three-dimensional spacer fabrics.

Recent developments include conductive textiles (Möhring, 2005; Möhring *et al.*, 2005). This new generation of 'smart textiles' is not yet used in car seat heatings, but offers great opportunities for the future, also for sensor technology, illumination and design.

Apart from the materials used for the heating, fundamental investigations of Hänel and co-workers (Hänel *et al.*, 1995, 1997) showed that their placement is as important. Only if the heating elements are near enough to the passenger and their distribution in the seat leads to an entire warming of the whole seat surface in contact with the person can a pleasant sensation soon be achieved.

If all components of seat and heating are attuned to each other, the heating can lead to a significantly faster warming of the car seat. In Fig. 8.12 the time dependent initial heat fluxes H_{ci} as measured with the seat comfort tester are given for a car seat with a jacquard woven fabric on top and (a) the heating switched off and (b) the heating turned on. After an identical starting point at $t = 0$ with a maximum heat flux of $H_{c\ max} \approx 330\ W/m^2$, the heating leads to a much faster decrease of the heat flux, i.e. warming of the seat. If the heating is switched off, it takes 180 s until a heat flux is achieved, which is perceived as slightly cool, and 275 s until a pleasant heat flux is reached. If the heating is turned on, it only takes 43 or 50 s, respectively.

In modern 'climate seats' active elements are included which not only improve comfort in the cold, but also in the heat. These seats employ a

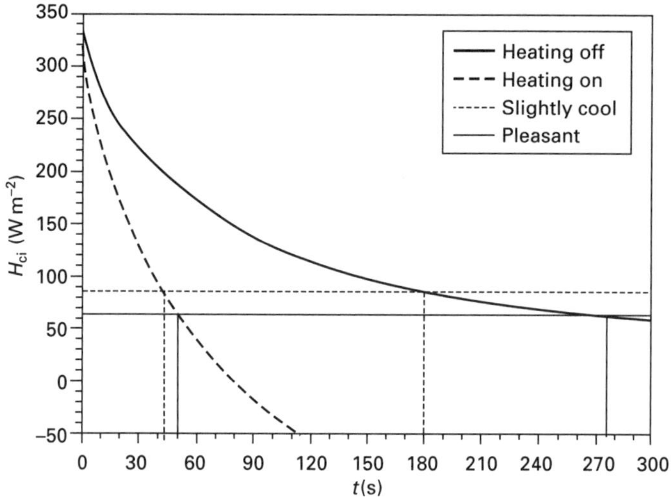

8.12 Influence of seat heating on the initial sensation of warmth. Initial heat flux H_{ci} as a function of time t, monitored by means of the seat comfort tester in a cool environment. The seat heating is either switched on or off.

Table 8.2 Steady state heat flux H_c as measured by means of the seat comfort tester after 2 hours of testing for two car seats with a different cushion design. High values are advantageous

Cushion construction	H_c
Forced ventilation	64 ± 13
Foam	16.0 ± 0.2

ventilation layer made of a three-dimensional spacer fabric, in which a forced air stream driven by a fan exchanges warm air with the cooler ambience. In Table 8.2 the great benefit of a cushion construction comprising a forced ventilation is shown with the example of two seats with a Jacquard cover (Bartels and Umbach, 1997). The steady state heat flux H_c, as measured by means of the seat comfort tester after two hours, is about 16 W/m^2 for a seat with a foam cushion. But the seat with forced ventilation is able to dissipate 64 W/m^2, i.e. a factor of four.

In most of today's cars air conditioning is built in, leading to a significant reduction of the thermal stress of the driver and the passengers in the summer time. However, forced ventilation in a seat is still of great benefit. Owing to their thickness, which is necessary to be able to carry the weight of the passenger and ensure ergonomic comfort, all cushions and back rests have a high thermal insulation, which is comparable or even higher than that of warm winter clothing. Thus, even in a pleasant cabin environment, a ventilated seat guaranteeing a higher heat flux is more comfortable.

8.5 Conclusions

Today, comfort is one of the most important properties of a car seat. Modern technical textiles used as covering, lining and cushion materials can significantly improve climatic comfort, especially in comparison to non-textile materials like leather or foam:

- Textile covering materials show better water vapour resistance ('breathability') and feel less cold in the winter time than leather.
- Three-dimensional spacer fabrics as linings or cushions transport more heat and moisture than foams and, thus, reduce the accumulation of sweat. They can be combined with a fan, which enforces the ventilation, to further improve the dissipation of heat and moisture.
- Linings made of textiles including super-absorbent polymers are able to absorb high amounts of perspired moisture, even during heavier sweating.

The seat comfort can be tested via the physiological laboratory apparatus skin model and seat comfort tester, which lead to reproducible results, which are in good agreement with practice, and are much faster and cheaper and, thus, highly efficient for a product development process. Seat trials with test persons are best positioned as a validation procedure at the end of such a process. If seat trials are performed in a climatic chamber, test conditions can be kept reproducible.

8.6 Future trends

It is foreseeable that in the future the importance of comfort in the car will be even more significant as one of the main factors in the competition between car manufacturers. The car seat is that component in the car, which has the largest contact area with the passenger, and it stays in direct touch throughout a journey. Thus, the car seat is one of the main components for comfort. Consequently, modern climate seats are increasingly desired by the consumer – and ordered.

Owing to their physiological benefits in comparison to concurrent materials, the demand for technical textiles for use in car seats will increase. As approximately 3.5 kg or 10 square metres are used per car just for the upholstery (Erth *et al.*, 2005), textiles for car seats possess interesting market potential.

8.7 Acknowledgements

We are grateful to the Forschungskuratorium Textil for the financial support of the research project (AiF-No. 13495 N), which was funded by the German Ministry of Economy and Technology via grants of the Arbeitsgemeinschaft industrieller Forschungsvereinigungen 'Otto-von-Guericke'.

8.8 References

5I3M (2003), *5th International Meeting on Thermal Manikin and Modelling*, Strasbourg, September 29–30.

ATA (1997), *4th International Conference and Exhibition on Comfort in the Automotive Industry – Recent Developments and Achievements*, Bologna, Italy.

Bartels V T (2003), 'Thermal comfort of aeroplane seats: influence of different seat materials and the use of laboratory test methods', *Applied Ergonomics*, 34, 393–399.

Bartels V T and Umbach K H (1997), 'Laboratory tests of thermophysiological seat comfort', *4th International Conference and Exhibition on Comfort in the Automotive Industry – Recent Developments and Achievements*, Bologna, Italy.

Bartels V T and Umbach K H (2003), 'Messverfahren zur Beurteilung der Atmungsaktivität von Schutzbekleidung und Fahrzeugsitzen', *Technische Textilien*, 46(1), 58–61.

Bartels V T and Umbach K H (2005), 'Physiologically optimised car seats – latest findings and trends', *13th Techtextil Symposium*, Frankfurt a.M., Germany.

Bauer B and Bartels V T (2005), 'Entwicklung von textilen Bezugsstoffen für Fahrzeugsitze und Polsterungen mit optimiertem Initialwärmeempfinden', technical report No. AiF 13495 N, Hohenstein Institutes, Bönnigheim, Germany.

Bauer B, Bartels V T and Umbach K H (2005), 'A study of the initial perception of warmth of car seats – the important first impression', *Kettenwirk-Praxis* 39(4), 16–18.

BPI 1.2, 'Measurement of the buffering capacity of textiles with the thermoregulatory model of human skin (Skin Model)', Hohenstein Institutes, March 1994.

Erth H, Gulich B and Fuchs H (2005). 'Mobiltextilien – Entwicklungstendenzen bis 2010', *44th International Man-Made Fibres Congress*, Dornbirn, Austria.

Faust E and Umbach K H (1984a), 'Climatic comfort in car seats – Part 1', *Automobiltechnische Zeitschrift*, 86, 99–102.

Faust E and Umbach K H (1984b), 'Climatic comfort in car seats – Part 2', *Automobiltechnische Zeitschrift*, 86, 399–405.

Frankowsky B and Umbach K H (1995), 'Quantitative Erfassung des klimatischen Sitzkomforts von KFZ-Sitzen', *Hohensteiner Forschungsbericht*, July.

Hänel S E, Dartman T and Shishoo R (1995), 'A new method for measuring mechanical and physiological comfort in car seats', *34th International Man-Made Fibres Congress*, Dornbirn, Austria.

Hänel S E, Dartman T and Shishoo R (1996), 'A comparison of measuring methods for comfort rating of beds', *Advances in Occupational Ergonomics and Safety I*, 2, 783–788.

Hänel S E, Dartman T and Shishoo R (1997), 'Measuring methods for comfort rating of seats and beds', *International Journal of Industrial Ergonomics*, 20, 163–172.

Helbig F (2006), 'Neue Dimensionen für funktionelle, reguläre 3D-Gewirke', *Melliand Textilberichte*, 87, 718–720.

ISO 8996, 'Determination of metabolic heat production', 1990.

ISO 10551, 'Assessment of the influence of the thermal environment using subjective judgement scales', 1995.

ISO 11092, EN 31092, 'Measurement of thermal and water-vapour resistance under steady-state conditions (sweating guarded-hotplate test)', 1993.

Mecheels J (1998), *Körper – Klima – Kleidung: Wie funktioniert unsere Kleidung?*, Berlin, Schiele & Schön.

Möhring U (2005), 'Conductivity in innovative textiles for automotive applications', *13th Techtextil Symposium*, Frankfurt a.M., Germany.

Möhring U, Neudeck A and Gimpel S (2005), 'Textile innovation in the autocar by conductivity, sensory and lucent structures', *44th International Man-Made Fibres Congress*, Dornbirn, Austria.

Thill U and Krause U (2005), 'Absorber comfort automotive seat', *13th Techtextil Symposium*, Frankfurt a.M., Germany.

Umbach K H (1986), 'Physiological aspects of seat comfort', *Symposium for Airlines and for the Aircraft Industry*, Frankfurt a.M., Germany.

Umbach K H (1999), 'Parameters for the physiological comfort on car seats', *38th International Man-Made Fibres Congress*, Dornbirn, Austria.

Umbach K H (2000a), 'Physiologischer Sitzkomfort im Kfz', *Kettenwirk-Praxis*, 34(1), 34–40.

Umbach K H (2000b), 'Climatic comfort of car seats', *Technical Usage Textiles*, 3(35), 34–38.

Umbach K H (2004), 'Messung und Beurteilung des klimatischen Komforts von Kfz-Sitzen', *Forschungskuratorium Textil e.V. Jahrestagung 2004 – Technische Textilien im Fahrzeugbau*, Eisenach, Germany.

Weiss M and Ulbrich T (2003), 'Automobil-Sitzheizungstechnologie im Wandel', *42nd International Man-Made Fibres Congress*, Dornbirn, Austria.

9 Smart textiles in automotive interiors

F BOUSSU, C COCHRANE, M LEWANDOWSKI and V KONCAR, ENSAIT, l'Ecole Nationale Supérieure des Arts et Industries Textiles, France

Abstract: In this chapter different smart textile structures suitable for use in car interiors are presented. One of the main interests of smart textile integration in car interiors is also related to the 'customization' trend. Designing very low weight car seats with heating fabric and achieving space saving and comfort thanks to a more breathable structure are now possible. However, only few applications can be found today involving smart textile structures comprising sensors, actuators, and computing and storage devices integrated into internal car elements. On the other side, car interiors contain a range of textile surfaces that may host textile-based sensors and actuators adapted to this specific space. They may be classified in function of the measured parameters and effects they are able to generate. This classification is given below:

- Sensors: temperature, humidity, strain, UV radiation, acceleration, light intensity, etc.
- Actuators: heating, cooling, different kinds of alert signals, flexible screens and displays, security systems, etc.

Flexible devices integrated into textile structures of a car interior could also be used as physiological sensors for various vital parameters. Heating textiles can be developed to be suitable for flexible structures as car seats, steering wheel, etc.

The potential use of shape memory alloys in car seats, so that they always have the same 'new' look, is also very interesting. Their price is currently the main problem. Nevertheless, they may be used in small quantities and mixed with traditional textile threads to achieve costs attractive for luxury cars

Keywords: actuators, sensors, shape memory alloys.

9.1 Introduction

9.1.1 Textile structures in car interiors

During the recent 'Futurotextiles' exhibition,[1] various car makers and most of the automotive suppliers made a brief review of the existing textile solutions used in car design and fabrication. Figure 9.1 indicates the textile parts that may be found in a car made at the present time.

- The first logo identifies non-woven parts such as: bonnet cover, fuel filter, oil filter, air filter, carpets, petrol tank, sun visor, roof panel, window shelf, silencer and driveshaft tunnel.

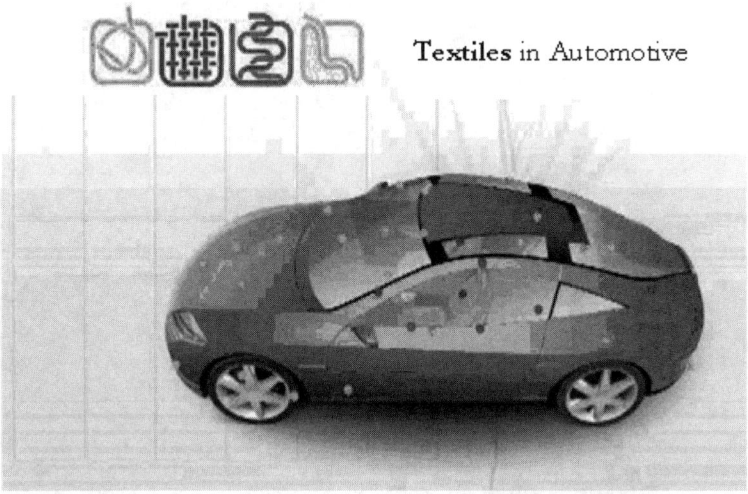

9.1 Textiles in automotive.

- The second logo is used to show woven parts such as: airbags, safety belts, bucket seat fabrics, convertible roof.
- The third logo indicates knitted parts such as: bucket seat covering, headrest covering and storage shelf net.
- Finally, the fourth logo is used to show composite parts such as: car body, tyres, brakes, bumpers, belts, hoses, dashboard and armrests.

Moreover, the actual bucket seat of the car can be recovered with a knitted or woven fabric, or with leather upholstery. The material used for seat recovering needs to meet several criteria. First, it must match the car interior's style and coordinate with different parts such as the dashboard and roof-panel colours. Second, there are numerous constraints resulting from the use of the bucket seat, such as: tensile strength, good drape behaviour, abrasion and pilling resistance, wet and dry resistance, UV resistance, ease of cleaning and recycling. Finally, costs must be kept as low as possible.

One of the future trends for bucket seats is 'customization' like the Happy attitude concept (see Fig. 9.2) described on the Faurecia web site (http://www.faurecia.fr/pages/happy/home.asp). Car interior themes can be swapped quickly and easily by the customer and a different theme can be chosen to match the current mood of the driver.[2]

Thanks to this new concept, the dashboard can be printed on demand and can also be made of the same textile material as the bucket seat covering. This customized material can be easily removed for cleaning.

Another future trend for bucket seats is to replace the foam by a textile material able to match various constraints such as resilience and resistance

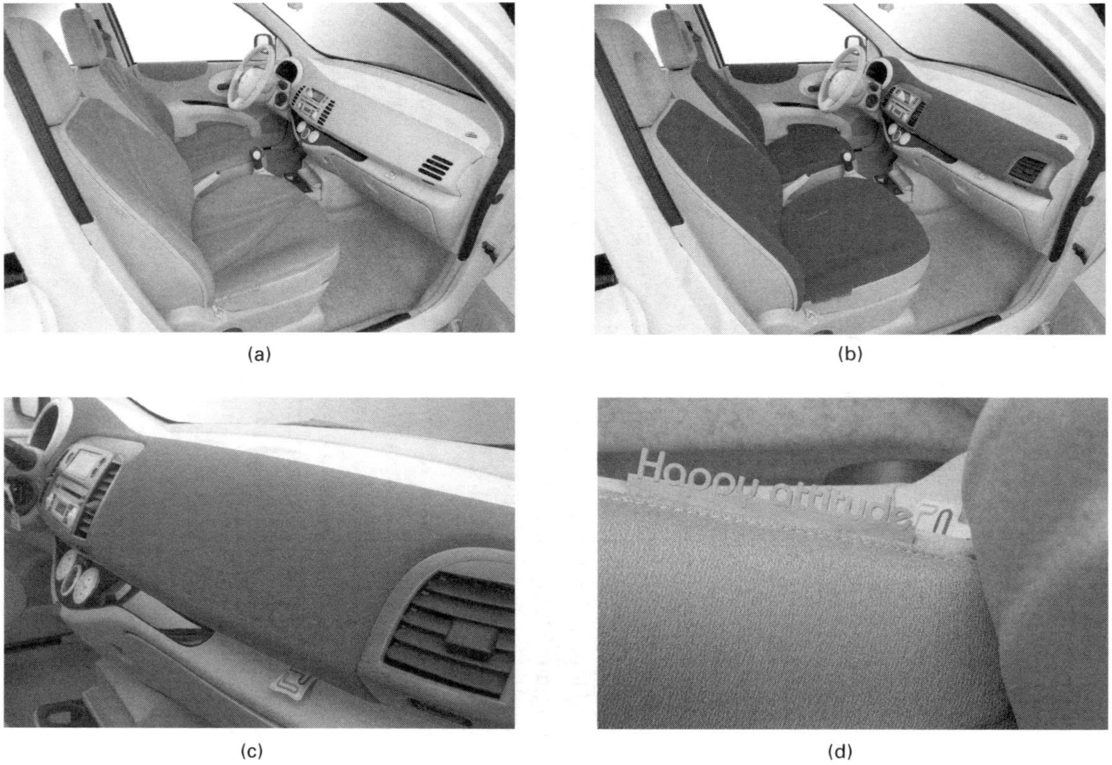

9.2 Faurecia Happy attitude concept.

to fatigue. As a result, the chemical agents used in foam production and the different solvents used during the production process are totally avoided, which may lead to a more easily recycled material such as the Banex ™ seat[3] (see Fig. 9.3).

Banex is an automotive seat fabric that achieves the cushioning effect of springs through machine-made warp-knitting and finishing using special polyester yarns. This makes it possible to design a seat with greater weight reduction, space saving and comfort than ever before, since a single sheet of fabric provides the same elastic function as the urethane foam and springs. The ability to add an anti-bacterial treatment to the fabric is another advantage.

Another solution using leno fabric has been produced by the quantum group (www.quantum5280.com) (see Fig. 9.4). The advantages of this method are numerous, including: polyurethane foam replacement, weight reduction of the car seat, a more breathable structure and space reduction of the entire car seat.

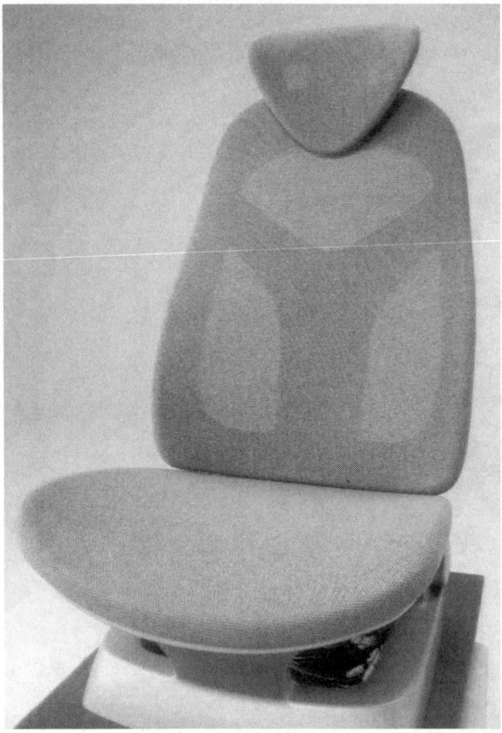

9.3 Banex car seat solution.

9.4 Leno fabric for car seat fabric fitted for baby.

9.5 Double face warp-knitted solution.

Better ageing of a car seat can be achieved by the use of a double-face warp-knitted fabric called spacer fabric (see Fig. 9.5). This fabric, coupled with reducing the height of the foam inserted in the final car seat structure, may help achieve new functionalities such as cooling or heating, breathable performance and foam fatigue reduction.

On the other hand, only a few applications can be found involving smart textile structures comprising sensors, actuators, and computing and storage devices integrated into the car interior. This is probably due to the fact that smart textiles are relatively new, insufficiently known and mostly applied to garments. For example, a smart shirt or wearable motherboard developed in the 1990s by S. Jayaraman from GaTech, was one of the first applications of smart textiles in garments. On the other hand, it is well known that a car interior contains various textile surfaces that are possible hosting structures for sensors and actuators. As an illustration, airbags are made from textiles and, in the case of an accident, need to be controlled in a smarter way than they are at present. Car seats and carpets are very large areas that are in permanent contact with occupants and they are excellent interfaces for the integration of sensors and actuators. There are already some applications such as heated seats or sensors that detect passenger presence in order to suggest seatbelt fastening. However, in all these cases the devices are not really integrated into the textile structures, they are more or less hidden behind them.

Nevertheless, it is possible to find an example of fully integrated smart textiles in an airbag control system comprising a restraining apparatus adapted to be connected to the occupant side of an airbag.[4] This extension/retraction device is used to adjust the degree to which the restraining apparatus can be extended, and a control device directs the extension and retraction of the device. In operation, the restraining apparatus can be retracted by the extension/retraction device in response to a signal sent by the control device to limit the physical extent to which the airbag can deploy to prevent injury to a vehicle occupant adjacent to the airbag.

The aforementioned airbag control system contains sensors and actuators interconnected via the apparatus for controlling airbag deployment. Generally, sensors and actuators adapted to a car interior may be classified as a function of the values they measure and the effects they generate:

- Sensors: temperature, humidity, strain, UV radiation, acceleration, light intensity, etc.
- Actuators: heating, cooling, different kinds of alert signals, flexible screens and displays, security systems actions, etc.

Different applications of smart fabrics inside the car seat can be envisaged[5] for discriminating between adult and child occupants and being able to make safety adjustments to the actual load to be supported by seatbelt actuators during an accident. From an aesthetical point of view, smart fabrics can be inserted in the car interior to improve illumination and replace the present switch systems.

Smart textile systems are defined in a more structured way in the following sections.

9.1.2 Smart textiles

The term 'smart textile' describes a class of apparel that has active functions in addition to the traditional properties of clothing. These novel functions or properties are obtained by utilizing special textiles or electronic devices, or a combination of the two. Thus, a sweater that changes colour under the effect of heat could be regarded as a smart textile.

Finally, all smart textiles can operate in manual or automatic mode. In the case of manual operation, the person who uses the textile can trigger these additional, intelligent functions, while in automatic mode the textile structure can react autonomously to external environmental parameters (temperature, humidity, light, etc.).

The informal taxonomy proposed by Swallow and Thompson[5] may help to define the combinations available by considering the association of different components of a smart fabric, such as being active or passive, sensor or effector.

9.2 Technical elements enabling the production of smart textiles

In this section, the building blocks for integration into smart textile structures are further described.

9.2.1 Peripherals

The main peripherals usually found in smart textiles are mentioned and briefly analyzed in the next few paragraphs.

9.2.2 Control interfaces – near the 'human interfaces'

The use of textiles in a car interior as supports for control interfaces is interesting because the control interfaces can then be close to the appropriate parts of the body.

In contrast to certain miniaturized communication devices, textile surfaces in a car interior have a greater surface area, which enables them to offer more functionality. For example, the small keyboard of a mobile phone that fits in the palm of one's hand becomes much more readable when transposed to the surface of a piece of fabric that is three times larger. New properties guaranteeing resistance to use and to washing should also be taken into account.

9.2.3 Sensors

Car interior textiles could be used to detect different actions, in particular recognizing the posture of the driver and passengers and their gestures, in order to facilitate certain commands that are intuitive or to adapt and prepare security systems such as airbag deployment or seat belt pretensioning and positioning. Moreover, when these sensors are connected to computing facilities and the control unit, they may allow the recognition of a situation and its context for a better interpretation of reality.

Sensors integrated into the textile structures of car interiors could also be used as physiological sensors for various parameters. This term refers to the sensors used to record the health or personal parameters of a driver and passengers in a broad sense and to raise an alert if necessary. The applications arising from the use of these sensors are numerous. We can, for example, use sensors to provide a physical performance analysis of a car driver or even a racing car driver, or to conduct a medical check-up for passengers in real time.

9.2.4 Interfaces for displaying information and actuators

For many applications, it is necessary to display or reproduce the information supplied by the communication systems that are integrated into a car interior's textile structures. Therefore traditional interfaces, such as displays, screens and loudspeakers, have to satisfy the same ergonomic and mechanical resistance criteria as those quoted for control interfaces. For example, the current characteristics of colour liquid crystal displays, rigidity, weight and power consumption, have to be adapted. Solutions such as micro-screens embedded in glasses and the use of technologies including flexible supports, such as flexible displays and optical fibres, have begun to appear.

In addition, the proximity of these textile structures to the natural human senses opens up new possibilities for the transmission of information. Visual and auditory ways of collecting information (such as screens and loudspeakers), which at present are largely developed the way they are because they do not require direct contact with the user, could soon be joined by tactile and olfactory methods. The seat fabric that changes the environment by diffusing a combination of perfumes is about to leave the realm of science fiction.

9.2.5 Data processing

The hardware elements of memory, computation and data processing (RAM, hard disks and processors) will certainly not evolve much in the short term unless they do so in the direction of miniaturization. Even if progress is made on developing flexible substrates, they remain fragile and require partly rigid protection in order to be integrated into smart textile structures. However, their integration has become entirely possible, as seen by the incorporation of a micro PC into the loop of a belt. It is also conceivable that only a small quantity of information would need to be processed locally in smart textiles, and that more complex functions and more significant memory requirements could be handled by higher-powered remote servers. This difference between local and remote processing requires the development of specific algorithms, as for the case of intelligent vehicles.

9.2.6 Connectors

Connection problems are another major issue for state-of-the-art smart textiles. The principal question is how to transport information and energy among the various components of the electronic system with optimal efficiency. The concepts of weight distribution and ergonomics must be taken into account in distributing the different components around various parts of the vehicle.

In addition, it would seem to be an advantage to have only one energy source distributing power to the disparate electronic interfaces, thus allowing better energy management. On the other hand, each electronic interface could have its own computation and storage capacities, which would allow resources to be allocated and weight to be distributed in a better way.

It is important to examine the problem of control and the centralization of information distribution. In fact, to be able to manage all of the functions of a complex communicating device, it is necessary to combine output control and information input on a single interface. This means that accessing emails or directions on a map web site, for example, must be done on a single screen.

9.2.7 Energy

Energy autonomy is still a major handicap for the majority of mobile electronic devices. Many users of wireless devices have no doubt dreamt of never having to recharge their mobile phones. Even if electronic circuits now require increasingly less power, new features keep being developed and create an additional need for energy (a larger screen size implies a need for greater power consumption).

An interesting alternative seems to be the use of renewable energy sources. Solar energy and wind are relatively difficult to adapt for use with vehicles because they require large surface areas to be truly effective. On the other hand, many studies have been carried out on techniques that will make it possible to recover the energy released by vehicle movement.

9.3 Textile sensors

9.3.1 Conductive composite

A sensor based on a thermoplastic elastomer (Evoprene)/carbon black nanoparticle composite has been developed during this study. This sensor displays general mechanical properties that are strongly compatible with any flexible and soft material such as a textile structure and offers great potential for use in car interiors.

To improve the precision and reliability of a sensor, it is important to know the influence of external parameters such as rate of strain deformation, temperature and humidity on the sensor's response and on its mechanical behaviour.

The design and calibration of the sensor on a nylon fabric have been demonstrated in previous papers.[6,7] This present study will describe more particularly the behaviour of the sensor in various climatic conditions of different temperatures and relative humidity, as well as some results on the sensor's response at different strain rates.

Full details of the sensor's fabrication and optimization can be found in our previous work.[7] The conductive polymer composite that the sensor material is made from is prepared via a solvent process. Carbon black particles (Printex L6, from Degussa) are dispersed with a thermoplastic elastomer (Evoprene 007 from Alpha Gary, which is a Styrene-Butadiene-Styrene (SBS) co-polymer) in chloroform (Aldrich). After mixing for the necessary time at 55°C, the blend obtained is applied as a coating on the textile substrate and left to dry at room temperature. The final concentration of carbon black filler particles in the Evoprene polymer matrix is 27.6 vol.%, after total evaporation of the chloroform solvent. This concentration is greater than the percolation threshold so that there is a good conducting network throughout the composite. Finally, a protective latex film is applied on the sensor. Figure 9.6 shows schematically the dimensions of the sensor on the fabric, which is a light nylon fabric with a mass per unit area of 42 g.m^2. The linear density of the weft and warp yarns is equal to 3.3 tex.

The complete sensor system (conductive polymer composite + latex film) has the following dimensions: 10 mm wide, 110 mm long and only 46 μm thick. The sensor track is parallel to the weft direction of the fabric (sample dimensions: 50 mm × 300 mm). Stainless steel yarns, used as electrical

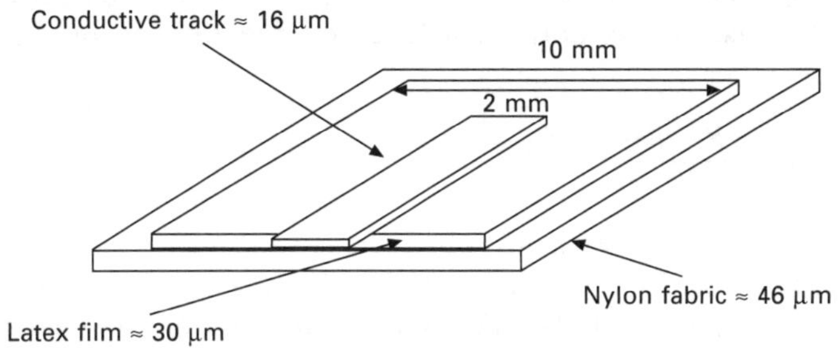

9.6 Schematic representation of the structure and dimensions of the sensor integrated on the fabric.

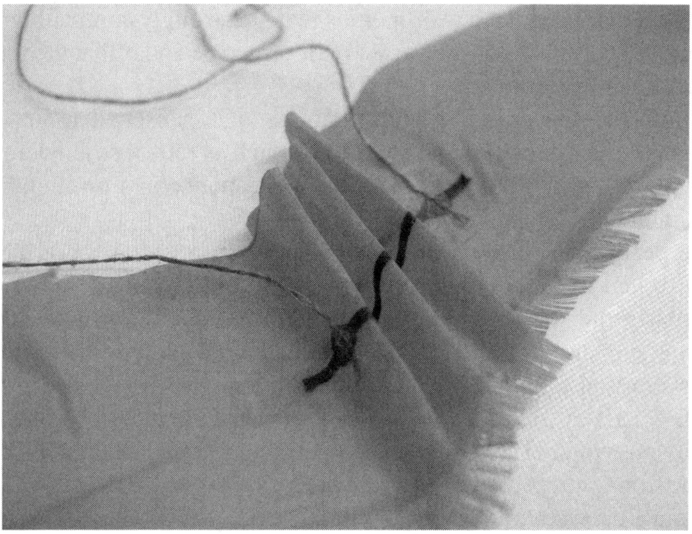

9.7 Set-up of sensor on fabric, with electrical connections.

connections, are added at the ends of the sensor track in order to measure its electrical response (Fig. 9.7).

9.3.2 Experimental results

Tensile characterization

The sample was set up on a tensile testing machine (MTS 2/M), with the sensor track positioned parallel to the direction of fabric extension and at the centre of the test specimen (fabric test length = 200 mm). Different rates of extension, 16, 200 and 500 mm/min, were applied to different samples, and the electrical resistance of the sensor was recorded during elongation until the sample broke. Since the different sensor specimens will necessarily present slight variations in their intrinsic resistance values, a normalized relative electrical resistivity is defined to characterize the sensor's electrical property:

$$Rr = \frac{(R - R_i)}{R_i} \Delta R / R \qquad 9.1$$

where R_i is the initial resistance of the sensor measured before extension, and R the resistance at a certain length l of the sample. The sample was left at rest for about 1 minute before beginning the test, and R_i represents the mean value of resistance measured during this lapse of time. The electrical resistance was measured until the fabric broke up, which occurred at around

45% elongation. In general textile applications such as weaving, the elongation zone of interest is within the 0 to 10% range. An example of the electrical behaviour of the sensor in this zone is given in Fig. 9.8.

To be able to compare our gauge sensitivity to existing strain gauges, a first assumption of a linear variation was made. In this case, the gauge factor (K) is defined as follows:

$$K = (\Delta R/R)/\varepsilon \qquad 9.2$$

In the elongation range (0–10%), K is greater than 30. In comparison, the coefficient K of a classic metal gauge (copper-nickel) is 2.1. Moreover, a classic metal gauge has a range of elongation between 0.1 and 0.5%.[8]

The linear relationship (Eq. 9.2) applied to our sensor is, however, not very satisfying. In fact, the electro-mechanical behaviour of the sensor can be better modelled by the following equation:

$$\Delta R/R = a\varepsilon^b \qquad 9.3$$

where a represents the sensitivity of the sensor, while b relates to the deviation from the linear model, in which for the ideal case $b = 1$ and $a = K$.

The good fitting of this model is illustrated on Fig. 9.8 by the solid line. In this case, $a = 99.9$ and $b = 1.51$. The two parameters depend not only on the intrinsic characteristics of the sensor (initial resistance, length to width ratio, composite quality, etc.), but also on the elongation rate. Table 9.1 gives the values of a and b obtained at other rates of deformation. In all cases, there is a good match between the experimental data and the model (Eq. 9.3).

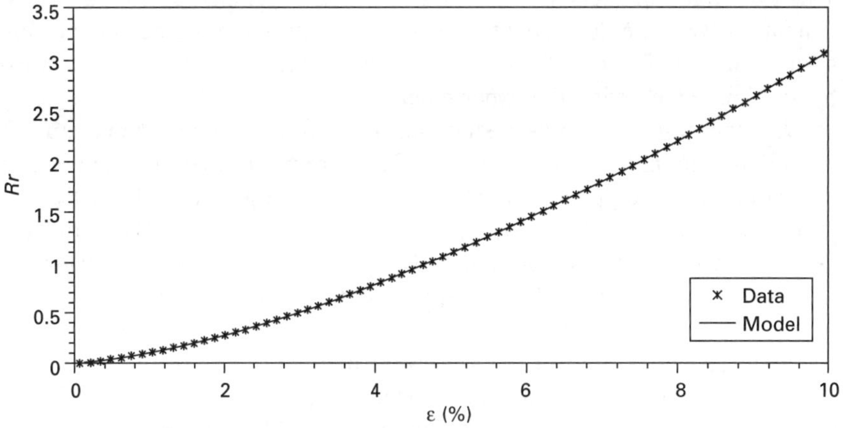

9.8 Dependence of relative electrical resistivity of sensor on elongation ε, at an extension rate of 16 mm/min.

Table 9.1 Elongation rates

	Elongation rate (mm/min)		
	16	200	500
a	99.9	200.8	264.2
b	1.51	1.69	1.79

9.9 Set-up of samples in climatic chamber.

Influence of climatic conditions

This investigation is carried out with a climatic chamber (Excal 2221-HA, Climats®). The samples, mounted on a metallic grid, are placed at the centre of the chamber (Fig. 9.9). The electrical resistance *R* of each sensor is recorded by an ohmmeter during the experiment.

With the aim of separately studying the two parameters, temperature *T* and relative humidity *RH*, two types of test are performed in which one of the parameters is kept constant while the other undergoes a cycle. In the iso-humidity cycle, *RH* is kept constant, and *T* is increased from 10 to 50°C and decreased back to 50°C, in successive 10°C steps. At each step, the temperature is kept constant for a period of 150 minutes, during which time the resistance of the sensor is recorded. In the isothermal cycles, *T* is kept constant while *RH* undergoes a cycle from 20 to 90% in incremental steps of 10%, again with a 150 min equilibrium time for each step. At the beginning of each cycle, a reference value of electrical resistance R_0 is measured at 20°C/40%*RH*. As in the electromechanical characterization (Eq. 9.1), the resistance is expressed as a normalized relative resistance *Rrc*:

$$Rrc = \frac{(R - R_0)}{R_0} \qquad 9.4$$

For each cycle experiment, two values of *Rrc* are measured for each *T*/*RH* pair, one during the upward direction, and the second during the downward direction of the cycle. It was found that there was little difference between the two values, so to illustrate the results, a mean value was calculated. The results of the iso-humidity cycles are given in Fig. 9.10, and those of the isothermal cycles are given in Fig. 9.11.

In Fig. 9.10, the curves can be divided in two categories. For $RH < 60\%$, similar curves are obtained: relative resistance first decreases and reaches a constant value when *T* increases. For $RH = 60\%$, *Rrc* undergoes different variations when *T* increases. In all cases, for constant *RH*, the variations

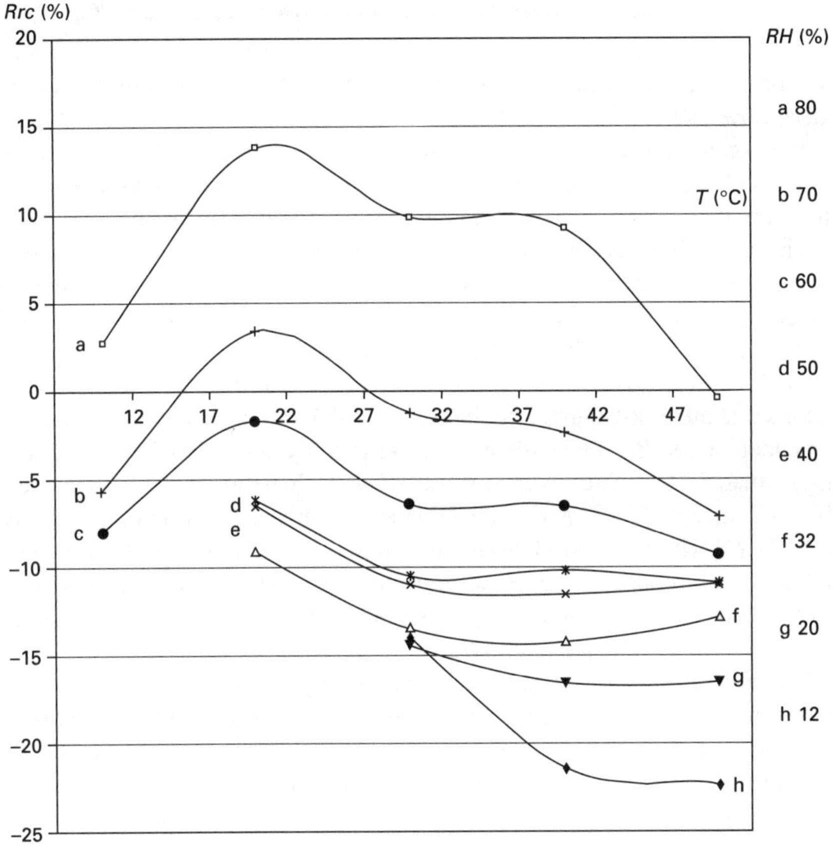

9.10 Iso-humidity cycles for temperature varying between 10 and 50°C.

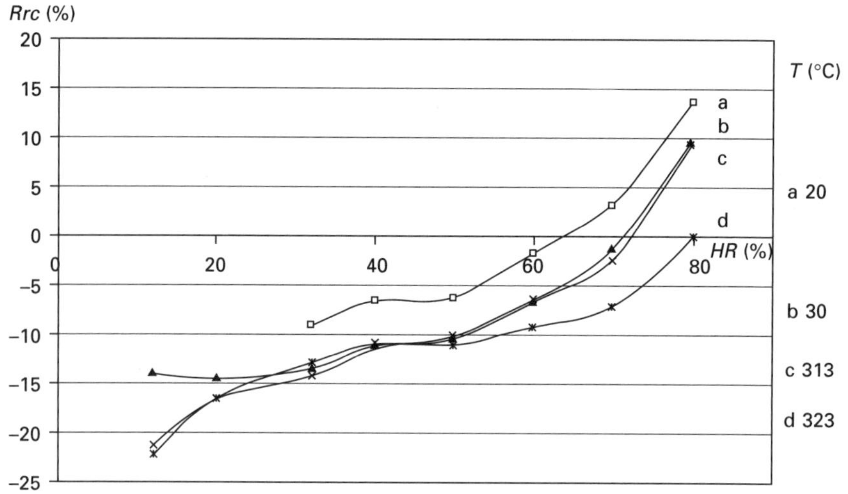

9.11 Isothermal cycles for RH varying between 20°C and 50°C.

nevertheless remain relatively low, since ΔRrc is at a maximum of 9% for the curve 80% *RH*.

It is remarkable and unusual that the electrical resistance decreases when *T* increases for this type of conductor, carbon black, which displays a metallic-like conduction.[9] However, although carbon behaves in a 'metallic' way when conducting electricity in the sensor, the carbon particles are surrounded by a polymer matrix, which is an insulator. *RH* and *T* probably have an impact on the molecular arrangement in the polymer, leading to a modification in the arrangement of the conductive carbon particles.

With regard to the isothermal curves (Fig 9.11), the same behaviour is obtained at all temperatures. *Rrc* increases with *RH*, with a somewhat sharper dependence for *RH* lower than 35% (sensitivity around 0.5%/%*RH*) and higher than 55% (sensitivity around 1.2%/%*RH*). In between these two values, the resistance does not vary much, and the sensor sensitivity amounts to only 0.2%/%*RH*. There are very few studies found in the literature on the influence of humidity on conductive polymer resistance. The only study is about sensors used to measure hygrometry and in that particular application, the electrical resistance varies linearly with *RH*.[10,11]

The increase in resistance with *RH* can be attributed to the absorption of water molecules by the carbon particles. These particles have a complex structure and generally are in an agglomerated form in the composite, hence presenting a porous network with numerous pores which are accessible to water molecules by capillarity. The absorption of these water molecules may lead to an 'un-percolation' of the conducting network of carbon particles, leading to a loss in conductivity of the system.

The sensitivity of carbon black particles to moisture is confirmed in Fig. 9.12.

The study of water content in the particles at different *RH* shows the same behaviour as observed in Fig. 9.11: at about the same value of relative humidity (40–50%), there is a sharp increase in water absorption in carbon particles, and resistivity can be observed in the sensor.

9.3.3 Conclusion

The electrical conduction of insulating polymers charged with conductive particles inducing a metallic behaviour (see *I/V* curve) is due only to physical phenomena. If the number of conductive particles is significant (concentration above percolation threshold), the conductive network is continuous, guaranteeing electric conduction.

In order to create a sufficiently sensitive sensor, the conductive particle concentration has to be higher than the percolation rate. Indeed, when the composite undergoes a strain deformation, the number of electrical contacts between particles will decrease, and the number of contacts has to remain high so as not to affect the general electrical conductivity. A fall in the number of 'electrical contacts' or 'conductive paths' (and thus an increase in electrical resistance) can also result from temperature and/or moisture. This sensitivity to these parameters could possibly lead to the use of this material as a moisture and/or temperature sensor.

The results regarding electrical and mechanical characteristics are encouraging. Below a 10% strain, the sensitivity of the textile-based sensor is 14 times higher than that of traditional metal gauges. However, the sensor response is dependent on the strain rate, and thus it has to be calibrated

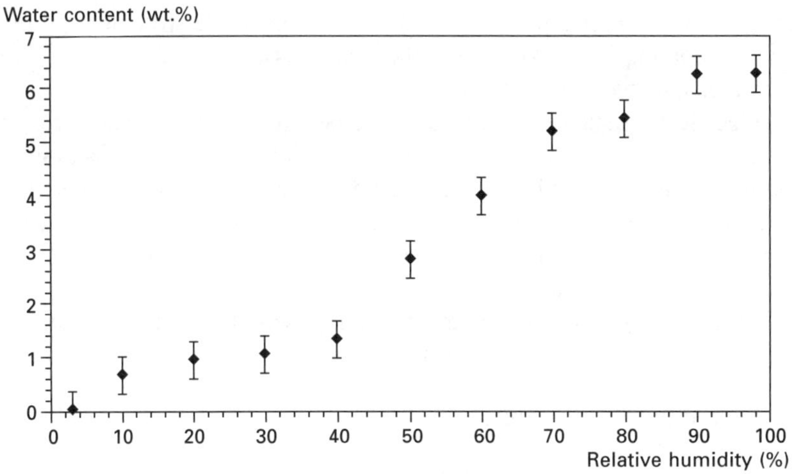

9.12 Water content in carbon black particles at different *RH* (T = 80°C).

according to the strain rate found in the envisaged application. The conductive composite is also rather sensitive to moisture and temperature. Calibration curves can be plotted to correct the data according to these two climatic parameters. More measurements and tests have to be carried out in order to optimize and improve our textile sensor's properties with regard to the function of their application.

9.4 Textile actuators

9.4.1 Heating fabrics for car interior

Heating textiles can find applications in numerous and varied fields such as sports and leisure, medical, or automotive.[12–14] Non-textile conventional sensors and actuators based on wire heating technologies exist that may be integrated into textile structures, but they are often not appropriate because of their rigidity and cost. The situation is even more complex with actuators because they need large power supplies that are rarely flexible and lightweight.

The heating element developed and presented in this work is designed specifically to be suitable for flexible structures. It comprises a plane woven textile element in which are integrated electrodes composed of metallic yarns, woven or stranded directly in the fabric in a comb architecture arrangement. These electrodes are connected to a power supply. In order to ensure a uniform heat distribution, a thin conductive coating is applied on the fabric surface and electrode arrangement. The coating is based on a conductive polymer composite consisting of synthetic rubber and conductive carbon black nano-particles. The heating element can thus be shaped in any desired pattern by choosing specific dimensions for the electrode structure and the area of the coating layer. Moreover, the whole assembly can be placed in selected areas, depending on the application.

The electrical resistance of the heating element (electrodes and conductive coating, Fig. 9.13) can be calculated. The two stainless steel yarns parallel to the warp direction are used to supply the power that is necessary to heat the fabric and also create – together with the stainless steel yarns on the weft direction – a comb structure. This comb structure electrode design is made to create a parallel resistance (decreases the resistance).

Taking into account that the element is composed of a number of parallel resistances, and the distance between the comb teeth is constant:

$$\text{Number of parallel resistances} = \frac{L}{lp}$$

with L: conductive coating width
lp: distance between two adjacent teeth

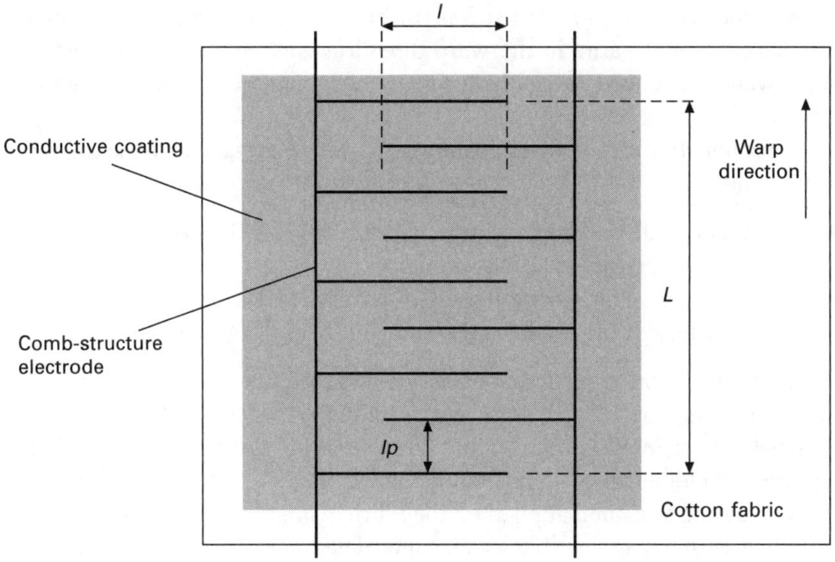

9.13 Pattern of heating element.

The total resistance is thus: $R_{tot} = \dfrac{R}{L/lp} = \dfrac{\rho \times lp/S}{L/lp} = \dfrac{\rho \times lp^2}{LS}$

where *l*: comb teeth superposition width
S: coating cross-section area= l/*e*, *e* is the coating thickness
ρ: coating resistivity

The coating resistivity depends on the amount of carbon black loading in the composite. It can be predicted by the following relation:

$$\rho = \rho 0\ (V - Cp) - t$$

where *ρ*0: constant
V: vol.% of carbon black
t: critical index
Cp: concentration of carbon black (vol.%) at percolation threshold.

Experimental: design of textile heating element

The textile fabric was made on a hand weaving machine (ARM, electronic control Selectron). A plain weave was chosen, and the warp and weft were made of cotton yarns with warp and weft densities of 27 yarns/cm and 10 yarns/cm, respectively. For the electrodes, stainless steel 2-ply yarns (Créafibres, France) were used, each ply consisting of 275 filaments. The overall yarn count was 500 tex, with a resistivity of 14 ohms/m. The distance between the teeth of the comb (*lp*, Fig. 9.13) was 0.7 cm, while the distance between the two parallel steel yarns in the warp direction was around 7 cm. The steel yarns were introduced manually during the weaving process according to the pattern in Fig. 9.13.

The conductive carbon black coating composite was made with the following ingredients:

- a synthetic rubber latex solution, Kraton IR-401 (Kraton Polymers), an anionic dispersion of polyisoprene
- a dispersing agent, Disperbyk-2010 (from SPCI)
- carbon black (CB) particles, Printex L6, from Degussa (Table 9.2).

The preparation procedure was as follows: the dispersing agent was put into water and the CB particles were gradually added while mixing continuously. A paste was obtained and left to rest for some time. The polymer was finally added while mixing gently in order to avoid a too strong shearing.

The coating was then applied on the fabric using a magnetic coating table equipped with a magnetic bar as a scraper. The thickness of the coating was, however, hard to adjust with this technique. Moreover, the quantity of coating deposited depended on the viscosity of the solution, which itself depended on the amount of water in the coating solution. If the viscosity was different, even if the same thickness was deposited on the fabric, the final quantity of composite would differ, since the absorption of the coating would be different. Hence, the final resistance would also be different. After drying at room temperature, the resistance of the actuator was measured with an ohmmeter, and was found to be equal to 12 ohm. The fabric with the electrodes is shown in Fig. 9.14, before and after the application of conductive coating. The final coated surface is approximately 5 × 7 cm.

Characterization of textile heating element

The electrical and heating properties of the element were studied. The electrical resistance of 3 × 3 cm coated samples with various carbon black contents was measured (Fig. 9.15). It was found to depend, of course, on the CB loading level: the higher the amount of CB, the lower the resistance.

The optimum CB loading is thus around 35 to 45% in weight. A coating containing 37.5 wt.% was used for the heating property study.

Table 9.2 Characteristics of carbon black particles

Particle average diameter (nm)	Structure (DBPA number)	Specific surface area (m^2/g)	Volatiles (%)
18	122	250	<1

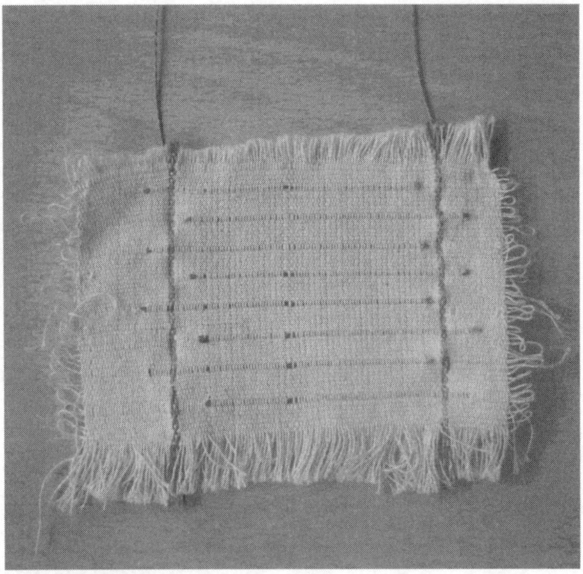

(a)

(b)

9.14 Fabric with heating element: before and after coating with conductive composite.

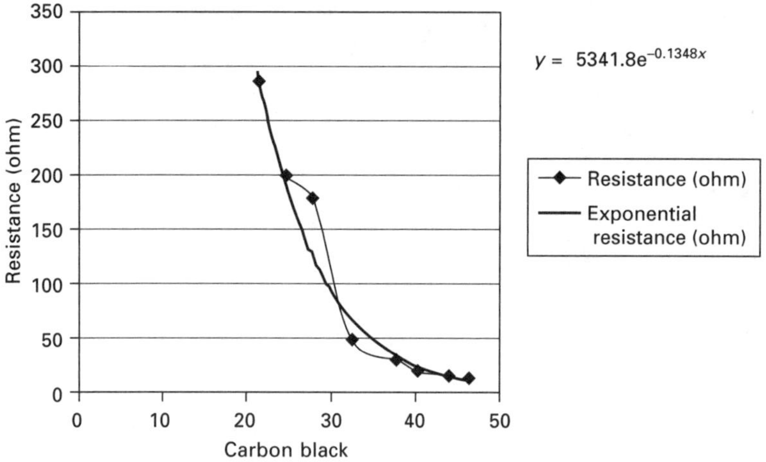

9.15 Variation of electrical resistance with carbon black content.

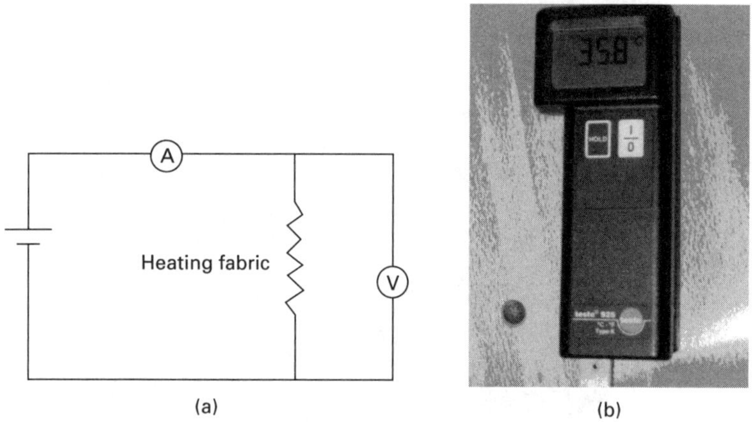

9.16 Power circuit with fabric heating element, and temperature sensor.

The coated fabric was connected to an ammeter and a power supply as shown in Fig. 9.16. It is important to note that the sample temperature may change as a function of time even when applying the same voltage. Because of this, the temperature of each sample was measured for 60–80 min for a given voltage. The temperature of the sample surface was recorded every five minutes with a temperature sensor (Testo® 925). The temperature was not the same for every point of the heating element, and the highest temperature was obtained at the centre of the sample, but the variation was not very important.

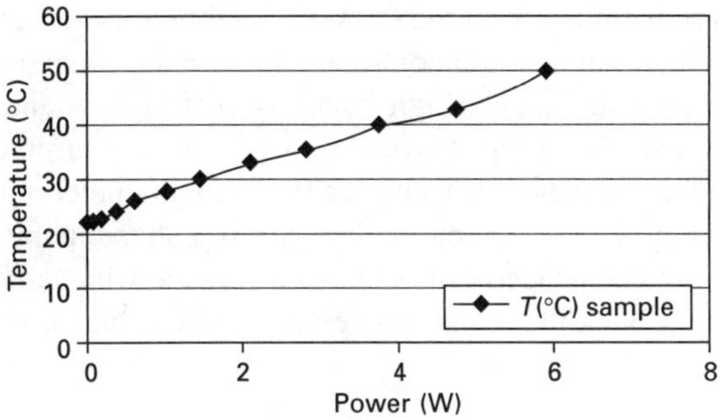

9.17 Temperature of heating fabric as a function of power supply.

The results are presented graphically in Fig. 9.17. The measurements were made at room temperature (22°C).

The increase in temperature was significant, about 30°C from an initial temperature of 22°C, for the highest power supply (6 W). The heating process was found to be quite rapid; the time taken to reach the highest temperature was about 10 minutes.

Conclusion

An increase in heating efficiency requires quite a large percentage of carbon black, more than 35 wt.%. It was found that the temperature was not homogeneous over the whole coated surface, the sample being hotter at the centre than at the corners. Therefore this characteristic has to be improved, maybe by modifying the pattern design of the electrodes. The heating was, moreover, limited to the coated area; it is possible that the use of intelligent design could result in the dissipation of the excess heat generated to other parts of the fabric.

After being coated with the CB solution, the woven samples became quite rigid. Other studies therefore need to be done to obtain a more flexible final structure, and our next work will be on knitted structures.[15] The steel yarns will be integrated inside the knitted fabric and, as this type of textile structure is intrinsically more deformable that woven structures, the final heating element will normally be more flexible.

9.4.2 Shape memory alloys for car seats

Introduction

The existing solutions presented in Section 9.1 do not fit all the requirements and future functionalities that need to be integrated inside car seat fabric.

Therefore, a research team based in the GEMTEX laboratory has attempted to develop a new fabric which could combine attractive fabric patterns with properties such as: stab resistance, high buckling resistance[16] and smooth heating of the car seat. The proposed solution is a woven structure with Nitinol (made of nickel and titanium) yarns of 0.25 mm diameter included in the weft direction (Fig. 9.18). This particular yarn is a shape memory alloy wire which has the notable property of being super elastic which allows it to be subjected to greater buckling than a conventional alloy such as steel (Fig. 9.19).[17]

Experimental

The main advantage in the future of using SMA wires inside the fabric could be controlled buckling as its elastic property increases with a particular applied temperature when close to the human body but it becomes more rigid under a different temperature. To illustrate this change of fabric shape under different temperatures, the sample has been educated to be flat at ambient temperature and bent during heating.

An interesting possibility for this fabric could be the design of 'moving' parts to ensure a surface motion which could help to avoid the inclination to sleep while driving.

9.18 View of the fabric made of polyamide yarns in the warp direction and SMA Nitinol yarn in the weft direction.

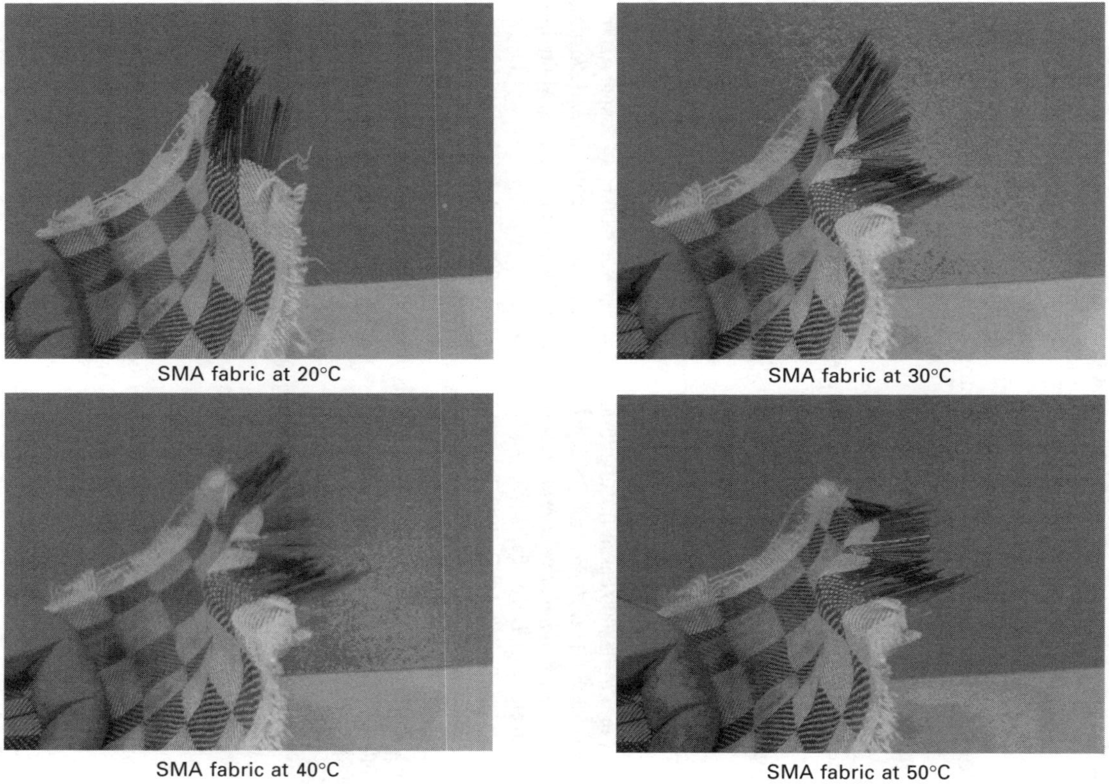

9.19 Shape memory alloy behaviour.

9.5 Conclusions

Different building blocks (sensors and actuators) have been presented in order to show the development and integration possibilities of smart textile structures designed for car interiors. Sensors based on conductive composites may be used to detect fabric lengthening or passenger presence, or to evaluate seat belt ageing. They may also be used to control air bag deployment as shown in the patent of Mr Jayaraman. Alternatively, heating fabrics are suitable for car seats. The advantage of our solution is that it is low cost, and has a long lifetime and is particularly reliable. The other advantage is that the same actuator may be used as a sensor, helping with temperature control and regulation.

The potential use of shape memory alloys in car seats, so that they always have the same 'new' look, is also very interesting. Their price is currently the main problem. Nevertheless, they may be used in small quantities and mixed with traditional textile threads to achieve costs attractive for luxury cars.

9.6 References

1. 'Futurotextiles': an exhibition on innovating textiles, lille3000: international cultural event http://lille3000.ensait.fr, Lille, France, 14 October 2006 to 14 January 2007.
2. Habert Cédric, The 'Happy attitude' concept, *Decautex conference Texile and Plastics: Innovation for automotive interior decoration*, Faurecia Interior Systems, Mulhouse, 30 November–1 December 2006.
3. Kawashima Textile MFG, 39th Tokyo Motor Show 2005.
4. Jayaraman *et al.*, Method and apparatus for controlling air bag deployment, US Patent 6,254,130 B1, Jul. 3, 2001.
5. Swallow S., Thompson A.P., *Applications for woven electrical fabrics*, Part IV Applications, Chapter 25, Intelligent textiles and clothing, Woodhead Publishing, 2005.
6. Cochrane C., Koncar V., Lewandowski M., Dufour C., 'Design and development of a flexible strain sensor for textile structures based on a conductive polymer composite', *Sensors*, 2007 **7** 473–92.
7. Cochrane C., Koncar V., Dufour C., 'Création d'un capteur d'allongement souple, compatible textile', *Matériaux 2006 Conference*, November 13–17, Dijon, France, 2006.
8. Asch G., *Les capteurs en instrumentation industrielle*, 5th edition, Dunod, Paris, 2000.
9. Cochrane C., Koncar V., Dufour C., 'Nanocomposite material sensors for textiles', *AUTEX 2006 World Textile Conference*, June 11–14, Raleigh, NC, USA, 2006.
10. Barkauskas J., 'Investigation of conductometric humidity sensors', *Talanta*, 1997 **44** 1107–12.
11. Barkauskas J., Vinslovaite A., 'Investigation of electroconductive films composed of polyvinyl alcohol and graphitized carbon black', *Materials Research Bulletin*, 2003 **38** 1437–47.
12. Droval G., Glouannec P., Feller J.F., Salagnac P., 'Simulation of electrical and thermal

behaviour of conductive polymer composites heating elements', *Journal of thermophysics and heat transfer*, 2005 **19** (3) 375–81.
13. El-Trantawy F., Kamada K., Ohnabe H. 'In situ network structure, electrical and thermal properties of conductive epoxy resin-carbon black composites for electrical heater applications', *Materials Letters*, 2002 **56** 112–26.
14. Kirkpatrick S., 'Percolation and conduction', *Reviews of Modern Physics*, 1973 **45** (4) 574–88.
15. Lewandowski M., Koncar V., Cochrane C., Giraud S., 'Electro-conductive heating textile elements based on a carbon black conductive polymer composite', *Intelligent Textiles & Mass Customisation International Conference*, Casablanca, Morocco, Nov. 2007.
16. Boussu F., 'Shape memory alloys used in fabrics for car seats', *2nd NEST conference Smart and intelligent materials*, Gothenburg, Sweden, 22–23 February 2007.
17. Boussu F., *Development of Shape Memory Alloy Fabrics for Composite Structure*, Part II Shape memory materials, Chapter 8, Intelligent textiles and clothing, Woodhead Publishing, 2005.

10

Reducing noise in automotive interiors

J Y CHEN, Louisiana State University Agricultural Center, USA

Abstract: Noise reduction is a big issue involved in the design and production of vehicle interior parts. This chapter focuses on the principles of noise control and engineering approaches for vehicle noise reduction using acoustic textile materials. Noise sources and noise paths in vehicles are first discussed. Mechanisms of textile materials for noise absorption and noise insulation in vehicle interiors are analyzed. Various noise control methods using different types of acoustic textile materials are illustrated by exhibiting some typical application cases. Finally, future trends in the new surface vehicle design and production are addressed, including new material requirements, novelty interior structures, and highly integrated acoustical interior systems.

Key words: automotive interior, acoustical material, textile material, nonwoven composite, vehicle noise, noise absorption, noise insulation, transmission loss, normal incidence noise wave, sound impedance tube, reverberant rooms.

10.1 Causes of noise generation in moving vehicles

10.1.1 Noise sources

The generation of vehicle interior noise comes from many sources (Fig. 10.1). These sources can be categorized into three types: vehicle noise sources such as engine noise, intake noise, exhaust noise, and brake noise; road and wind (aerodynamic) noise sources; and miscellaneous noise sources such as the noises of squeak, rattle, and tizz from interior components and ancillaries that are usually difficult to predict in many cases. These noise sources are briefly described below (Harrison, 2004).

Engine noise is a combination of combustion noise and mechanical noise. Combustion noise derives from fuel ignition in the cylinders that produce gas forces to move pistons. As a result, engine structure vibrates and radiates noise. For diesel engines, the combustion noise dominates the engine noise. During the conversion of gas forces into mechanical forces by a crank system composed of pistons, conrods, crankshaft, and bearings, the engine structure reacts to the crank movement and generates mechanical noise. For gasoline engines, the mechanical noise is the most significant contributor to the engine noise. When an engine runs, oxygen is supplied and combusted fuel is emitted

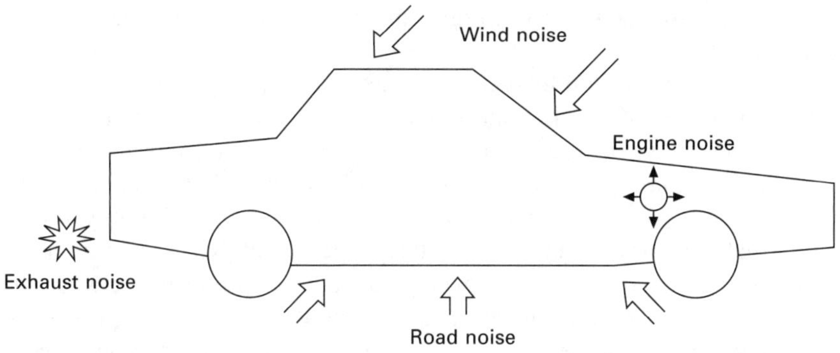

10.1 Vehicle noise sources.

during each piston stroke in a cylinder. Intake noise and exhaust noise occur in association with the vibration of the sucked air flow and exhausted gas flow. In addition, the engine timing drive, valve train, and fuel injection device also produce some levels of mechanical noise.

Brake noise is generated by the friction between pads and rotor disks. The type of brake noise depends on vehicle speed. At high speeds, brake squeal occurs within a noise frequency range well above 1000 Hz. All brake components (the disk, pads, and calliper assembly) contribute to this noise radiation. At moderate speeds, brake moan occurs with a noise frequency level around 100 Hz. Brake creep-groan happens at speeds less than walking pace and the noise frequency is also around 100 Hz. The brake noise with a frequency around 10 Hz is called brake judder. It also occurs when the vehicle speed is below walking pace. During the start-stop transition of a moving vehicle, the vehicle itself vibrates as a resonant system to develop those low-frequency brake noise sources.

The contact and friction between the tires and road generates road noise. The road noise becomes interior noise when it is transmitted to the interior by an interaction with vehicle structure. Auto interior acoustics must control this part of road noise. The road noise also produces noise impact on passing-by vehicles. This part of the road noise is often called tire noise. Tire manufacturers are interested in this noise source because all tire products need to pass the drive-pass noise test. As vehicles are driven at high speeds, wind noise becomes a significant source of interior noise. The wind noise paths that transmit sound waves into the interior can be observed through the following aspects. Airflow excites the vibration of the wind shield glass and roof panel thus causing the structure to radiate the noise inside a vehicle. Airflow passes over the underside of a vehicle and causes a noise leak into the vehicle (mainly through the wheel-arch areas). Wind noise is leaked into the interior through door and window seals or is radiated through door and window vibration. Airflow blowing in the vehicle through partially-open windows or sunroofs causes intensive low-frequency noise and buffeting.

The miscellaneous noises such as squeaks, rattles, and tizzes usually have a random and chaotic nature and in many cases they are easy to detect but hard to locate. When there is a relative motion between interior material pairs, squeaks may occur. These material pairs can be hidden deep inside the composite structure of an interior part such as a dashboard assembly. As the paired materials possess different values of static and dynamic coefficients of friction, the relative motion develops the stick-slip process between the materials and causes the squeak noise. Material types, surface properties, frequency and amplitude of excitation, normal loads, interference levels, temperature, and humidity are all factors that help provoke squeaks. Rattle noises often result from hard and loose-fitting interior trims. Analysis of this noise is not easy because objective measures are not very consistent with human responses to this noise. In contrast, tizz noises are often caused by high-frequency contact vibration among interior parts and are relatively easy to find and eliminate.

10.1.2 Noise propagation and noise control approaches

As described in acoustics, sound waves traveling in space take the forms of reflection, refraction, diffraction, and diffusion. In reality, there are multiple sound sources existing in the same space and at the same time. Complicated acoustic phenomena, such as room reverberation and resonance, sound concentrations, and sound standing points, can be observed. Considering a passenger car, its space is usually divided into three parts: engine compartment; passenger compartment; and trunk compartment. The vehicle interior noise we are discussing here means that all noise waves the driver and passengers receive are inside the passenger compartment. Among the noise sources described in the previous section, only squeaks, rattles, and tizzes are directly from inside the passenger compartment, all other noise sources occur outside and transmit into the passenger compartment by certain ways of interaction with the vehicle structure.

According to the automotive acoustics, there are two types of interacting mechanism between the noise sources and vehicle structure. One type is called airborne noise path that means outside noise waves directly leak or transmit into the passenger compartment to cause interior noise. The other type is called structure-borne noise path. From this path, outside noise waves excite the vibration of vehicle structure and cause interior part surfaces to radiate noise inside the passenger compartment. Figure 10.2 illustrates a typical system for vehicle interior noise control in which a noise source, a noise path, and a noise receiver are three components. Therefore, the principle

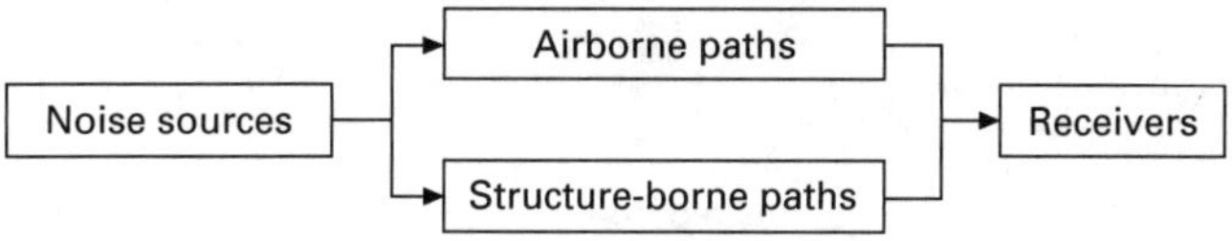

10.2 Vehicle noise paths.

of controlling the vehicle interior noise is to interrupt the airborne or structure-borne paths for noise propagation.

The major engineering approaches for the interruption of airborne paths include effective sealing of the passenger compartment to avoid direct noise leaking; high performance insulation of noise sources, such as the encapsulation of engine noise using panels or the encapsulation of road noise using heavy floor padding materials, which helps block noise waves transmitted from outside to inside; and efficient noise absorption to reduce reflected noise level inside the passenger compartment. For the interruption of structure-borne paths, isolating and damping the vibration of vehicle structure are two main methods. These methods are usually effective for the control of low-frequency noise, because the structure-borne noise is within the range of low noise frequency. In the range of high noise frequency, it is the noise received through airborne noise paths that usually dominates the interior noise levels.

10.2 Theory of noise-minimizing properties of textiles

To insure vehicles' acoustical performance, auto designers and manufacturers need to select and use different acoustical materials for maximum reduction of interior noise. Some materials are able to reduce airborne noise, while others help control structure-borne noise. Each material selected must possess a specific property that can contribute to at least one of these acoustical functions: noise absorption, noise insulation, vibration isolation, and vibration damping. The following sections discuss some primary theoretical issues regarding the effectiveness of textile materials on noise absorption and noise insulation, because the absorption and insulation materials account for a large percentage in the consumption of auto acoustical materials.

10.3 Textile materials for noise absorption

As a noise wave reaches the surface of an auto interior part during its propagation by an airborne path, it will become three different components: reflected part, transmitted part, and absorbed part (Fig. 10.3). A receiver on the same side as the sound source can receive both the incident and reflected sound waves. Absorbing the incident noise wave by an acoustic material is

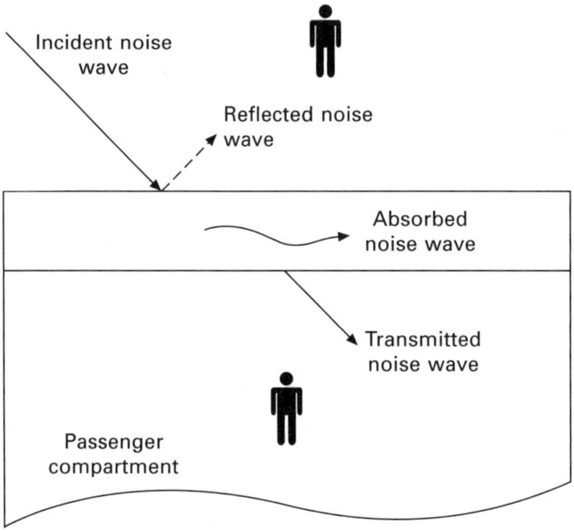

10.3 Airborne noise transmission.

an effective way to lower the reflected noise level in the interior noise control engineering. A common question is what types of material can be used as excellent noise absorption materials. These materials must function like air, being able to prevent noise reflection from incidence surface, and like an energy disperser capable of dissipating the noise energy entering the material.

Porous materials such as textiles and foams are ideal noise absorption materials. The feature of numerous tiny air pockets inside the material allows the materials to perform two important absorption functions described above. Textiles materials in the form of wovens, knits, and nonwovens are widely used as auto interior materials ranging from decor fabrics to substrates and composites. These different applications provide different levels of noise absorption in the vehicle passenger compartment. The principle of noise absorption by these textile materials is generally described by these mechanisms: (1) heat conduction converted from the sound energy encapsulated in porous areas in the material; (2) viscous losses because of the oscillating air flow entering the porous areas inside the material; and (3) material vibration mainly caused by the noise waves encapsulated in closed porous areas.

It is difficult to use theoretical methods for describing the relationship between material structures and noise absorption performance, because of diverse textile fiber materials and complex fabric geometric structures. In practice, the material physical properties, in terms of flow resistance, porosity, and structure factor, are often used to characterize the material performance for noise absorption.

10.3.1 Flow resistance

Assuming that a noise wave enters a porous material, there exists a steady air flow flux passing through the material. The flow resistance of the porous material, denoted as *r*, is defined as (Harrison, 2004):

$$\frac{\partial p}{\partial x} = -\, r u_{\mathrm{m}}, \qquad 10.1$$

where u_{m} is air volume velocity per unit cross-sectional area of the material (m/s); *p* is air pressure, and *x* is displacement of air flow in the material. The standard ISO 9053-1991 provides a protocol of instrumentally evaluating the flow resisitance of porous materials. To address this issue, an instrumental method of measuring the material flow resistance was proposed (Bies and Hansen, 1996). This method measures the pressure difference across a material specimen in a tube with a steady air flow to determine the flow resistance *r*. The equation for calculating *r* is expressed below:

$$r = \frac{\rho A \cdot \Delta p}{v_{\mathrm{m}} l}, \qquad 10.2$$

where ρ = gas density (kg/m^3); Δp = pressure drop (N/m^2); A = material cross-sectional area (m^2); v_{m} = air mass flow rate (kg/s); and l = material specimen thickness (m). It should be noted that the above equation is based on the assumption that the material performs the same resistance to both oscillatory acoustic flow and steady air flow. Overall, this equation is only suitable for measuring low speed internal flows (Fahy and Walker, 1998).

Another instrumental method was proposed to directly measure an oscillatory acoustic flow passing through a porous material installed in an impedance tube (Ingard and Dear, 1985). This device is composed of a tube, a noise source, two pressure sensors, and a specimen installed in the tube with a distance equivalent to an odd number of $\frac{1}{4}$ wavelengths from the closed end of the tube. The following equation is used for the calculation of flow resistance:

$$\frac{r}{\rho_0 c_0} \approx \left| \frac{p_1}{p_2} \right|, \qquad 10.3$$

where ρ_0 is air density and c_0 is sound speed in air. In the case of measuring fiber porous materials, the above equation is modified to include more specific parameters (Bies and Hansen, 1996). The new format becomes:

$$\frac{rd}{\rho c} = 27.3 \left(\frac{\rho_{\mathrm{m}}}{\rho_{\mathrm{f}}} \right)^{1.53} \left(\frac{\mu}{d \rho c} \right) \left(\frac{d}{b} \right), \qquad 10.4$$

where

ρ = gas density (kg/m^3);

ρ_m = porous material (fabric) density (kg/m^3);
ρ_f = fiber density (kg/m^3);
μ = gas viscosity (Ns/m);
b = fiber diameter (m);
c = sound speed in gas (m/s);
d = sample thickness (m).

A standard method for the determination of airflow resistance of acoustical materials is ASTM C 522. Acoustical materials suitable for this test include boards, mats, felts, fabrics, papers, and screens. Similarly, a specific test method for evaluating the textile porous property related to flow resistance is the air permeability test described in the standards ASTM D 461, ASTM D 737, and BS 5636. The air permeability of textile materials is determined by the volume of air passed through a fixed area of sample at a pressure difference in a time period. The Frazier Air Permeability Tester is an instrument commonly used in the industry.

10.3.2 Porosity

Material porosity is used to describe the fraction of porous space in a material. It is defined as:

$$\delta = \frac{V_0}{V}, \qquad 10.5$$

where V_0 is the volume of porous space of the material and V is the total volume of the material including the porous space (Harrison, 2004). The value of δ is between 0 and 1. For noise absorption materials, δ is usually high (above 0.90). In the discussion of a sound wave propagating in a tube with a fixed cross-sectional area, a linear equation of mass conservation in free air is obtained (Fahy and Walker, 1998), expressed as:

$$\frac{\partial u}{\partial x} = -\frac{1}{\rho_0}\frac{\partial \rho}{\partial t}, \qquad 10.6$$

where u is air flow (particle) velocity. Because the air bulk modulus is defined as:

$$k_0 = \rho_0 c_0^2 = \rho_0\left(\frac{\partial p}{\partial \rho}\right), \qquad 10.7$$

Equation 10.6 can be rewritten as:

$$\frac{\partial u}{\partial x} = -\frac{1}{k_0}\frac{\partial p}{\partial t}. \qquad 10.8$$

Now considering the situation that a sound wave passes through a porous material, the material bulk modulus k (defined as k_0/δ) and air volume flow rate u_m are used. Therefore, the above equation becomes:

$$\frac{\partial u_m}{\partial x} = -\frac{1}{k}\frac{\partial p}{\partial t}. \tag{10.9}$$

For textile materials, the formation of porous space is complicated because of diverse structures. In woven and knitted fabrics, the porous space consists of structural pores formed by yarns and yarn pores formed by fibers. If specialty fibers (such as hollow fiber or activated carbon fiber) are used for weaving and knitting, micropores inside the fiber will also be included. In nonwoven fabrics, the porous space is directly formed by randomly laid fiber. The nested fiber web structure makes nonwoven fabrics much bulkier than woven and knitted fabrics.

From Equation 10.5, the volume of porous space V_0 needs to be known in order to determine the porosity of textile materials. Some commercial instruments are available for characterizing pore volume, pore size, pore volume distribution, and other porous properties. The PMI porometers (Porous Materials, Inc.), the ASAP™ porosimetry system (Micromeritics Instrument Corporation), and the SA 3100 Surface Area and Pore Size Analyzer (Beckman Coulter Inc.) are among those instruments widely accepted for different industrial applications.

10.3.3 Structural factors

It is not enough to use porosity alone for describing materials' porous properties, because even if two porous materials indicate the same porosity, their noise absorption performance can still be totally different. Some other structural factors also must be taken into consideration. A theoretical fundamental used for analyzing material structural influence is the linear inviscid Euler equation (Kinsler *et al.*, 1982). This equation is expressed as:

$$\frac{\partial p}{\partial x} = -\rho_0\frac{\partial u}{\partial t} \quad \text{or} \quad u(x,t) = -\frac{1}{\rho_0}\int \frac{\partial p}{\partial x}\,dt. \tag{10.10}$$

In the case that there is an orientation of pore space in the material measured with an angle θ to the surface of the material, Harrison (2004) gives the following description:

$$\frac{\partial p}{\partial x}\cos\theta = -\rho_0\frac{\delta u_m}{\partial t}\frac{1}{\delta\cos\theta} \tag{10.11}$$

$$\frac{\partial p}{\partial x} = -\frac{s\rho_0}{\delta}\frac{\partial u_m}{\partial t} \quad \text{where} \quad s \propto \frac{\alpha}{\cos^2\theta}. \tag{10.12}$$

In Equation 10.12, s is called structure factor and α is an attenuation constant of the material. Overall, a random orientation of pores may help increase the flow density. In a more specific case of a plane wave propagating in a tube with a termination impedance, the linear inviscid Euler equation is modified as:

$$\frac{\partial p}{\partial x} = -\frac{s\rho_0}{\delta}\frac{\partial u_{\mathrm{m}}}{\partial t} - ru_{\mathrm{m}}. \qquad 10.13$$

For a simple harmonic motion with frequency ω, the above equation takes the following specific format:

$$\frac{\partial p}{\partial x} = -\left[\frac{s\rho_0}{\delta} + \frac{r}{i\omega}\right]\frac{\partial u_{\mathrm{m}}}{\partial t} = -\rho_{\mathrm{m}}\frac{\partial u_{\mathrm{m}}}{\partial t}. \qquad 10.14$$

The parameter ρ_{m}, because of including several structural effects, is used to represent the effective bulk density of porous materials. Ultimately, the introduction of ρ_{m} helps determine the characteristic acoustic impedance of air inside a porous material, defined as:

$$Z_c = \sqrt{k\rho_{\mathrm{m}}}. \qquad 10.15$$

It should be noted that besides those important material factors discussed above, there are some other structural components to which some attention must be paid in engineering practice. Irregularity of pore size and uniformity of pore space distribution inside the material body affects a steady flow volume speed inside the material. Side pockets outside a flow stream in the material tend to cause a reduced acoustic pressure along a given strain gradient in the pore. These structural features are also called the tortuosity of acoustical materials. This property indicates the effectiveness of preventing direct flow through the acoustical material. In contrast, porosity indicates the ease of noise waves entering the material. They both contribute to the flow resistance of acoustical materials.

Regarding the textile materials applied for the noise absorption, fabric structures, yarn structures, and finish methods may influence the acoustic property of absorption. Fiber orientation in fabric materials, particularly in nonwovens, is a special parameter closely related to the pore orientation. In addition, fabric weight (mass per unit area), fabric thickness, fabric yarn count, fiber type, fiber linear density, fiber cross-sectional shape, and fiber surface property are all influential components that may play a role in the determination of the material capacity for noise absorption. In general, textile materials with the use of finer fibers and with smaller pore sizes tend to produce higher viscous losses and to increase the material flow resistance. Figure 10.4 gives a brief summary of the discussion in this section (Wyerman, 2003).

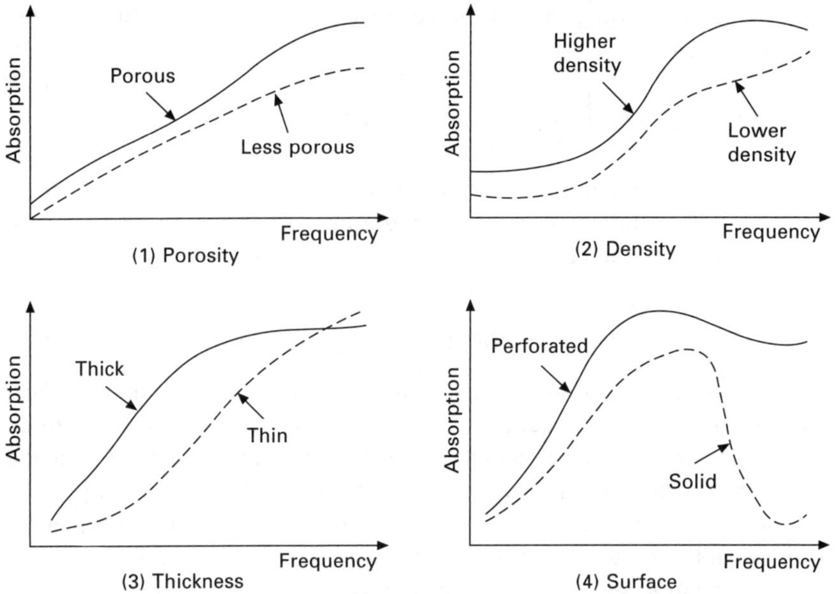

10.4 Effect of material properties on noise absorption.

10.4 Textile materials for noise insulation

10.4.1 Introduction

A passenger sitting inside a sealed passenger compartment of a vehicle can still receive external noise waves (Fig. 10.1). But these noise waves are transmitted noise waves with a certain level of noise intensity reduction, because of the insulation material existing between the noise source and receiver. The acoustical materials used for blocking the transmission of noise from one side to the other side of the material are called barriers. In vehicle engineering, acoustic barriers are important interior materials to reduce airborne noise. For best acoustic performance, a barrier material should have these features: a smooth and high density surface to maximally reflect noise waves, a non-porous structure to effectively encapsulate incident noise waves, and a certain material stiffness to minimize excited vibration. In a wide range of textile materials, natural fiber and synthetic fiber nonwovens (felts) with sufficient fabric weight and thickness are suitable for the noise insulation application in auto manufacturing. They are usually incorporated together with decor fabrics (such as wovens, knits, and genuine or artificial leather), interlinings (foams, films, etc.), and substrates into interior parts as a composite backbone after a hot-pressing process.

The insulation performance of a barrier material is assessed by sound transmission loss (TL). This parameter is determined by the ratio of incident noise intensities (w/m^2) to transmitted noise intensities, as defined below (Harrison, 2004):

$$\text{TL} = 10\log_{10}\left[\frac{I_i}{I_t}\right]. \qquad 10.16$$

The unit of TL is dB and its value varies corresponding to the noise frequency. The TL mass law indicates that the sound transmission loss increases as the mass or frequency increases.

10.4.2 Noise transmission through thin panels

Most nonwoven felts for interior noise insulation form a thin panel after hot-pressing. In the range of audible sound frequencies, thin panels are often excited by noise to radiate bending waves that cause a panel deflection in the plane perpendicular to the panel surface. This deflection involves the deformations of extension, compression, and shear. Thus, the mechanical properties and dimensional parameters of the panel are used in the determination of the propagation speed of bending waves. The phase velocity of bending waves, denoted by C_B, is given by:

$$C_B = \left(\frac{B}{\rho S}\,\omega^2\right)^{\frac{1}{4}}, \qquad 10.17$$

where B is bending rigidity (N·m); ρ is panel density, S is panel area; and ω is radial frequency (=$2\pi f$). For a thin plate with thickness d, the above equation becomes:

$$C_B = \sqrt{1.8 c_L d f} \qquad 10.18$$

$$c_L = \sqrt{\frac{E}{\rho(1-\nu)}}, \qquad 10.19$$

where E is the Young's modulus (N/m^2), ν is the Poisson's ratio, and f is frequency (Hz).

When the bending wavelength from the panel is equal to the noise wavelength in the air, a strong coupling between the panel bending waves and noise waves occurs. This means that the noise impinging on the panel from any angle of incidence will induce a strong responding bending wave in the panel. This phenomenon is called coincidence. The frequency needed to make this coincidence happen is called the critical frequency. For the isotropic panel, this critical frequency is determined by (Harrison, 2004):

$$f_c = \frac{c_0^2}{2\pi}\left(\frac{m}{B'}\right)^{\frac{1}{2}}, \qquad 10.20$$

where m is the panel surface density (kg/m^2) and B' is the modified bending rigidity expressed as:

$$B' = \frac{d^3}{12} \cdot \frac{E}{1 - \nu^2}. \qquad 10.21$$

The TL-frequency curve given in Fig. 10.5 illustrates the sound transmission property of an isotropic thin panel responding to the increase of sound frequency (Bies and Hansen, 1996; Wyerman, 2003). This curve exhibits three transmission property regions. In the range of low frequencies, the TL values are controlled by panel stiffness and resonance. At the frequency where the first panel resonance occurs, the sudden decrease of TL is partially controlled by the system damping. The medium frequency range is mass law region. In this region, TL complies with the mass law relationship determined by Green (2005):

$$\text{TL} = 20\log_{10}(mf) - 47.2. \qquad 10.22$$

The high frequency range is called coincidence region, because this range is around the critical frequency. At the frequencies below the critical frequency, the wavelength in air is longer than the wavelength in the panel. As a result, the panel is coupled poorly to the airborne noise wave. This poor coupling makes the panel have good TL performance. In contrast, at the frequencies above the critical frequency, the panel is coupled well to the airborne noise wave with a certain incident angle. The good coupling causes a poor TL property. Therefore, the panel TL depends on damping control and can be increased at a rate of 9 dB per octave.

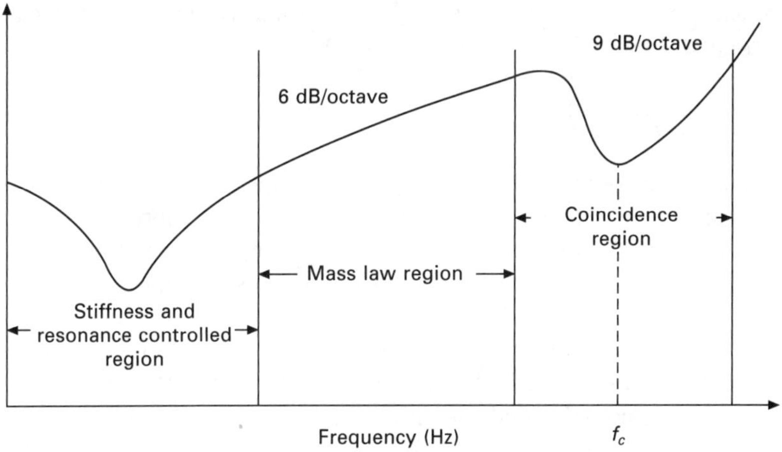

10.5 Noise insulation property of thin panel.

10.5 Testing methods for acoustical properties of textile materials

Vehicle acoustic performance has become an important invisible factor to improve passenger comfort, in contrast to visible factors (esthetic attributes) for vehicle interiors, such as color and style. From an engineering point of view, acoustic performance of a vehicle can be evaluated at three levels: material level, component level, and system level (Khambete, 2003). At the material level, a major task is to evalute the noise absorption and noise insulation of flat (2D) materials subject to strikes of any incident noise waves. At the component level, the effectiveness of 3D interior parts on noise attenuation in a reverberation room is of interest to investigate. This method is also called buck testing. At the system level, focus is on the dynamic overall acoustic performance of vehicles on roadside driving. Considering the focus of this book, the scope of present discussion is limited to the material level evaluation of noise absorption and noise insulation for textile materials.

10.5.1 Measurement of noise absorption

There are two instrumental approaches for assessing the material ability to absorb noise: one is the test of normal incidence noise absorption and the other is the test of random incidence noise absorption. The standard test methods are listed in Table 10.1.

ASTM C 384/ISO 10534-1. These methods use a standing wave impedance tube to allow a plane wave traveling in the tube. After hitting the test sample, the plane wave is reflected to produce a standing wave. Maximum pressure (P_{max}) and minimum pressure (P_{min}) in the tube are measured by a movable microphone, so that the absorption coefficient of normal incident noise can be determined:

$$\alpha_n = \frac{4 P_{max} P_{min}}{(P_{max} + P_{min})^2}. \qquad 10.23$$

The standing wave method is not only a good educational model but also a good practical instrumental approach for testing absorption coefficient, reflection coefficient, impedance, and admittance of the normal incidence noise. One disadvantage of this method is its slow test procedure.

ASTM E 1050/ISO 10534-2. These methods use a two-microphone impedance tube to determine the ability of acoustical materials for absorbing normal incidence noise waves. The Brüel and Kjær instrument of impedance tube system is commonly used for these standard test methods within the frequency range 0–6400 Hz. This instrument system includes Type 4206 Impedance Tube, PULSE Analyzer Type 3560, and Type 7758 Material Test

Table 10.1 Standard test methods for sound absorption

Noise source	Standard	Title
Normal incidence	ASTM E 384-98	Standard Test Method for Impedance and Absorption of Acoustical Materials by Impedance Tube Method
	ISO 10534-1: 1996	Acoustics – Determination of sound absorption coefficient and impedance in impedance tubes – Part 1: Method using standing wave ratio
	ASTM E 1050-98	Standard Test Method for Impedance and Absorption of Acoustical Materials Using a Tube, Two Microphones and a Digital Frequency Analysis System
	ISO 10534-2: 1998	Acoustics – Determination of sound absorption coefficient and impedance in impedance tubes – Part 2: Transfer-function method
Random incidence	ASTM C 423-99a	Standard Test Method for Sound Absorption and Sound Absorption Coefficients by the Reverberation Room Method
	ISO 354: 1985	Acoustics – Measurement of sound absorption in a reverberation room. Amendment (1998)

Software. The testing principle of this system is illustrated in Fig. 10.6. A sound source is mounted at one end of the impedance tube and the material sample is placed at the other end. The loudspeaker generates broadband, stationary random sound waves. These incident sound signals propagate as plane waves in the tube and hit the sample surface. The reflected wave signals are picked up and compared to the incident sound wave. The frequency range to be tested depends on the diameter of the tube. A large tube (100 mm diameter) is set up for measuring the nonwoven sound absorption in the low frequency range from 50 Hz to 1600 Hz. A small tube (29 mm diameter) is set up for testing the material sound absorption in the high frequency range of 500–6400 Hz.

The major parameters to be measured are corrected transfer function H fallen into the real part H_r and imaginary part H_i, complex reflection coefficient R determined by the real part R_r and imaginary part R_i, and normal incidence sound absorption coefficient α (taking values between 0 and 1). These parameters are described below:

$$H = \frac{\overline{H}}{\overline{H}_c} = H_r + jH_i \qquad 10.24$$

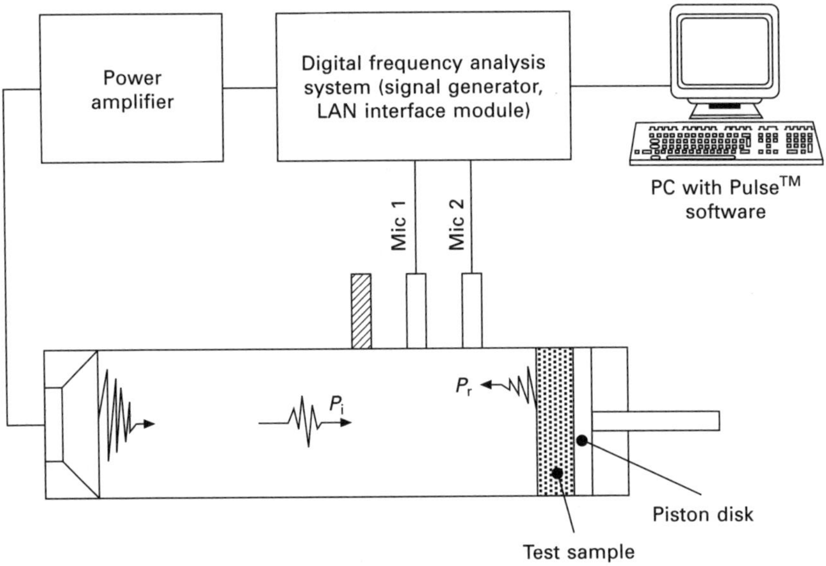

10.6 Measuring system configuration.

$$R = \frac{H - e^{-jks}}{e^{jks} - H} e^{j2k(l+s)} = R_r + jR_i \quad 10.25$$

$$\alpha = 1 - |R|^2 = 1 - R_r^2 - r_i^2 \quad 10.26$$

$$\frac{z}{\rho c} = \frac{1 + R}{1 - R} = \frac{r}{\rho c} + j\frac{x}{\rho c}, \quad 10.27$$

where

$\overline{H}$ = measured transfer function;
$\overline{H}_c$ = microphone calibrated factor;
$j = \sqrt{-1}$, indicating an imaginary unit in the equation;
c = speed of sound (m/s);
ρ = density of air (kg/m^3);
f = sound frequency (Hz);
$k = 2\pi f/c$ (m^{-1});
l = distance from the test specimen to the center of the nearest microphone (m);
s = center-to-center spacing between the two microphones (m);
$r/\rho c = \alpha/[2(1 - R_r) - \alpha]$, acoustic resistance ratio;
$x/\rho c = 2R_i[2(1 - R_r) - \alpha]$, acoustic reactance ratio;
$z/\rho c$ = acoustic impedance ratio.

Figure 10.7 exhibits a tested result of the normal incidence sound absorption coefficients (α) as a function of the sound frequency (f) for the cotton/cotton

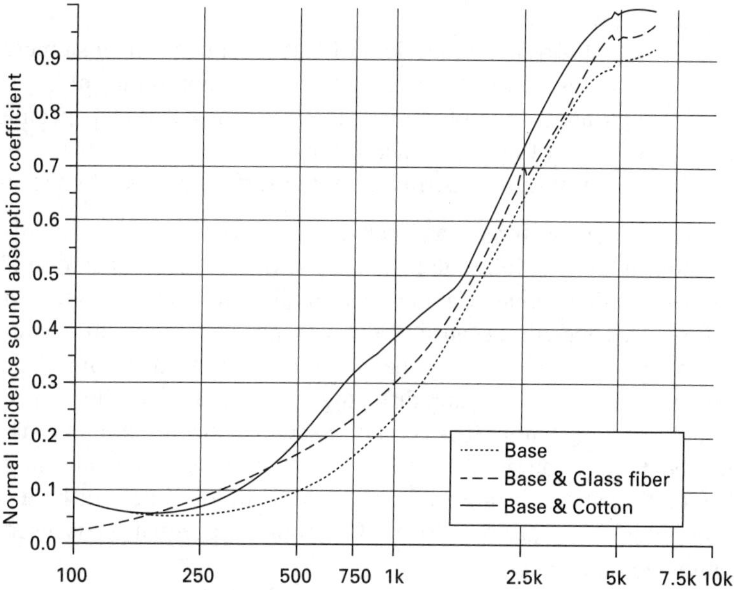

10.7 Noise absorption of cotton nonwoven composites.

and cotton/fiberglass nonwoven composites, using the above instrument. The curves combine the low frequency data and the high frequency data together to indicate a whole bandwidth of the 1/3 octave band frequency. The x-axis uses a log scale. The two-microphone method is a very fast and accurate method with a broad band of normal incidence noise frequencies. It only needs small sample sizes for testing.

ASTM C 423/ISO 354. These standard methods use a reverberation room to test materials for random incidence noise absorption. A major procedure is to measure decay rate ($\Delta dB/\Delta t$) that can be determined by turning off a steady noise source and measuring sound pressure decrease over time elapse. The sound absorption of the reverberation room, denoted as A in m^2 or Sab, is determined by the Sabine formula:

$$A = 0.9210\,\frac{Vd}{c}, \qquad 10.28$$

where

V = volume of reverberation room (m^3);
c = sound speed (m/s);
d = decay rate (dB/s).

For testing an absorption material, the noise absorption coefficient of the measured sample is calculated by the following equation:

$$\alpha = \frac{A_2 - A_1}{S} + \alpha_1, \qquad 10.29$$

where

α = absorption coefficient of test specimen (dimensionless);
A_1 = absorption of the empty reverberation room (m^2);
A_2 = absorption of the room with the installed test specimen (m^2);
S = area of test specimen (m^2);
α_1 = absorption coefficient of the surface covered by the specimen.

Alpha Cabin Method is another reverberation method commonly used in the auto industry (Müller and Krobjilowski, 2003). Compared to the above standard method, Alpha Cabin uses a small reverberation room with a sound diffuse environment, because of the presence of a sound diffuse area inside the room. Figure 10.8 shows the structure of an Alpha Cabin, composed of three sound sources, a sound diffusing area, and a microphone. The band of random noise is within 0.4–10 kHz. Specimen size used in tests is between 0.6 and 2.4 m^2. The standard size is 1.2 m^2. The measurement is taken at five different positions in the cabin and reverberation time will be determined after integrating decay curves. The following equation is used for the computation of measurement:

$$\overline{\alpha} = 0.805\left(\frac{1}{t} - \frac{1}{t'}\right) \qquad 10.30$$

$$\overline{\alpha}_E = S \cdot \overline{\alpha}, \qquad 10.31$$

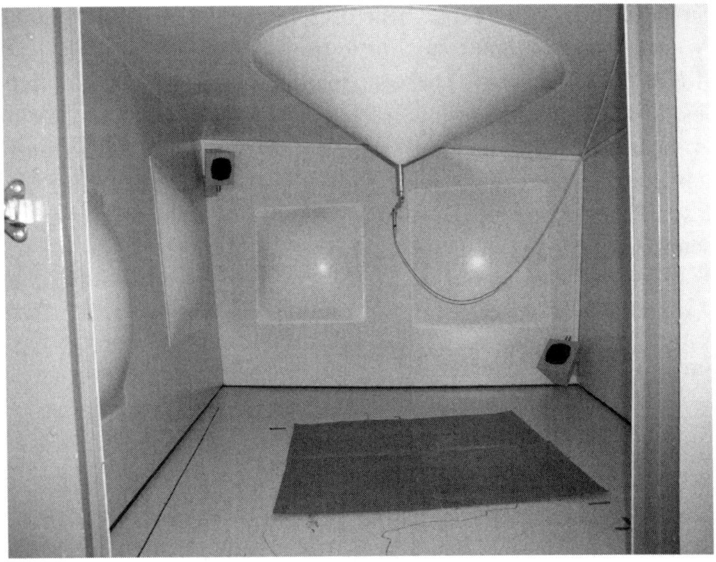

10.8 Alpha Cabin in the BIK Research Laboratory at Univeristy of Bremen, Germany.

where, $\overline{\alpha}$ is absorption coefficient when standard-size specimen is used; t is time of reverberation without specimen; t' is time of reverberation with specimen; $\overline{\alpha}_E$ is equivalent absorption when non-standard size specimen is used (m^2); and S is surface area of specimen (m^2).

In summary, the reverberation room methods allow the use of large samples for the measurement of materials exposed to random incident noise waves, and therefore are suitable for testing inhomogeneous materials. The material acoustical properties of reflection coefficient, impedance, and admittance cannot be determined through these room methods. Also, the value of noise absorption coefficent can be larger than one, meaning that the numerical numbers of this absorption coefficient are not comparable with those obtained from the impedance tube methods. Because a dedicated test room is needed, the implemention of the reverberation room methods is overall costly and inconvenient.

10.5.2 Measurement of noise transmission loss

The evaluation of material ability for noise insulation also has two instrumental approaches: the reverberation suite method dealing with random incidence noise and the impedance tube method coping with normal incidence noise. These measurement approaches are described in different test standards. It is essential to understand test scopes and physical definitions for each standard method, so that measured data resulting from different standards can be interpreted properly. The measuring methods discussed below are three most important standard methods widely adopted by vehicle OEMs.

ISO 140-1 and 140-3 / ASTM E 90. These test standards are designed for laboratory measurement of airborne noise transmission loss through a building partition or partition element. The measurement method uses two reverberant rooms (as a suite), one for noise source room and the other for noise receiver room. A test panel specimen is mounted between these two rooms (Fig. 10.9). The noise transmission loss for the test panel is determined by the ratio of noise power incident on the panel (W_i) to the noise power transmitted through the panel (W_t), expressed as:

$$\text{TL} = 10\log_{10}\frac{W_i}{W_t}. \qquad 10.32$$

The noise power impinged on the panel is calculated by:

$$W_i = I_i S, \qquad 10.33$$

where S is panel area. The noise power transmitted through and absorbed by the receiver room is determined by:

$$W_t = I_t A, \qquad 10.34$$

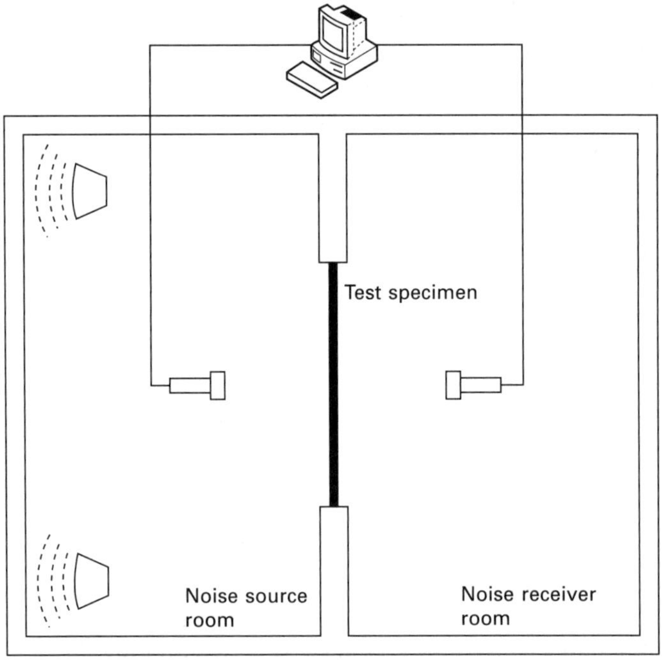

10.9 Reverberant room method for measurement of noise transmission loss.

where A is noise absorption of the receiver room. Rewriting Equation 10.32 obtains:

$$\mathrm{TL} = 10\log_{10}\left(\frac{I_i S}{I_t A}\right) = 10\log_{10}\left(\frac{I_i}{I_t}\right) + 10\log_{10}\left(\frac{S}{A}\right). \qquad 10.35$$

Because the measured noise pressure and energy in a diffuse room are comparable, the term $10\log_{10}(I_i/I_t)$ is the noise reduction (NR) indicating the difference of the noise pressure levels between the two reverberant rooms $(SPL_i - SPL_t)$. Therefore,

$$\mathrm{TL} = \mathrm{NR} + 10\log_{10}\left(\frac{S}{A}\right) = SPL_i - SPL_t + 10\log_{10}\left(\frac{S}{A}\right). \qquad 10.36$$

SAE J1400. This standard test method, entitled 'Laboratory Measurement of the Airborne Sound Barrier Performance of Automotive Materials and Assemblies,' is established by the Society of Automobile Engineers (SAE) with the purpose of providing a test procedure that can determine the airborne noise barrier performance of materials and composite assemblies widely used for making vehicle interior products. The measurement set-up includes a reverberant source room, a hemianechoic receiver room, and a signal analyzer.

A test sample is mounted between the two rooms. This test also needs a reference sample (usually a homogeneous limp material like lead) for the characterization of test condition. Random incidence noise is generated in the verberant source room. The sound pressure levels in both the source room and receiver room are measured. The TL of test samples is calculated according to the TL mass law. For the reference specimen, the measured transmission loss is expressed as:

$$TL_0 = 20 \log_{10} (mf) - 47.2. \tag{10.37}$$

For the test specimen, the TL is calculated by:

$$TL = NR - CF, \tag{10.38}$$

where NR is noise reduction measured by the test specimen and CF is called correlation factor determined by:

$$CF = NR_0 - TL_0, \tag{10.39}$$

in which NR_0 is noise reduction measured by the reference specimen (Hoult, 2003).

ASTM Work Item 5285. This test method, entitled 'Measurement of normal incidence sound transmission of acoustical materials based on the transfer function method,' has been proposed as Draft 050415 by an ASTM technical committee, and will be designated as ASTM E xxxx-04. This in-progress standard method describes the use of an impedance tube, four microphones, and a digital frequency analyzer for measuring material transmission loss. The sound source to impinge test materials is normal incidence noise wave. The Brüel and Kjær transmission loss tube Type 4206T is designed for the TL measurement. This tube set is actually an extension of the Brüel and Kjær impedance tube Type 4206, including an additional pair of microphones and two extended tubes, a large tube (diameter 100 mm) for measuring sound frequencies within 50–1600 Hz and a small tube (diameter 29 mm) for measuring sound frequencies within 500–6400 Hz.

The measuring principle is illustrated in Fig. 10.10, in which A, B, C, D are coefficients that represent complex amplitudes of the sound waves in the field of normal incidence sound wave tube. These coefficients are determined by the following equations (Song and Bolton, 2000):

$$A = \frac{j(P_1 e^{jkx_2} - P_2 e^{jkx_1})}{2 \sin k(x_1 - x_2)}, \tag{10.40}$$

$$B = \frac{j(P_2 e^{-jkx_1} - P_1 e^{-jkx_2})}{2 \sin k(x_1 - x_2)}, \tag{10.41}$$

$$C = \frac{j(P_3 e^{jkx_4} - P_4 e^{jkx_3})}{2 \sin k(x_3 - x_4)}, \tag{10.42}$$

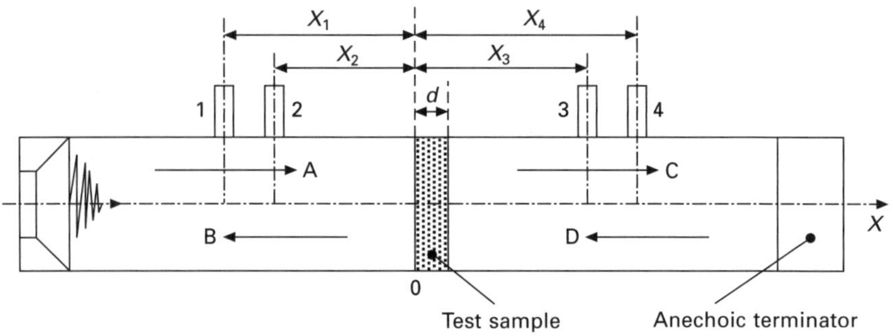

10.10 Impedance tube method for transmission loss test.

$$D = \frac{j(P_4 e^{-jkx_3} - P_3 e^{-jkx_4})}{2 \sin k(x_3 - x_4)}, \qquad 10.43$$

where P_1, P_2, P_3, and P_4 are transfer functions measured by four microphones located at 1, 2, 3, and 4; k is wave number in the ambient fluid. The transmission loss is defined as:

$$\text{TL (dB)} = 20 \log_{10} \left| \frac{C}{A} \right|. \qquad 10.44$$

Figure 10.11 exhibits a test result of three acoustic nonwovens using the Brüel and Kjær TL tube: cotton nonwoven (Base) with weight 62 g/m^2 and thickness 35 mm; glass fiber-surfaced cotton nonwoven composite (Base & Glass fiber) with weight 189 g/m^2 and thickness 38 mm; and cotton-surfaced cotton nonwoven composite (Base & Cotton) with weight 336 g/m^2 and thickness 38 mm.

10.6 Engineering of acoustic textiles for noise control in vehicles

10.6.1 Type and function of auto interior materials

To achieve the best acoustic performance for a vehicle, four different types of acoustical materials are needed for use in vehicle interiors. Noise absorbers and barriers are mainly for the control of airborne noise. Vibration isolators and dampers are primarily for the control of structure-borne noise. These materials are integrated into interior parts (composite assemblies). Each material may become an important component in these composite assemblies, such as soft-touch skin or rigid 3D structure. Figure 10.12 illustrates all interior applications of these acoustical materials (Wyerman, 2003). It can be seen that the barrier materials are heavily used to insulate engine noise, road noise, and exhaust noise. Effective noise barrier materials should feature

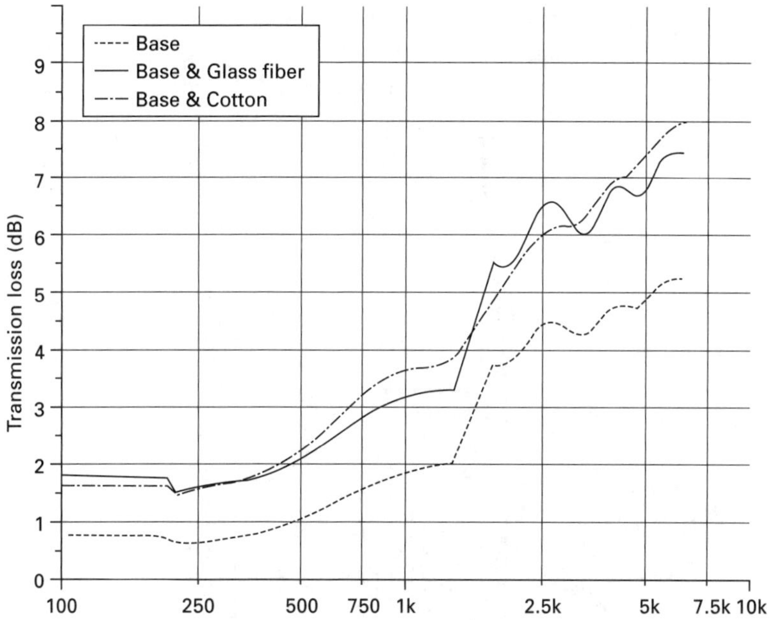

10.11 Transmission loss curves of cotton nonwoven composites.

non-porosity, high surface density, flexibility, and double-wall panel structure. Inside the passenger compartment, headliner, floor carpet, door trims, pillar trims, and seats become major absorbing and insulating components for reducing interior noise. 'Absorber everywhere' is a good interior design philosophy that helps achieve the excellent acoustical performance in vehicle manufacture. Ideal noise absorption materials should have unique porosity and tortuosity in structure and high flow resistance in the interaction with noise waves.

It is estimated that the consumption of textile materials in each car is about 20 kg. This includes interior trims, tires, seat belts, airbags, and composites (World Textile Publications Ltd, 1999). Major textile and foam materials used in vehicle acoustical applications are listed in Table 10.2. Woven, knitted, and nonwoven fabrics are commonly used as stylish decor layers of interior parts. Fibers selected for these fabrics are man-made fibers like PET, PA, and PP. The polypropylene-blended natural fiber nonwovens have become a core layer in the interior composites. They can be either bulky format or rigid format (panel) depending on whether they are required for acoustic absorption or for acoustic insulation. Foam materials are prominently used in two interior structures: surface components for soft touch and noise absorption and padding components for noise insulation. These foams can be made from PUR, PP, PE, or PVC.

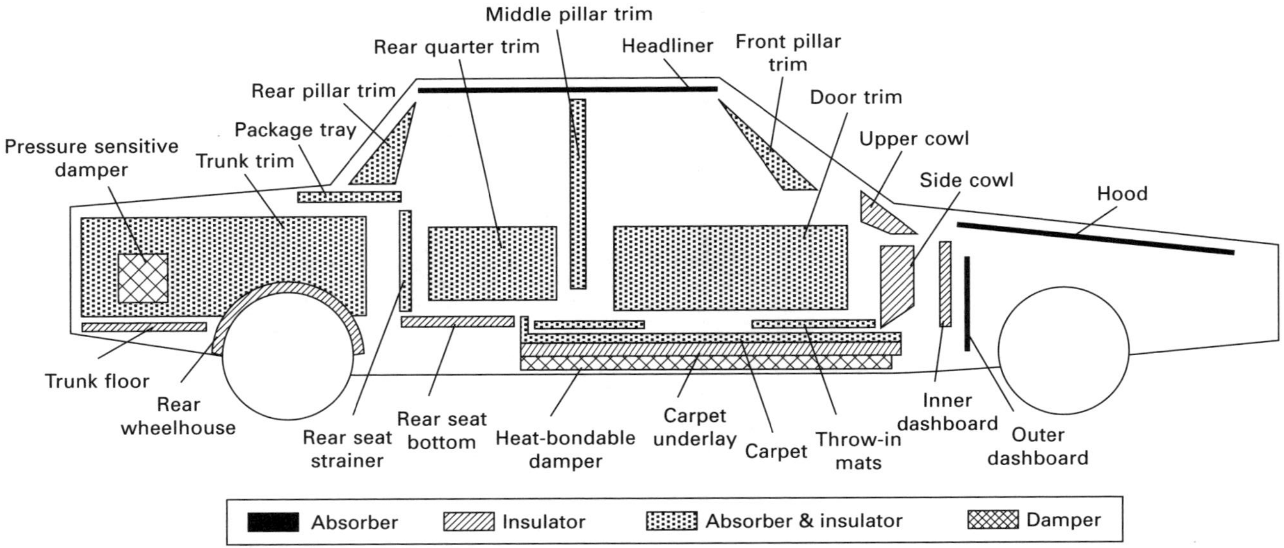

10.12 Acoustical materials in vehicle interior.

Table 10.2 Major textile and foam acoustical materials in auto interiors

Area	Interior part	Component	Material
Engine	Hood absorber Outer dash absorber	Single layer	Foams: polyurethane, melamin Felts: cotton/phenolic resin, glass/thermoset resin
Passenger	Floor carpet	Decor	Needlepunched PET nonwovens; PA or PET tufts
		Insulator Heavy layer	Foams, thermoplastic-bonded felts
	Inner dash insulator	Single layer Heavy layer	Foams, thermoplastic-bonded felts
	Headliner	Decor	Nonwovens, warp knitted fabrics
		Insulator	Phenolic resin bonded felts, natural fiber/PP, PUR foam and glassmat sandwich, polyester felts
		Retainer	Spunbond nonwovens
	Door panel Door trim	Decor	Wovens, warp and weft knits, nonwovens, PVC skins, leathers
	Door insert	Soft touch	Foams
		Substrate	Thermoset resins, natural fiber/PP
		Retainer	Spunbond nonwovens
	Instrument panel	Decor	PVC skins, leathers
		Soft touch	Foams
	Seat	Cover	Wovens, warp and weft knits, PVC skins, leathers
		Pad	Foams
Trunk	Side trim	Decor	Nonwovens
		Insulator	Natural fiber/PP, trilaminate
	Liner (floor cover)	Decor	Nonwovens
		Insulator	Recycled fiber felts
	Load floor	Single wall	Natural fiber/PP, wood/resin, baypreg sandwich
		Double wall	Twinwall (glass/PP, natural fiber/PP)
	Package tray/ parcel shelf	Decor	Nonwovens
		Rigid layer	Natural fiber/PP, wood/resin, baypreg sandwich

10.6.2 Interior noise control through different acoustical materials

There are two engineering routes for reducing interior noise levels in vehicles. One route is to eliminate noise sources or to interrupt any noise path that allows noise waves to enter the vehicle interior. The other route is to use acoustical absorption materials to absorb noise waves already transmitted into the passenger compartment. The first route is essential and is closely related to the vehicle quality rendered by design and installation. Also, some noise sources such as wind noise or road noise are simply unavoidable. The second route is very practical in engineering. But under certain circumstances, the noise reduction efficiency may not be high enough because the selection and application of acoustical materials are influenced by other factors such as cost, weight, and recyclability. To address the question of how to use high performance acoustical materials for leveraging efficient interior noise reduction, some typical application cases are selected and analyzed below.

Material approaches for controlling noise absorption

Issue of surface material influence. Use of interior surface materials is choosy and highly dynamic, because of customers' need for new vehicle design and for more choices of vehicle models. Interior noise levels may vary with the difference of surface materials or with the change of surface material area (e.g., with or without riders). It is considered that the passenger compartment is neither like a truly reverberant room nor an anechoic room. In the study of this semi-diffuse acoustic environment, a parameter called room constant R is defined (Harrison, 2004):

$$R = \frac{S\bar{a}}{1 - \bar{a}}, \qquad 10.45$$

where S is total room area (m^2) and $\bar{a}$ is average Sabine absorption in the room. Accordingly, the sound pressure difference in the reverberant field is determined by:

$$\Delta SPL = 10\log_{10}\left(\frac{4}{R}\right). \qquad 10.46$$

Table 10.3 provides the sound absorption data in terms of fabric seats and leather seats in a typical European passenger wagon (Harrison, 2004). Using the data from this table, the calculated R is 14.98 for fabric seats and 11.53 for leather seats. Applying Equation 10.46, the difference of noise reduction between the use of fabric seat cover and the use of leather seat cover can be calculated by:

$$\text{NR} = 10\log_{10}\left(\frac{14.98}{11.53}\right) = 1.1\ (\text{dB}). \qquad 10.47$$

This is a comparative value, indicating that the fabric seat surface may result in a slightly greater effect on the noise reduction (here 1.1 dB extra) compared to the leather seat surface.

Biobased nonwoven sandwich composites for enhancing noise absorption. Biobased nonwovens using natural fibers (cotton, ramie, flax, hemp, kenaf, etc.) and polypropylene fiber as binder have been increasingly used for making auto interior parts. These biobased nonwovens feature lightweight,

Table 10.3 Interior surface area and noise absorption of a typical European wagon

Interior	Surface area S (m^2)	Fabric seat		Leather seat	
		$\bar{a}$	$S\bar{a}$	$\bar{a}$	$S\bar{a}$
Headliner	3.50	0.60	2.10	0.60	2.10
Carpet	1.47	0.42	0.62	0.42	0.62
Footwell	1.05	0.42	0.44	0.42	0.44
Trunk	1.12	0.42	0.47	0.42	0.47
Front screen	1.05	0.03	0.03	0.03	0.03
Rear screen	1.05	0.03	0.03	0.03	0.04
Seat squabs	1.96	0.89	1.74	0.61	1.20
Seat backs	2.80	0.89	2.49	0.61	1.71
Side glass	3.00	0.03	0.09	0.03	0.09
Side trim	3.00	0.18*	0.54	0.18*	0.54
Dash	0.84	0.18*	0.15	0.18*	0.15
Rear trim	0.84	0.18*	0.15	0.18*	0.15
Total	21.68		8.86		7.53

*Estimated.

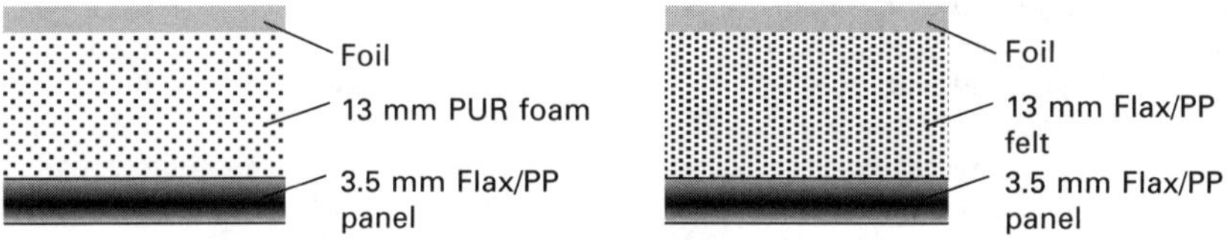

10.13 Two nonwoven sandwich acoustical composites.

enhanced acoustical performance, and high capacity of impact energy absorption. A typical engineering use of these nonwovens in the vehicle interior applications is a sandwich structure nonwoven composite in place of foam insulators. The research at the University of Bremen (Müller and Nick, 2005) has conducted a comparative study on a nonwoven sandwich composite and a foam sandwich composite (Fig. 10.13). The sound absorption property of these composites in response to normal incident noise was measured using the Brüel and Kjær impedance tubes (Fig. 10.14). The result indicates that in the tested freuquency range, the nonwoven sandwich has a higher average value of absorption coefficient than the foam sandwich. When using the nonwoven sandwich to replace the foam sandwich, the material weight is reduced and the ability to absorb noise is enhanced, particularly in the frequency range above 3000 Hz.

Material approaches for controlling noise insulation

At the range of low frequencies, the TL performance of noise barrier materials is usually low, mainly because the barrier materials tend to vibrate and

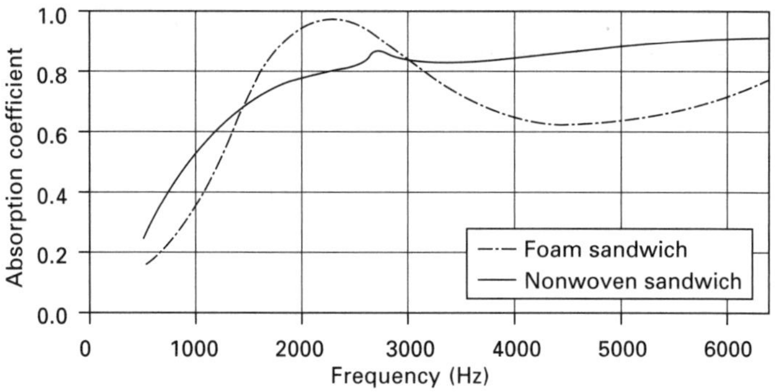

10.14 Comparison of noise absorption: nonwoven sandwich vs. foam sandwich.

Table 10.4 Sample specifications of biobased nonwovens and foams

Sample	Description	Thickness (mm)	Weight (g/m^2)	Density (kg/m^3)
Cotton	Blended with 30% PP	11	1024	93.1
Kenaf	Blended with 30% PP	11	997	90.6
Hemp	Blended with 30% PP	11	940	85.5
Foam #5	Recycled foam with scrim surface	12.7 (1/2″)	1265	99.6
Foam #6	Recycled foam with coated surface	9.5 (3/8″)	1777	187.0
Foam #7	Recycled foam with foil surface	9.5 (3/8″)	1590	167.4

resonate more than at the high frequency range. The improvement of the interior material property for low frequency noise insulation becomes an important technical issue. This issue can be seen in the design of car flooring systems, particularly the flooring system for diesel engines. Heavy padding layers (foams) are often required for enhancing the barrier performance to block the engine and road noise transmitting into the passenger compartment. However, a pay-off of this design is an increase of total vehicle interior weight and production cost. The other concern of using heavy foam materials for floor insulation is the disposal issue of ELV interior materials.

As an alternative, natural fiber or synthetic fiber nonwoven felts can be used as carpet underlayers for improving the efficiency of noise insulation. A recent study reported the TL evaluation of three natural fiber nonwoven felts in comparison with three commercial foam materials (Chen *et al.*, 2006). The material specifications are listed in Table 10.4. The TL property was measured using the Brüel and Kjær impedance tube. As shown in Fig. 10.15, the cotton/PP felt has superior TL performance, compared to the commercial

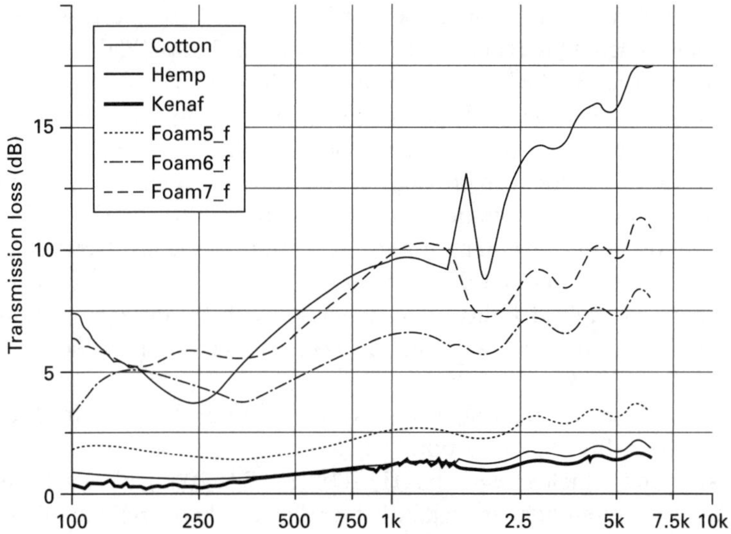

10.15 TL comparison: biobased nonwovens vs. foams.

foam products and the kenaf/PP and hemp/PP felts. It can aslo be seen that the type of foam surface influences the TL significantly. Here, Foam #7 (foil surface) is better than Foam #6 (coating surface) and Foam #5 (scrim surface).

10.7 Future trends

Future vehicle visions are energy efficiency, intelligent interior functions, and ecological performance. Because today's automotive industry is growing globally, the competition for auto interior design and production is fiercer than ever before (Weiser, 2005). Although the rivalry theme is still winning customers, vehicle OEMs are increasingly using merging technologies in all stages of manufacture, in order to keep their products safer, quieter, and more cost-effective. Consequently, vehicle interior producers (Tier 1 and Tier 2 suppliers) are facing these challenges: strong price competition, increasing raw material price, and shorter development time period. Vehicle acoustics will be a leverage for the consumer automobile estates. Future trends in the development of innovative interior materials are envisaged below.

In the material aspect, use of lightweight interior materials that contribute to the reduction of overall vehicle weight is a priority in responding to OEMs' urgent need for improving vehicle energy efficiency. Natural fiber-based nonwoven composites will become a major type of lightweight auto interior material because of their performance in noise reduction, impact energy absorption, and renewability. A typical application case is twin wall (or twin sheet) panels for trunk interiors, e.g. 50/50 fiberglass/PP such as 'Sommold' from Faurecia (Nick and Duval, 2005), or 50/50 natural-fiber/PP such as 'KENBOARD air™' from R+S Technik GmbH (Fischer, 2005). Other new textile materials particularly innovated by nanotechnologies will also gain attention in the interior acoustical applications. For example, micro- and nano-fiber fabrics can provide a highly porous surface for noise absorption. Nano finishing technologies can be used to modify fabric surface micro texture for an improvement of noise reduction efficiency. New technologies for the acoustic material development also include tunable noise absorbers and barriers, hybrid lightweight materials with balanced noise absorption and transmission loss, and active noise and vibration reduction systems (Wyerman, 2003).

In terms of the material design aspect, 'features designed in' and 'design for disassembly' have become a new design approach for today's sound package and vehicle integration (He 2005; NASA, 2003). New vehicle interior features are required to design in, rather than to add on, right at the beginning stage of the product development. This will allow interior parts to be defined and refined earlier so that a balanced solution for whole interior system can be made. The concept of design for disassembly reflects an OEM's focus on vehicle recyclability. The EU has issued an ELV regulation that 95% of a vehicle by weight has to be recyclable by 2015 (Kemper and Hobi, 2003). Under this guideline, European vehicle designers have to insure the recyclability of all newly produced interior parts to be accountable. Recycling automotive interior parts is very difficult because most of them are petrobased polymer materials with mixed composite structures. Current disposal methods for these auto interior parts are either garbage dumps or burning. A possible solution for this problem is to make interior parts either using one single polymer or using biodegradable materials. The EU-funded project Controlled Close Loop Recycling (CONCLORE), headed by the University of Bremen in Germany, is focusing on a concept car with all PET interiors. The most significant impact of this research is the establishment of an industrial ally that brings the university research lab, automaker, interior manufacturer, PET producer, nonwoven producer, and RFID technology provider to work together for implementing the EU ELV regulation.

Regarding the design tools for interior acoustical materials, poro-elastic modeling (PEM) is a major simulation method for single and multilayer 2D panel materials (Dinsmore, 2005). A typical frequency range for this method is 0.1–10 kHz. This method can provide a quick solution with minimal computational needs; enabling simulation of different material combinations; and allowing inverse computation to determine material parameters. Finite element analysis (FEA) and statistical energy analysis (SEA) are commonly used for simulating interior parts, interior systems, and complete vehicles (Manning, 2003). FEA is powerful at low frequencies (below 200 Hz) but has difficulty modeling damping and absorption. SEA is suitable for analyzing high frequency response. The establishment of the FEA and SEA models enables the performance of vehicle sound packages to be predicted without costly vehicle testing. Many CAD software tools for simulating interior sound packages or for sound mapping are commercially available, such as FEWaves, VNoise, AutoSEA, SEA-XP, and Comsol Multiphysics™.

10.8 References

Bies, D. A. and Hansen, C. H. (1996), *Engineering noise control – theory and practice*, Second edn, E&FN Spon.

Chen, Y., Negulescu, I., Sun, L., Jiang, N., Zhang, X. and Parikh, D. V. (2006), 'Acoustical and Flame Retardant Performance of Biobased Nonwovens as Auto Interior Materials', in *Techtextil Symposium North America*, Messe Frankfurt, Atlanta, GA.

Dinsmore, M. (2005), Design tools for acoustical materials, http://www.sae.org/events/nvc/dinsmore_final.pdf.

Fahy, F. and Walker, J. (1998), *Fundamental of Noise and Vibration*, E&FN Spon.

Fischer, D. (2005), 'IMG vs. "One-Shot" Process ThermoGrain,' in *IQPC Workshop – Interior Surfaces and Materials*, International Quality and Productivity Center, Frankfurt, Germany.

Green, E. R. (2005), The principles of acoustical attenuation, http://www.sae.org/events/nvc/green_final.pdf.

Harrison, M. (2004), *Vehicle refinement: controlling noise and vibration in road vehicles*, SAE International, Warrendale, PA.

He, H. (2005), Sound package and vehicle integration, http://www.sae.org/events/nvc/he_final.pdf.

Hoult, P. (2003), Test methods for determining performance of acoustical materials, http://www.sae.org/events/nvc/ws-mon-presentation.pdf.

Ingard, K. U. and Dear, T. A. (1985), 'Measurement of Acoustic Flow Resistance', *Journal of Sound and Vibration*, 103 (4), pp. 567–572.

Kemper, P. and Hobi, G. (2003), 'Hot Melt Adhesives Combine with Nonwoven Fabrics', *Adhesives and Sealants Industry*, pp. 22–24.

Khambete, S. (2003), 'Acoustic Influence of Nonwoven Fiber Products in Automotive', in *2003 Auto Interior Show/Conference*, The Auto Interior Show, Detroit, MI.

Kinsler, L. E., Frey, A. R., Coppens, A. B. and Sanders, J. V. (1982), *Fundamentals of acoustics*, Third edn, John Wiley & Sons Inc., New York.

Manning, J. E. (2003), Predicting the performance of acoustical materials, http://www.sae.org/events/nvc/ws-mon-presentation.pdf.

Müller, D. H. and Krobjilowski, A. (2003), 'Processing, Applications and New Developments of Natural Fiber', in *Beltwide Cotton conferences*, National Cotton Council of America, Memphis, TN, pp. 2406–2420.

Müller, D. H. and Nick, A. (2005), 'Noise, Vibration, Harshness, Lightweight, Costs, Recycling: Too Many Requirements?', in *IQPC Workshop – Interior Surfaces and Materials*, International Quality and Productivity Center, Frankfurt, Germany.

NASA (2003), 'Designing for Disassembly', *NASA Tech Briefs*, pp. 22–24.

Nick, A. and Duval, A. (2005), 'Twinwall sommold innovative mechanical and acoustical felt material', in *13th International Techtextil Symposium for Technical Textiles, Nonwovens and Textile-Reinforced Materials*, Messe Frankfurt, Frankfurt, Germany.

Song, B. H. and Bolton, J. S. (2000), 'A transfer-matrix approach for estimating the characteristic impedance and wave numbers of limp and rigid porous materials', *Journal of the Acoustical Society of America*, 107 (3), pp. 1131–1152.
Weiser, T. (2005), 'Requirements and potentials of innovative interior trim materials', in *IQPC Workshop – Interior Surfaces and Materials*, International Quality and Productivity Center, Frankfurt, Germany.
World Textile Publications Ltd. (1999), 'Changing landscape: notable developments in the automotive sector', *Nonwovens Report International*, July 1999, (340), pp. 34–37.
Wyerman, B. R. (2003), Basics of acoustical materials, http://www.sae.org/events/nvc/ws-mon-presentation.pdf.

11

Leather and coated textiles in automotive interiors

M MEYER, H SCHULZ and M STOLL, Research Institute of Leather and Plastic Sheeting (FILK), Germany

Abstract: Besides textiles, leather and leather substitutes are important upholstery materials in the automotive industry. Compared to textiles they are easier to clean and more durable which causes the use of synthetic leathers in the public transport sector and the use of leather in private automobiles.

The selection of the different leather types such as aniline leather, semi-aniline leather and leathers which are pigmented and finished depends on their intended use and the required quality. Nevertheless, all these leathers are manufactured by a sequence of processing steps. The first steps from raw hide to crust include wet operations performed in drums complemented by mechanical operations mainly to purify the material and to adjust the thickness. The final operations comprise drying and finishing by several lacquer coatings to improve fastness.

With the beginning of the series production of automobiles the need for leather substitutes became obvious as the demand for interior material grew very fast. There was a need for heavy-duty, easy to process and cost-efficient material with leather-like properties. This chapter deals with the development and evolution of polymer coated textiles, commonly summarised under artificial leather, in the automotive industry. It will give an overview of the state of the art regarding textile substrates, coating mediums, coating technologies, material properties and application areas of the various materials.

In the automotive industry the requirements for materials are specified in great detail. This results in extensive testing standards. The EN and ISO standards available today are summarised at the end of this chapter.

Key words: leather, synthetic leather, polymer-coated textiles, automotive interior, producing technologies, physical testing methods, chemical testing methods.

11.1 Introduction

Leather and its substitutes are used in the automotive industry as upholstery material for seats and headrests, as covering material for dashboards, back shelves, door linings, centre consoles, gear shifts, steering wheels and ceilings, and also for accessories. Compared to textile materials, leather and synthetic leather are easier to clean and more durable. Synthetic leather, therefore, is used in the public transport sector (in taxicabs, ship cabins, buses and trams). Leather, however, is more expensive than synthetic materials. It represents exclusivity and comfort. It is used in private cars according to the rule that the more expensive the car, the more leather it will contain.

Leather has been used as an interior material for cars since the invention of automobiles. In the early years of car manufacturing, automobiles were luxury goods finished with luxury material, i.e. leather. From the 1920s–1940s, manufacturers used heavyweight vegetable-tanned and pigmented leathers, finished with an acrylic base coat and nitrocellulose. The 1960s saw the introduction of chromium-tanned nappa leathers, which were at first finished using polyurethane coatings, and therefore showed better fastness properties. Chromium-free leathers with improved dry heat behaviour have now been on the market for 15 years, for use as dashboard coverings, for instance. They are combination-tanned with glutaraldehyde and synthetic tannins.

Today, most modern automotive leathers used in upholstery are chromium-tanned nappa leathers with polyurethane/acrylate finishing. They have a matt-finished appearance and cannot be polished (Wachsmann, 2005). In recent years, however, chromium-free tanned leathers have become increasingly important. Automotive leathers are high-end products with properties adjusted for special use in car interiors, reflected in the standardised requirements described in Section 11.5.

However, the beginning of mass car production in the 1930s required cheaper substitutes. The first leather substitute used in cars was based on nitrocellulose-coated textiles. Then, at the beginning of the 1950s, the first PVC-coated textiles came onto the market. PVC had already been invented a hundred years earlier, but technological difficulties had delayed its technical use (Lueger *et al.*, 1931). In parallel to leather finishing, polyurethane-based substitutes were invented as upholstery materials in the 1960s. These PVC and polyurethane-based substitutes with optimised polymeric systems are still in use today.

The next section will describe the technologies for making leather and leather substitutes, their special qualities and the necessary requirements when they are used as automotive interior materials. A short section will also deal with leather board, a leather waste-based substitute.

A compendium describing the use of leather in the automotive industry has been compiled by Buirski *et al.* (1999).

11.2 Technologies for producing leather for automobiles

Leather is made from the naturally grown collagen fibre network of hides and skins by suitable chemical and physical processes (Reich and Taeger, 2007). The properties of automotive leathers are usually similar to those of upholstery leathers for furniture, with some improved features relating to better abrasive behaviour, improved fastness parameters and minimised gaseous emissions.

The types of leather used in the automotive industry are:

- Aniline leather: full grain leather which is fully penetrated and aniline-dyed; it has no pigmented finish, but is oil and water resistant. Because of its natural appearance, this type of leather shows poorer light fastness and abrasive properties than finished leathers.
- Semi-aniline leather: full grain leather, with a light pigmented finish allowed.
- Pigmented and finished leathers: full grain or light/medium corrected grain, with slight or heavy buffing; with a pigmented finish consisting of several coating layers, embossed with a brand-typical grain or fancy pattern, and showing improved light fastness.
- Coated split leathers: flesh splits with a pigmented finish of several layers, embossed with brand-typical grain structures or pattern.
- Nubuck: leather with heavily buffed grain.

Many automotive manufacturers insist upon full grain leathers, especially in the case of more expensive leather qualities. Buffing is not allowed during the manufacture of these leathers, although they are finished with pigmented lacquers.

As shown in Fig. 11.1, tanning of hides for the manufacture of leather is a complex procedure consisting of a dozen technological steps. These steps can be divided into a sequence of four technological sections: beamhouse, tanning, retanning/wet end, and finishing. Each of these sections comprises both chemical and mechanical operations. The chemical operations of the first three stages are performed in rotating drums or mixers. The volumetric capacity of these drums ranges from 100 kg up to several tons, to contain raw material and float. Water is used as the transport medium for chemicals, and as float and solvent. After working in the raw materials, the hides are not dried until the finishing stage. Therefore, an efficient work system and suitable preservation of the material and floats are necessary.

Amongst others, the most important difference between making leather and processing textiles is the handling of pieces, and the fact that leather is a naturally grown raw material with all its visible defects.

The principles of the various tanning technologies have been described in detail elsewhere (see references in Section 11.6).

11.2.1 Raw material

Leather is made from animal hides. For automotive leathers, bovine hides are used almost exclusively. Trading companies buy hides from abattoirs,

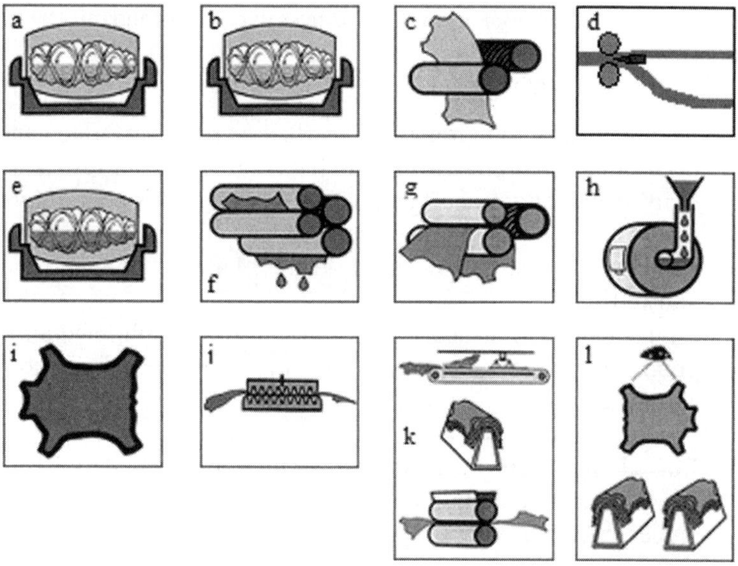

11.1 The process steps of leather manufacturing are soaking (a), liming (b), fleshing (c), splitting (d), deliming – bating – pickling – tanning (e), sammying (f), shaving (g), wet end (retanning, dying, fat liquoring) (h), drying (i), milling – staking (j), finishing (k), final sorting (l) (with permission of German Leather Association VDL).

preserve them by salting or cooling, and trade them worldwide. The quality of the raw material strongly affects the final leather quality, and therefore the final price.

On the one hand, the quality of the hides depends on the slaughtering itself and also the preservation after slaughtering. On the other hand, the livestock husbandry conditions are responsible for the final quality of the leather. Parasites and insect bites, physical injuries inflicted by the farmer, by barbed wire or other animals lead to individual defects of the grain, which reduce the leather's quality. The tanners are unable to notice or control many properties of the raw hides because of the pelage – and the tanner may buy a pig in a poke.

11.2.2 Beamhouse

The aim of the beamhouse operations is the unhairing, purification, bleaching and opening up of the structure of the hides. The beamhouse processes begin with soaking and washing the hides several times in order to remove dung and non-collagenous proteins, and to rehydrate the salted hides. During soaking and washing, chemicals such as surfactants, enzymes and soaking auxiliaries are added to improve and speed up rehydration and cleaning.

The purpose of the subsequent liming step at pH ~12.8 is to remove the epidermis and hair (keratin) and open up the collagen fibre structure. To do this, the hides are soaked overnight in solutions of sodium and calcium hydroxide, sharpened by the addition of sodium sulfide. The addition of sulfide is necessary to reduce the disulfide bridges of the keratin, making it susceptible to alkaline deterioration. After washing again, the now hairless pelts are neutralised by deliming, using organic acids in combination with ammonium sulfate. For the manufacture of soft leathers, the pelts are bated in an enzyme bath to further open up the structure.

11.2.3 Tanning

Before tanning, the pelts have to be adjusted to acidic pH. However, pure collagen swells in acid at pH < 5 with a maximum swelling at pH 2.8. Swelling, therefore, has to be prevented during leather manufacture and the pelts are acidified in the presence of high amounts of sodium chloride (5–10%) (pickling), which strongly suppresses the swelling in order to prepare the collagen structure for tannage.

Today, around 80–90% of leathers in the world are tanned with chromium-III salts, in combination with synthetic tannins as retanning agents. Only small amounts of leather are tanned with combinations of aldehydes, synthetic and vegetable tannins (free of chromium leather, FOC).

The resulting wet products are sammied, that is, pressed mechanically to obtain the slightly wet semi-finished wet-blue (Cr-III salts) and wet-white (aldehyde-synthetic), respectively. The semi-finished products are traded worldwide.

11.2.4 Wet end: retanning, dying, fat liquoring

Tanneries which do not cover the complete technology buy semi-finished leather (wet-blue, wet-white) on the worldwide market place for further processing – comprising wet end and finishing. Wet-end technologies include retanning, e.g. by synthetic tannins and/or vegetable tanning agents, followed by dying and fat liquoring. In special cases, the resulting crust leathers are subject to after-treatment with special hydrophobic and fire protective agents. All these steps are again performed in rotating drums. Finally, the crust leathers are dried, either on a toggle drier, by vacuum drying or hang drying.

11.2.5 Finishing

Finishing is the final coating of the leather's surface. Today, finishing is one of the most important steps in leather manufacturing, especially for the automotive sector. The finish strongly affects many properties of the leather, such as fastness to light, sweat, water and oil, improved abrasive and soiling behaviour, and these properties facilitate application in automobiles. The finish is the face of the leather, printable with artificial brand grain structures. An overview of the techniques and systems used can be found in Reich (2006). Requirements differ between the US, European and Asian markets. In the US, there is a demand for higher abrasion properties, whereas in Europe higher flex performance is needed. The Japanese market tends to require the combination of both of these properties (Tomkin, 2005).

With regard to finishing techniques, leather is a complicated substrate because of its flexibility, softness, rigidity, high porosity, hydrophilicity, and above all its thermal sensitivity. Many coating techniques used for other materials, such as foils of synthetic polymers, textiles or paper, are also suitable for leather finishing. In principle, the finish is composed of different layers: prefinish, impregnation, base coat and top coat. The layers increase in hardness from prefinish to top coat. Finishing recipes consist of polymer components (polyacrylates, polybutadienes, polyurethanes), colouring pigments, softeners, hardeners/cross-linkers, UV protective agents, and further additives. Spraying is the most common application technology, though to an increasing degree reverse roller-coating, and on a limited scale, brush-coating, casting and screen process, are also in use. Between these coating operations, mechanical operations such as staking and milling are suitable to increase softness. Over the last twenty years, finishing systems have changed from solvent-based to water-based systems (Reich, 2006).

In Europe, finishes are built up by a larger numbers of lighter coatings, whereas in the US a minimum number of heavier coats is applied. Finishing lines in the US are long multi-stage plants designed to obtain high throughput. European tanneries, in contrast, often combine different techniques, which gives more flexibility when changing colours or products. American leathers may show higher pigment-to-resin ratio than leathers produced in Europe, thus interfering with the flex properties (Tomkin, 2005).

11.2.6 Mechanical operations

Chemical operations in leather manufacture are also accompanied by a set of mechanical operations. In contrast to textile or synthetics technologies, the tanner has to handle the pieces, and so all mechanical operations are oriented towards this requirement. The limed hides are fleshed by fleshing machines. Fleshing is usually combined in one production line, in which a splitting machine splits the pelts with a sharp band knife, mostly into two parts – grain split and flesh split. The tanned hides are shaved to unify the thickness, then sammied before storage. These semi-finished wet-blues or wet-whites are traded around the world. Tanneries operating wet-end technologies often have to shave their wet semi-finishes again. At the end of the wet processes, the leathers are dried either on a toggle dryer, by vacuum drying or hang drying. During the finishing operations, the leathers are adjusted and unified in softness by milling and staking. If corrected grain is allowed, the surfaces of the crust leather are buffed, and the finished leathers are often embossed to produce a unified brand-typical appearance.

11.2.7 Sorting

Each leather is unique as each animal was an individual – with visible defects. The price to be obtained for leather depends dramatically on its visual appearance and so leathers are sorted several times during the manufacturing process. The first sorting step is usually carried out at the semi-finished stage, but sometimes the pelts have already been sorted before this. Furthermore, crust leathers are sorted at the end of the wet-end process as well as at the finished stage, and especially in the automotive sector, the cut pieces are sorted before the coverings are sewn. Sorting is always performed by trained employees. Complete automation of sorting has so far only been realised at a rudimentary level.

11.2.8 Further treatment of the leather

The subsequent processing steps comprise a multitude of further stages, which in rare cases are performed at the car manufacturing plant itself, although more often at independent factories, such as punching, sewing and upholstery factories and component suppliers. The tannery usually receives a release note for a leather batch from the car manufacturer. Leather for certain purposes is frequently developed specifically by the car manufacturers and tanneries. This development comprises, e.g., colour, design, grain, thickness, firmness, climate properties, odour, and even the permitted visual characteristics. Reference samples of these specially developed leathers are produced and deposited at the car manufacturer's and the tannery. The car manufacturer then buys batches of this leather and has it punched and sewn, partly by the tannery itself, partly by affiliated factories. Thereafter, it is further processed by component suppliers, and the components themselves are delivered at the necessary time to the car manufacturer's production lines.

Punching is one of the most important steps where money is made or lost, because the yield of punched area in relation to the required area per lot is determined as a result of negotiation between supplier and customer before the leathers are produced. Nearly all leathers show certain visible defects which are often of natural origin – such as scars, insect bites, coarse patches – and only in some cases may derive from technological sources. However, most of these characteristics are not accepted by customers. During punching, therefore, attention must be paid to circumvent these patches or hide them in non-visible areas. Furthermore, leather for automotive applications is divided into different regions (ABC), depending on its intended use. For seat upholstery, only the core region of the butt is allowed to be used, whereas coverings of gear shifts or door linings can also be made from the belly. Cutting skill determines profit or loss.

11.2.9 Technological trends

Technological trends are heading in two directions, one relating to the whole leather industry and the second pertaining especially to automotive leather. There is generally a tendency to eliminate (aqua)toxic, mutagenic and teratogenic chemicals from the whole process. Furthermore, automotive leather is being optimised with regard to its utilisation properties, such as haptics, abrasion and soiling behaviour, fastness and water vapour permeability.

With regard to beamhouse processing, the most important aim is to reduce, or completely eliminate, sulfide application for unhairing. Oxidative techniques and enzymatic processes are alternatives discussed nowadays (Heidemann, 1993; Trommer, 2002; Bronco *et al.*, 2005; Brady *et al.*, 1990; Alexander, 1988). However, neither of these techniques could completely displace the established technologies.

The use and handling of chromium salts are not risk-free, and therefore many attempts have been made to substitute broadly used chromium tannage. Glutaraldehyde tannage was introduced in the 1980s. Nevertheless, technologically and economically, chromium tannage is still more advantageous than wet-white tannage, which moreover did not produce the expected ecological benefits (Trommer and Kellert, 1999; Wilford, 1999).

The manufacture of chromium-free leather leads to the blocking of the cationic side chains of collagen as a result of aldehyde tannage. However, established auxiliary systems are optimised for chromium tanned leathers, where amino groups are still available for further reactions. Therefore, new and more flexible wet-end auxiliary systems are needed for the retanning, dying and fat liquoring of FOC leathers (Günther, 2002).

During retanning, leather can be further upgraded by the addition of microspheres. Their distribution is based on diffusion, and they become deposited predominantly in loose areas and voids. After drying, the microspheres are activated by exposure to pressureless steam, which does not wet the leather but softens the shell of the microspheres. The heat allows them to expand to form micro-cells. This results in a permanent filling effect, without making the leather heavy and hard (Tegtmeyer *et al.*, 2007).

As already described, the leather finish plays a very important role, especially in the automotive sector. Foam finish is currently in use to achieve softer surfaces. The technologically difficult technique of foaming has been completed by the addition of microspheres to the finishing layers. Using microspheres as a component of established finishing systems influences several parameters in a positive way, such as the masking of defects, better matting behaviour and the prevention of noise, soiling and hydrolysis (Simpson *et al.*, 2005).

The use of cool pigments is another very new technology in the finishing process, especially to prevent black and dark leathers from heating up in the sun. In one system, the pigments are transparent to IR radiation, but the leather surface reflects IR (Fennen *et al.*, 2006). The other system uses pigments which appear black in the visible spectrum, but reflect IR radiation (de Volder *et al.*, 2007). Compared to dark leathers without cool pigments, the resulting differences in surface temperature can reach 20°C.

In addition, the technique of ink jet printing is already used to individualise leather goods. Printing inks which are curable by UV radiation have now been developed, leading to printing with a high level of fastness (Rudolph, 2006).

Powder coating to avoid solvents, and radiation-curing lacquer systems to eliminate chemical hardening of the binders are among new technological trends summarised by Günther (2002) and Reich (2006). However, both principles are still under investigation.

11.3 Technologies for the production of coated textiles and artificial leather for the automotive industry

Coated textile substrates – summarised under the name artificial leather – compete with 'real leather' in the different fields of application, including of course for upholstery fabrics, although the structure of conventional artificial leathers differs basically from that of leather.

As described, leather consists of a meshwork of collagen fibres, which becomes continuously denser towards the grain side (Fig. 11.2). Artificial leathers are compound materials made of textile substrates, consisting mostly of several polymeric layers (Fig. 11.3).

Leather was the preferred material for the first motor-powered vehicles. The limited numbers of individual models, each produced laboriously by hand, were upholstered by the saddler using traditional technology. Supplying the demand for upholstery leather was not yet a problem, and in relation to the price of the automobile, costs played a subordinate role. This situation changed with the conversion of the automobile from a luxury car into a basic commodity, and when mass production started. To produce larger quantities, it was necessary to find alternative upholstery and covering materials to the expensive leather, which was also harder to process. There was a call for a high-quality substitute for leather.

11.2 SEM of a cross-section of a bovine leather.

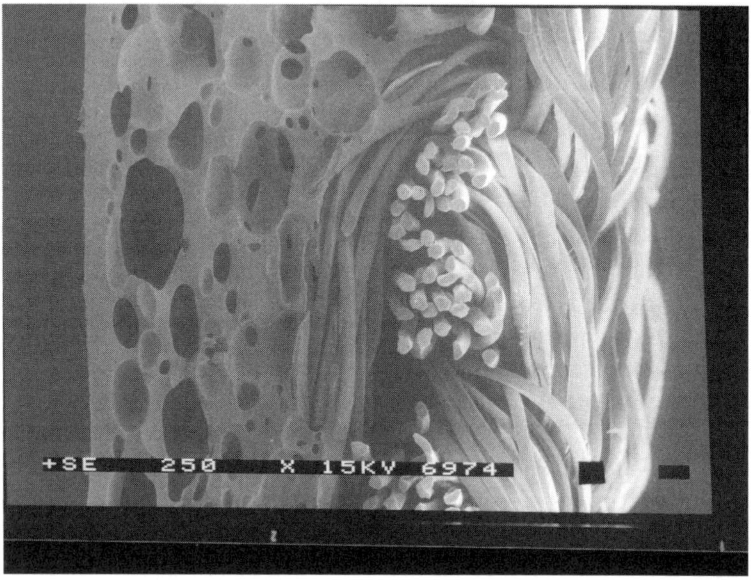

11.3 SEM of an artificial leather with a foamed interface layer.

The demand was for a material which looked nearly as expensive and aesthetically-pleasing as leather, which had a similar smooth surface feel (haptics) and similar useful properties. However, the processing needed to be less laborious than for leather, and much cheaper.

The first leather substitutes, used in automobiles since the turn of the century (1900), were nitrocellulose (NC) artificial leathers known since about 1880. They were first used as upholstery fabrics in series production in 1927 by Citroen, Fiat and Ford. These nitrocellulose artificial leathers remained in use until about 1940 in many areas, e.g. as upholstery fabrics for automobile seats. But a general breakthrough in the use of synthetic-based upholstery fabrics could not be achieved. The properties of NC artificial leathers were too different from those of natural leather.

Although the discovery of polyvinyl chloride (PVC) goes back as far as 1835 (Regnault), it was about 100 years before the first industrial production started (Lueger, 1894; Münzinger 1940).

Once the difficulties of controlling even polymerisation, stabilising and processing had mostly been resolved, PVC became one of the most important types of thermoplastics. Its variety and simplicity of processing, excellent technical properties and low price opened up a wide field of applications.

The production of plastified PVC pastes and their application in textile coatings was first described in DE 685 839 of the Kötitzer Ledertuch- und Wachstuchwerke AG (1937) and in 1938 in the patent of BASF. In automobiles, foils made of plastified PVC were first used on paperboard for textile coverings in 1943. The war period, along with the economic restrictions imposed by post-war occupying powers, led to a short interruption of this development, but in the following decades it was more dynamic. From 1949 onwards, when PVC artificial leathers were first used for mass production in automobiles, the different types of PVC-based artificial leather became the predominant cover fabrics for upholstery and covering and lining materials in motor vehicles. Both car manufacturers and customers considered them as cheap substitutes, and this view was held over a long period of time, which was disadvantageous for all kinds of plastics. Today, however, vehicle interiors would be unthinkable without plastics.

Despite, or maybe due to, the constant rise in the quantities of both leather and artificial leather processed for use as vehicle interior covering materials, the properties required of the material have continued to increase – and are still doing so. This is true for strength and elongation properties, and especially for visible and haptic appearance and comfort behaviour, particularly temperature and humidity management.

The impregnation, coating or other finishing of textile substances normally produces impermeable, compact materials with unfavourable properties, which cannot substitute for leather. Only more functional system solutions, both in the development of the substrates (microfibres) and the formation of a porous polymer coating, provide better comfort. In addition to elongation and strength properties, which are particularly found in textiles, the so-called breathability of the polymer compounds plays a large role. Thus, the permeability, absorption and humidity releasing capabilities are influenced. For synthetic materials this is achieved by microporosity and/or hydrophilicity, and it is described by the water vapour permeability (WVP) or the absorption of vapour capability (AVC) (Träubel, 1999).

11.3.1 Textile substrates

The textile substrate used influences how the goods will turn out with regard to the chosen fabrics (woven fabrics, knitted fabrics, non-woven material), and also with regard to the chosen fibres (natural or synthetic fibres, fibre yarn count), for properties such as visual appearance, grip, comfort, and in particular the mechanical properties of the compound material.

Processing non-woven material using micro-fibril fibres achieved a considerable improvement in the material's qualities (Anon., 1988). That is why these products are called leather materials (Nagoshi, 1987). Products such as Alcantara, Clarino, Sofrina and Lorica have become well known. These micro fibres with a fibre diameter <1 μm are generated by forming matrix fibril fibres. Due to this matrix fibril structure, these fibres can also be processed in conventional plants for non-woven material production. To create excellent material properties (strength, stress-strain behaviour, softness), flexibility of the micro fibre fibrils is necessary, which is implemented by the extraction of the matrix (Fig. 11.4) However, this process step is both time- and energy-consuming and results in high prices for the leather material.

Based on their high requirement profiles, especially with regard to strength, substrates consist mainly of a high proportion of synthetic fibres (mostly

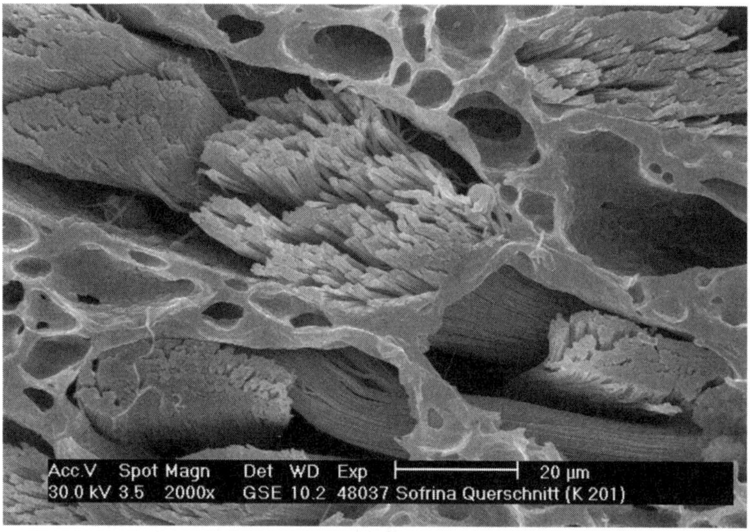

11.4 SEM of a micro fibre bundle (Sofrina).

polyesters). These substrates thus do not meet, or do not meet satisfactorily, the requirements for compound materials based on natural raw materials with high comfort levels.

The use of natural fibres for the production of coating substrates is restricted. Non-woven materials can be made from viscose fibres by needle-punching, and also by water-jet interlooping, but this shows particularly low wet strength, and therefore certain coating techniques, such as coagulation procedures, are not possible.

Some fibre manufacturers have modified viscose fibres. The so-called Lyocell procedure produces fibres which show an increased wet strength and can thus be used to process non-woven materials, which are coated with the help of the coagulation procedure (Mädler, 1998).

The fundamental idea behind the use of non-woven materials in textile compound structures is based on the symbiosis of the properties and the advantages of the non-woven material in including functional fibres (such as super absorbent fibres), or in combining with other area-measured material, such as woven fabrics, knitted fabrics, clutch, foils, etc., to meet the often very heterogenic requirements made of the textile backing material (Schäfer, 1992) (Fig. 11.5)

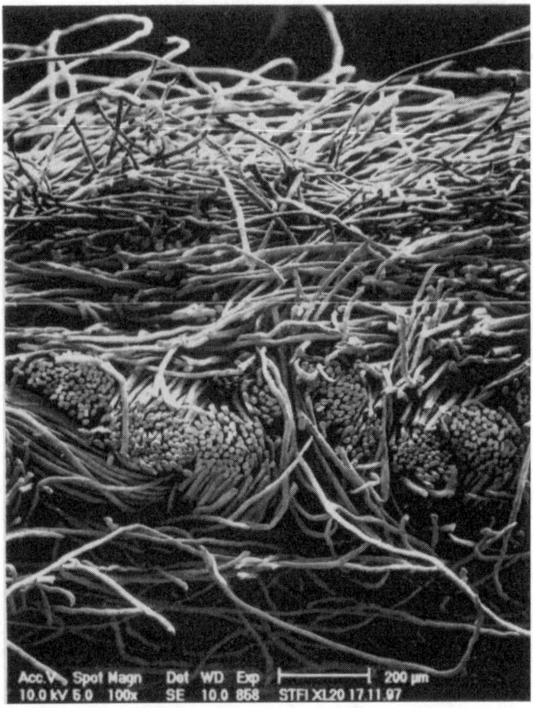

11.5 SEM of a fabric-needleflet composite.

11.3.2 Porous polymer layers

The known chemical and physical foaming procedures lead mainly to closed-cell structures with pore sizes between 20 and 200 μm and above (Schürer, 1993; Günter, 1993). Chemical foaming agents such as azodicarbonamide are used for foaming PVC pastes or plastic melts. The physical foaming procedures are based on the use of combustible hydrocarbons, such as pentane and isobutene, as well as so-called microspheres (micro-hollow balls made of plastics with enclosed gases) or the mechanical integration of air in liquid plastics or melts.

Like compact coatings, closed-cell structures only have low permeability and are thus not suitable as leather substitutes. The formed cell walls can be partly destroyed by subsequent mechanical destruction or thermally-initiated expansion of the micro-hollow balls, and the foam can be made permeable.

A polymer layer of pores communicating with each other can only generally be achieved by the precipitation of a polymer solution. The advantage of a layer generated in that way is that the sorption capability is allowed for, independent of the thickness of the formed layer, and thus layer thicknesses of over 100 μm can be achieved. Depending on the conditions, micro and/or macro pores (0.1 μm to 50 μm in size, in exceptional cases 100 μm) can be produced.

In 1958, a process for producing microporous coatings from polyurethane solutions by pre-coagulation and subsequent coagulation had already been published. In this process, a film applied to a polyurethane sheet material was pre-coagulated in the organic solvent under climate control (humidity of air >90%) and coagulated in a precipitation bath consisting of a solvent/non-solvent mixture (dimethylformamide/water) (Fig. 11.6). The remaining solvent was subsequently washed out (Wittke, 1983).

With constant further development, the quality of this process was improved and production technologies rationalised (such as the elimination of the laborious pre-coagulation phase) (Träubel, 1999; Merkle and Tackenberg, 1984; Freitag, 1983).

At the end of the 1970s/beginning of the 1980s, the range of poromeric materials was extended with impregnated woven materials (so-called immersed coagulates) for light upper materials and garments (Lomax, 1984).

A procedure was developed at the then Forschungsinstitut für Leder- und Kunststofftechnologie (Research Institute for Leather and Plastics Technology) to use a mechanically foamed polyurethane solution in dimethylformamide for the coagulation procedure (Stoll, 1994). By forming a so-called double structure, as seen in Fig. 11.7, the production of a solid porous layer with high water vapour permeability was successfully achieved. The technology described requires special systems engineering, consisting of a coat application unit, precipitation and washing baths, and a dryer.

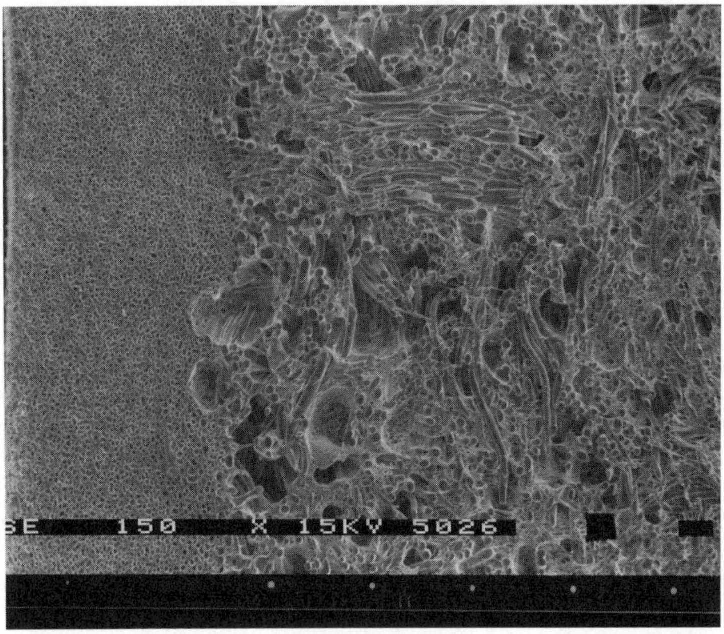

11.6 SEM of an artificial leather on the base of a non-woven with a coagulated top coat.

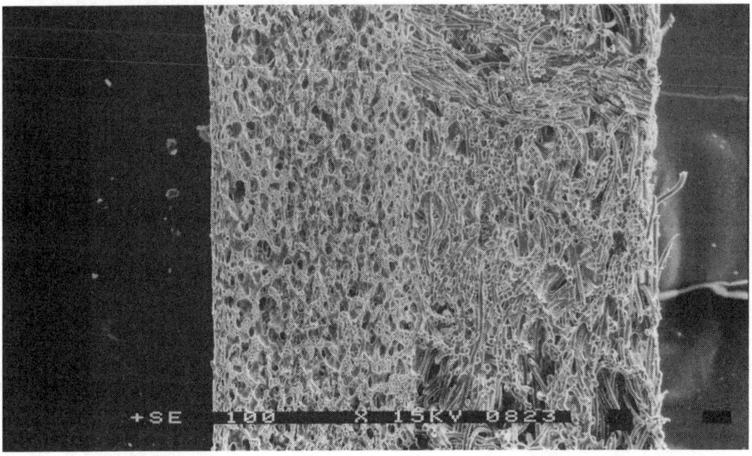

11.7 SEM of a porous artificial leather on the base of a non-woven with a microporous foamed coat.

Especially designed polyurethanes, dissolved in low-boiling solvents and capable of absorbing a large amount of water, can be applied to suitable substrates in conventional coating machinery plants. In graded temperature zones, the solvent vaporises first so that the polyurethane in the non-solvent gradually coagulates, followed by the evaporation of the water. In this way, very uniform pores can be achieved, up to a maximum of 3 to 7 μm, which allow the water vapour molecules to pass unobstructed. However, only layers with a thickness of < 100 μm can be formed by this so-called vaporisation coagulation (Loogk *et al.*, 1983; UCB–Firmenschrift, 1995; Träubel *et al.*, 1994).

An alternative way of forming permeable layers is with hydrophilic coatings. By integrating hydrophilic components into the polymeric chains (polyurethanes or polyesters), the transport of water molecules is possible without the existence of pores. The hydrophilic groups are able to absorb the water molecules and transmit them to an adjacent group. By using a different water vapour partial pressure at each side of the coating, water molecules are absorbed on the wet side and transmitted via the hydrophilic groups to the dry side. There they will be released into the environment. The polyurethane or polyetherester membranes used (such as Sympatex) are very thin (10–25 μm), as the WVP strongly reduces with thickness (Träubel, 1999; Hürten and Spijkers, 1997)

Over the past years, an improved procedure for forming a polymer layer with a porous capillary structure has been developed at the Research Institute of Leather and Plastic Sheeting, in cooperation with the Institute for Applied Polymers (IAP) in Golm. The procedural principle applied is based on the precipitation of a polyurethane (PUR)–solution in dimethylformamide (DMF), using the controlled release of a means of precipitation from a thermally-sensitive hydrogel, which is added to the solution before application.

When a certain specific temperature for this material has been reached, water will be released from the hydrogel. This additional water exceeds the stability limit of the ternary system DMF/PUR/water and by nucleation starts the precipitation process.

The hydrogels suitable for the new procedure are based on the development of a linear poly-n-isopropylacrylamid (PNIPAM) copolymer with a reproducible phase transformation at about 50°C, a so-called LCST (lower critical solution temperature) procedure. In order to avoid spontaneous coagulation when adding the hydrogel to the PUR solution, a casing of the gel particles based on styrene/divinylbenzene (DVB) is necessary (Mädler, 2005; Bohrisch *et al.*, 2007).

The feasibility of the new coagulation procedure has been proven and three main advantages were achieved compared to the conventional procedure:

(a) a 2- to 3-fold increase of process speed,
(b) variable possibilities of texturing/capillary structure and
(c) improved sorption properties of the polymer layers generated.

11.4 A special case: leather board

Leather board is made up of an irregularly laid mixture of fibres, with different leather fibres as the skeleton substance. This fabric is bonded using mainly natural binders to obtain a firm board-like material.

Leather board is produced in accordance with paper technologies. Shavings which accrue during thickness-machining and waste from leather-cutting are milled to fine fibres and powder. These natural fibres are pulped with water, mixed with natural rubber or other synthetic-based lattices, emulgators and further additives, and poured out onto a sieving belt. The wet film is pressed continuously between roller presses to remove the water mechanically and is then dried in a heating channel.

Leather board is based mainly on natural materials; it smells like leather, feels board-like, can be imprinted with many structures and finished with lacquers. It is used in the shoe industry as insole material and for stiffeners, as well as to manufacture, for example, bags and belts.

Leather board has also been tested for use as automotive interior material (Richter, 2005). With regard to mechanical parameters such as tensile strength, elongation or tear strength, the leather board has poorer properties than leather. Its use for specific applications, such as injection-moulded door interiors or back shelf covering, seems possible. However, light fastness, and especially emission behaviour, proved to be problematic parameters (Richter and Schulz, 2006). Lubricants were identified as one reason for the high gaseous emissions. Both the fat liquors used in the leather board manufacturing process and those brought into the process through the raw materials, e.g. the cutting waste, seem to be a problem. Leather board can, therefore, be manufactured for automotive interior materials. However, it is important to select pure raw materials to keep within the specification limits.

11.5 Testing of leather and coated textiles for automotives: physical and chemical requirements

The specifications for automotive leather and coated textiles contain many different requirements depending on their use in the car interior. Over the last ten years, the various activities of different CEN and ISO working groups have resulted in many new ISO and EN methods. For many standard parameters, EN and ISO standards are now available (ISO = International Organisation for Standardisation; CEN = European Committee for Standadisation; EN = European Standard).

The most important specified parameters and their corresponding test methods are summarised in Tables 11.1 and 11.2.

Table 11.1 Important specified parameters for automotive leather and harmonised test methods used

Property	Test method
Sampling and conditioning	ISO 2418, ISO 2419
Apparent density, mass per unit area	ISO 2420
Thickness	ISO 2589
Surface coating thickness	ISO 17186, VDA 230-204
Tensile strength, percentage elongation, elongation at break	ISO 3376
Tear load	ISO 3377-1
Permanent elongation	DIN 53360
Adhesion of finish	ISO 11644
Stitch tear resistance	ISO 23910
Seam fatigue	Different internal tests, no harmonised method
Stiffness in bending	VDA 230-209
Flex resistance	ISO 5402
Abrasion	ISO 17076, VDA 230–210, many different test methods are used
Scratch resistance	Different internal tests, no harmonised method
Cold flexibility	ISO 5402 at low temperatures
Cold impact	Different internal tests, no harmonised method
Heat storage, climate change test	ISO 17288 and many different internal tests, evaluation of colour change (ISO 105-A02 or -A05), appearance and shrinkage
Hydrolic ageing	VDA 230-208, ISO 17288
Light fastness and hot light ageing	ISO 105-B06, VDA 75202, SAE J 1885
Colour fastness to and for rubbing	ISO 11640
Colour fastness to perspiration	ISO 11641
Colour fastness to migration	ISO 15701
Resistance to chemicals, cleaning agents, sun lotions, insect repellents and more	Many different internal test methods
Fogging	DIN 75201, ISO 14288 VDA 277
Emission behaviour (VOC and SVOC)	VDA 278 VDA 270
Smell	VDA 275
Formaldehyde emission	ISO 17074, FMVSS 302, DIN 75200
Burning behaviour	ISO 14268
Water vapour permeability	VDA 230-206 part 1 and 2
Stick-slip behaviour	ISO 5398 part 1, 2, 3, 4
Total chromium content	ISO 17075
Chromium-VI content	ISO 4048
Fat, matter soluble in dichloromethane	ISO 17070
PCP	ISO 4684
Water content, volatile matter	ISO 4045
pH value	

German car manufacturers are very committed to harmonising methods for testing automotive leather, textiles and coated textiles. Representatives of all German car manufacturers are working in different groups of the Association of the German Automotive Industry, VDA, creating new test methods as required and adjusting common methods for application in the automotive industry. The aim of this work is to enhance the quality of leather and coated textiles and reduce testing costs. Examples of newly developed testing methods are those for stick-slip behaviour, determination of loose grain effect of leather and determination of stiffness in bending.

Many other activities are still in progress to develop objective and realistic test methods, e.g. to describe the soiling behaviour of material surfaces, evaluate surface feel or measure the cooling effects of leathers treated with special colours and pigments.

11.6 Sources of further information and advice

Over the last century, hundreds of books about leather chemistry and technology have been published. The chemical and technological background has been described in detail, e.g. in Stather, *Gerbereichemie und Gerbereitechnologie* (1957), in Herfeld, *Bibliothek des Leders*, published in several volumes (1990), in Heidemann's *The Fundamentals of Leather Manufacturing* (1993) and the anthology of Reich (2003). The most recent work was published with Reich as the main author on behalf of Taeger (2007). This book gives an excellent overview, especially of the chemical background to leather manufacture.

The most popular scientific journals in this field are the *Journal of the American Leather Chemists Association* (JALCA, USA), the *Journal of the Society of Leather Technologists & Chemists* (JSLTC, UK) and *World Leather*, in which market trends are discussed. The International Union of Leather Technologists and Chemists Societies (IULTCS; www.iultcs.org) is the umbrella organisation for the various national leather chemistry and technology societies. This organisation is responsible for the biennial congresses, which are the most important scientific events in this sector.

The materials and technologies for manufacturing compact and microporous leather substitutes are described in detail by Träubel in his book *New Materials Permeable to Water Vapor* (1999). The manufacturing and material details of poromerics are described in the books of Albrecht *et al.* (2000) and in Günter (1993). PVC-coated fabrics are described in Felger (1986). The machinery, equipment and technologies for manufacturing coated fabrics are described in Hufnagl *et al.* (1981) and Giessmann (2003).

In addition, short reviews describing leather-like materials are published in the following encyclopedia entries: 'Leather Imitates', Ullmanns *Encyclopedia of Industrial Chemistry* and 'Man-made leather', Kirk-Othmer,

Table 11.2 Important specified parameters for automotive coated textiles and harmonised test methods used

Property	Test method
Sampling and conditioning	DIN 16909, ISO 2231-B
Total mass per unit area	ISO 2286-2
Thickness	ISO 2286-3
Tensile strength, percentage elongation, elongation at break	ISO 1421-1
Tear resistance	ISO 4674-1
Permanent elongation	DIN 53360
Coating adhesion	ISO 2411
Stitch tear resistance	DIN 54301
Seam fatigue	Different internal tests, no harmonised method
Stiffness in bending	VDA 230-209
Flex resistance	DIN 53351, DIN 53359
Abrasion resistance	ISO 5470 part 1 and 2, VDA 230–210, many different test methods are used
Scratch resistance	Different internal tests, no harmonised method
Cold flexibility	DIN 53351 and DIN 53359 at low temperatures
Cold impact	VDA 237-101, EN 1876 part 1 and 2, ISO 4646, ISO 4675, different internal tests, no harmonised method
Scratch resistance	Different internal tests, no harmonised method
Heat ageing, climate change tests	EN 12280-1 and many different internal tests, evaluation of colour change (ISO 105-A02 or -A05), appearance and shrinkage
Hydrolic ageing	VDA 230-208
Light fastness and hot light ageing	ISO 105-B06, VDA 75202, SAE J 1885
Colour fastness to rubbing	ISO 105-X12
Resistance to amines	Some internal test methods, VDA method in preparation
Resistance to chemicals, cleaning agents, sun lotions, insect repellents, etc.	Many different internal test methods
Burning behaviour	FMVSS 302, DIN 75200
Stick-slip behaviour	VDA 230-206 part 1 and 3
Coefficient of friction	ISO 8295
Fogging	DIN 75201, ISO 6452
Emission behaviour (VOC and SVOC)	VDA 277 VDA 278
Smell	VDA 270
Formaldehyde emission	VDA 275

Encyclopedia of Chemical Technology. The most important journals in this field are the *Journal of Coated Fabrics*, the *Journal of Industrial Textiles*, and *Coating*.

11.7 References

Albrecht, W.; Fuchs, H.; Kittelmann, W.: *Vliesstoffe*; Wiley-VCH, Weinheim, (2000) – ISBN-10: 3-527-29535-6.

Alexander, K. T. W.: The 1988 John Artur Wilson Memorial Lecture: Enzymes in the Tannery – Catalysts for Progress. In: *JALCA* 83 (1988) 9, pp. 287–316.

Anon.: What is a Microfibre? In: *Textile Horizons* (1988) 4, pp. 49–50.

Bohrisch, J.; Hahn, M.; Mädler, A.; Stoll, M.: Synthesis and properties of new thermosensitive polymer systems. In: *Journal of Applied Polymer Science* 104 (2007) 4, pp. 2122–2129.

Brady, D.; Duncan, J. R.; Russell, A. E.: Model for Proteolytic Depilation of Skins. In: *JALCA* 85 (1990) 9, pp. 334–343.

Bronco, S.; Castiello, D.; D'Elia, G.; Salvadori, M.; Seggiani, M.; Vitolo, S.: Oxidative Unhairing with Hydrogen Peroxide: Development of an Industrial Scale Process for High-Quality Upper Leather. In: *JALCA* 100 (2005) 2, pp. 45–53.

Buirski, D.; Wachsmann, H.; Daniels, R.; Landmann, W. (eds.): *Sitting comfortability, A world leather publication*, Worlds Trades Publishing, Liverpool (1999).

de Volder, N.; Hallmanns, M.; Levy, J.: Annoying Energy II. Solar Heating of Dark Leather, XXIX. Congress of the IULTCS and the 103rd Annual Convention of the ALCA, Washington DC, USA, 20–24 June 2007.

Felger H. K.: *Kunststoff Handbuch 2/2 Polyvinylchlorid*, 2. Auflage, (1986) – ISBN-10: 3-446-14161-8.

Fennen, J.; Döppert, F.; Hess, M.: Solar Reflective Leather; IULTCS II. Eurocongress Istanbul, 24–27 May 2006.

Freitag, H.: Über die Herstellung von synthetischem Leder – Vlieskunstleder. In: *Coating* 16 (1983) 8, pp. 210–214.

Giessmann, A.: *Substrat- und Textilbeschichtung*, Springer Verlag, Berlin (2003) – ISBN-10: 3-540-43426-7.

Günter, O. (Ed.): Kunststoff-Handbuch, Vol. 7 Polyurethane, Ch. 10, 3rd edn. München, Wien: Carl Hanser Verlag (1993) ISBN 3–446–16263–1.

Günther, E.: Innovationen und Trends bei der Oberflächenveredlung von Automobilleder. In: *Leder und Häutemarkt* (2002) pp. 25–30.

Heidemann, E.: *The Fundamentals of Leather Manufacturing*. Darmstadt: Eduard Roether KG und Verlag (1993). – ISBN 3–7929–0206–0.

Herfeld, H.: *Bibliothek des Leders*. 2. Aufl. Frankfurt a. M.: Umschau Verlag (1990).

Hufnagl, W. *et al.*: *Kunstledertechnik*, 1. Aufl. Hothing Fachbuchverlag, Heidelberg, (1981).

Hürten, J.; Spijkers, J.: Laminierung von Sympatex Membranen. Teil 1 Laminierung von Sympatex Membranen. Teil 2 Adhäsion – *Kleben & Dichten* 41 (1997) 1–2, pp. 34–36; Adhäsion – *Kleben & Dichten* 41 (1997) 3, pp. 34–37.

Lomax, R.: Recent Developments in Coated Apparel. In: *Journal of Coated Fabrics* 14 Oct (1984), pp. 91–99.

Loogk, M.; Döring, E.; Reich, G.: Vergleichende Untersuchungen zur Beurteilung der

Abrieb und Scheuerbeständigkeit von Leder und Kunstleder Leder Schuhe. *Lederwaren* 18 (1983) 6, pp. 277–279.

Lueger O.: *Otto Luegers Lexikon der gesamten Technik und ihrer Hilfswissenschaften.* Stuttgart: Deutsche Verlags-Anstalt, 1894–1899 (1894).

Lueger, O. *et al.*: *Otto Luegers Lexikon der gesamten Technik*, DVA, Stuttgart, 3. Auflage, (1931).

Mädler A.: *Schlussbericht PUR-Beschichtung von Lyocellträgern*, FILK GmbH Freiberg (1998).

Mädler, A.: *Ausbildung und Charakterisierung von, permeablen Werkstoffverbunden durch Fällung von Polymerstrukturen*/FILK GmbH. Freiberg (2005). Forschungsbericht.

Merkle, R.; Tackenberg, W.: Alternatives to Leather: Laif Poromerics. In: *Journal of Coated Fabrics* 13 Apr (1984) 4, pp. 228–238.

Münziger, Walter M.: *Kunstleder-Handbuch.* Berlin-Wilmersdorf: Pansegrau (1940).

Nagoshi K.: Leatherlike Materials. In: *European Polymer Journal* (1987) 8, pp. 677–697.

Reich, G.: *Vom Kollagen zum Leder – Gesammelte Abhandlungen*, Verein für Gerberei-Chemie und Technik e.V., Reutlingen (2003).

Reich, G.: Der Finish macht. In: *Leder & Häute Markt* (2006) 01/02, pp. 32–40.

Reich, G.; Taeger, T.: *From collagen to leather – the theoretical background.* Ludwigshafen: BASF AG (2007).

Richter, R.: Lefa für Sohlen, Böden & Autos – Entwicklung und Einsatzmöglichkeiten von lederfasergefüllten Kunststoffen. In: *Leder & Häute Markt* (2005) 10, pp. 20–22.

Richter, R.; Schulz, H.: Kniffelig: LEFA im Fahrzeug – Emissionsarmes und alterungs beständiges LEFA für den Automobilinnenraum. In: *Leder & Häute Markt* (2006) 03/04, pp. 30–34.

Rudolph, H.: Individuelles Design auf Leder. In: *Leder & Häute Markt* (2006) 11, pp. 16–18.

Schäfer, W.: Vliesverbundprodukte für Anwendungen in der Elektrotechnik. In: *Melliand Textilberichte* 73 (1992) 11, pp. 866–869.

Schürer, M.: Vortrag zum 1. *Freiberger Polymertag* (1993), FILK GmbH Freiberg.

Simpson, P.; Hörner, K. D.; Dreissigacker, G.; Salgado, J.: Innovations for the upgrading of automotive leathers. In: *World Leather* 18 (2005) 5, pp. 31–35.

Stather, F.: *Gerbereichemie und Gerbereitechnologie.* Berlin: Akadamie-Verlag, (1957).

Stoll, M.: Verfahren zur Herstellung poröser Polymerschichten nach dem Koagulations verfahren. In: *Coating* 27 (1994) 1, pp. 9–11.

Tegtmeyer, D.; Voerlaender, O.; Zeyen, W.; Silberkuhl, F.; Tysoe, C.: New Micro-sphere Application in the Leather Industry, XXIX. Congress of the IULTCS and the 103rd Annual Convention of the ALCA, Washington DC, USA, 20–24 June 2007.

Tomkin, M.: Technical Differences Between US and European Automotive Leathers. In: *JALCA* 100 (2005) 4, pp.133–136.

Träubel H.: *New Materials Permeable to Water Vapor.* Berlin: Springer (1999) – ISBN 3-540-64946-8.

Träubel, H.; Schütze, D. I.; Pedain, J.: Bessere Tragehygiene durch wasserdampfdurchlässige Materialien. In: *Textilveredlung* 29 (1994) 6, pp. 171–177.

Trommer, B.: Ein neuer katalytisch-enzymatischer Oxidativäscher. In: *Leder & Häute Markt* (2002) 1, pp. 28–35.

Trommer, B.; Kellert, J.: Ökologischer Vergleich verschiedener Gerbarten. In: *Leder & Häute Markt* (1999) 6, pp. 25–36.

UCB–Firmenschrift *Breathable coatings* (1995).

Wachsmann, H. M.: Autopolsterleder im Wandel der Zeit. In: *Leder & Häute Markt* (2005) 08/09, pp. 12–13.

Wilford, A.: Waste in the shoe industry – consequences for the leather industry. In: *Leather* (1999) 201, pp. 37–40.

Wittke, W.: Die Herstellung von Polyurethan-Kunstleder (PUR-Kunstleder) nach dem Koalugationsverfahren. In: *Coating* 16 (1983) 5, pp. 128–132.

Part III

Safety applications of automotive textiles

12

Technical developments and market trends of automotive airbags

S K MUKHOPADHYAY, Sans Fibres (Pty) Ltd, South Africa

Abstract: Airbags are used in modern cars as a passive safety system for drivers and passengers. The bags are made of closely woven fabric largely coated with an elastomer. Nylon 6.6 is the predominant fibre in making airbag fabric. An airbag is connected to an inflator that is activated by the impact from an accident. Airbags are folded with care and efficiency and stored in the steering wheel, glove compartment and in other appropriate places in a car to ensure smooth deployment in the event of an accident. The actual performance of an airbag is closely related to its deployment behaviour. Currently airbag markets are largely controlled by three global regions – Western Europe, Northern America and Japan. Approximately 220 million airbags are expected to be fitted in cars globally in 2008. It is expected that the future growth of the airbag markets would come from China, India and Korea. Pedestrian safety is becoming a big issue in Western Europe and Japan. It might open up new opportunities for automotive airbags in these geographical regions.

Key words: automotive, airbags, airbag yarns, airbag sewing threads, air permeability, coated fabric, cover factor, crash speed, deployment, hot gas, inflator, industrial textiles, nylon 6, nylon 6.6, nylon 4.6, polyester, safety system, side curtains, woven fabric.

12.1 Historical background

John Hetrick was the father of the automotive airbag. He was granted a patent in the USA in 1953 for developing an automotive safety cushion assembly. Hetrick's idea was executed in automobiles through commercial production of vehicles with airbags by General Motors in the USA in the early 1970s. Nonetheless, the concept did not successfully take off due to lack of public interest in an expensive complicated and less effective technology. But the concept of automotive safety cushion assembly was never discarded. Further research enabled scientists to come up with a much improved, effective and more economical airbag system. In 1984, the concept of airbags as a passive safety system was used as a standard item in Mercedes Benz S-class cars. From the late 1980s most European, American and Japanese automobile manufacturers widely started using airbags for drivers and front seat passengers in modern cars to reduce mortality and serious injuries in the event of head-on collision.

Figure 12.1 shows the inside view of a modern car made in 1997 with driver airbag deployed. From 1998 it became a legal requirement in the USA for all new vehicles to have driver's and passenger's airbags. Fitted side curtains have become more of a common safety feature in European and American cars for saving lives in the event of a rollover collision. Side curtains will become a legal requirement for cars and light trucks to be made in the USA from 2009. Today sales of new cars are bolstered by the inclusion of several airbags in a car and they appear to add extra value in the eyes of consumers for safety and security in the event of an accident. From early 2000s, Toyota Avensis started providing 9 airbags, Audi A8 10 airbags and BMW 7 series 12 airbags as standard.

Airbags are closely woven textile envelopes that are mostly made up from nylon 6.6 filament yarns. An attempt was made unsuccessfully in the past to use nylon 6 yarns to make airbag fabrics. Currently polyester (PET) and nylon 4.6 filament yarns are used in small quantity for airbag fabrics and airbag sewing threads in some parts of the world. Nylon 4.6 behaves very similarly to nylon 6.6 but it is much more expensive. Polyester (PET), however, is a cheaper substitute but technically inferior. Table 12.1 highlights the relative merits of nylon 6, nylon 6.6 an polyester yarns for airbag application.

Table 12.1 Characteristics of various polymers for airbag yarns

Polymer characteristics	Nylon 6.6	Nylon 6	Polyester (PET)
Density (kg/m^3)	1140	1140	1390
Specific heat capacity (kJ/kg °K)	1.67	1.67	1.3
Melting point (°C)	260	215	258
Softening point (°C)	220	170	220
Heat to melt (kJ/kg)	589	522	427

Sources: various

12.1 The interior design of a modern car with deployed airbag.

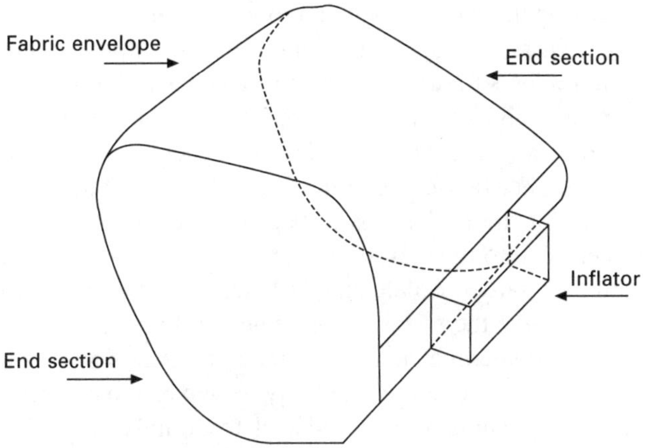

Source: US Patent 5375878, 1994

12.2 The schematic design of a passenger airbag attached with an inflator module. Source: US Patent 5375878, 1994.

An airbag is designed to open up from an impact (for most cars in the case of an impact in the front but for more advanced and expensive cars, appropriate airbags can be deployed on an impact on sides or at the back) in the car arising from a collision. Following a collision and bag deployment, an inflated airbag acts as a cushion or an energy absorber and prevents the occupants from striking any solid fixed parts of the car. An automotive airbag is connected to an inflator that is activated by the impact from an accident. Within a few milliseconds following an impact, the inflator fills the bag with nitrogen-rich gas or with another appropriate inert gas. On activation, the inflators produce mainly hot gas by activating a propellant to fill the bags. The textile envelopes (see Fig. 12.2) or airbags are made and very often appropriately coated to contain hot air produced by a small explosion arising from the activation of sodium azides or similar compounds (called propellants) while being impacted.

12.2 Materials and processes

The fabric used in making airbags needs to be strong, tough, hard wearing with low and controlled air/gas permeability behaviour. Nylon 6.6, as discussed in the previous section, is the material of choice for the yarn. Nylon 6.6 yarns made for airbags are strong, tough, resilient with good thermomechanical responses up to 200°C.

In the 1980s first generation airbags were largely made of 940 dtex and 700 dtex nylon 6.6 yarns with neoprene coating. The first generation materials were aimed at making drivers' and passengers' airbags meet requirements such as defined package volume for the storage compartment, good foldability, high softness, resilient, heat resistant and low air permeability. However, as more and more airbags were being fitted into cars, automotive manufacturers raised their requirements for technical performance of airbags. Amongst new performance criteria, high dynamic stress resistance to successfully withstand high crash speed, low fabric weight for lesser impact on the human body on bag deployment, higher fabric strength and greater air tightness for rollover protection. Hence in the mid-1990s airbag manufacturers brought second generation materials into the market. As a result 940 dtex yarn as an airbag material became virtually obsolete and 470 dtex and 700 dtex high tenacity nylon 6.6 yarns became the materials of choice. Also neoprene coating was mostly replaced with silicone coating for the bags. Also for curtain airbags used in rollover accidents, one-piece densely woven bag in combination with silicone coating for sealing was introduced to comply with air tightness requirements for rollover protection. As the market moved on and more and more automobiles in Western Europe, North America and Japan were fitted with airbags, a need was felt for third generation materials at the turn of the century. The requirements from the automotive manufacturers were raised to even stronger and lighter bags with lower and more controlled air/gas permeability. It was also expected that the new materials should be gentle on the skin during impact, with better packability and greater resistance to heat conductivity. In making third generation airbags, super high tenacity 470 dtex nylon 6.6 yarns were used with lighter fabric structure and more defined silicone coating. In the last 3 to 4 years, the market has been seriously evaluating 235 dtex nylon 6.6 high tenacity yarns as the next generation airbag material. Table 12.2 highlights the physical properties required in making good quality airbag fabrics.

In manufacturing airbag fabrics, the warp yarn is supplied on a beam. The yarns in the warp beam are normally sized to withstand loom-state friction and also to prevent yarns from rolling during drying and wind up. Various additives are used in the size recipe to reduce fabric fraying, yarn pull-out and distortion. Normally airbag fabrics are tightly woven with 1/1 structure and largely the fabrics are woven involving rapier and air-jet looms. However, airbag fabrics are now being woven on water-jet looms too. Today loom

Table 12.2 Airbag yarn characteristics

Yarn characteristics	Medium to high tenacity yarn	High to super high tenacity yarn
Tenacity (cN/tex)	72–78	80–85
Extensibility (%)	20–22	18–20
Hot air shrinkage @ 180°C (%)	8–9	6–7
Filament fineness (denier)	2.8–3.0	6.0–6.5

Sources: various

speeds in excess of 550 ppm (picks/minute) on rapier machines, 650 ppm on air-jet machines and 900 ppm on water-jet machines are achievable in airbag fabric construction. Water-jet looms are highly efficient and cost effective. Jacquard weaving machines are normally used to weave one-piece curtain bags. The fabric make-up and the actual size of driver and passenger airbags are quite different. American automotive airbags are somewhat larger than similar airbags for European or Japanese cars. Also irrespective of geographical location, passenger airbags are larger in size and consume more fabric than driver airbags. For example, depending on the size of a car and the country of origin, a driver's airbag consumes between 0.6 and 1.5 square metres of fabric, whereas a passenger airbag consumes between 3 and 4 square metres of fabric. Side curtain, a one-piece woven bag, normally consumes between 3.5 and 4.5 square metres of fabric.

After weaving, the airbag fabrics are normally scoured to remove size from the material. The complete scouring operation uses three to four wash boxes that are separated by a vacuum extraction system to capture the extracted size. Following scouring, fabrics are calendared and heat-set to achieve dimensional stability and a precise control of air permeability in downstream application. Finished fabric air permeability can be well engineered if a balance between temperature, tension, dwell time and pressure between the calendar rolls is appropriately maintained.

Currently the automotive industry uses both uncoated and coated fabrics in making airbags. For coated bags, silicone is the most common coating material. Perhaps it is worthwhile to mention here that the airbag industry has been debating for at least the last 10 years the relative merits of coated and uncoated fabrics for airbags. The coated airbag is considered to be environmentally unfriendly but it offers greater resistance to heat conductivity and tearing performance along with low air/gas permeability. However, the uncoated airbag is more environmentally friendly. It is easier to fold and pack into small spaces.

For coated fabric, materials need to go through a defined coating operation. Today's choice of an elastomer for coating is mainly silicone. In processing, silicone is applied as a single coat by blade technique. Depending on the type of fabric and finished fabric requirements, an amount of 50 to 80 g of silicone is consumed to coat a square metre of fabric. Once the elastomer is spread on the fabric it passes through an oven under tension to induce polymerisation and adhesion between the fabric substrate and the elastomer. The precise time and temperature settings of the oven depend upon the silicone formulation. When coating of an elastomer as a continuous layer is applied on a fabric, the pores of the fabric are blocked. Hence in the event of a coated fabric airbag deployment, airbags are expected to deflate through vents. High technology silicone coating is widely used to make lightweight airbag fabrics that are flexible, durable and less prone to degradation over

Table 12.3 Mechanical performance of coated airbag fabrics

Performance parameters	Values
Yarn	470 dtx
Weave	Plain 1/1
Construction (threads/cm)	Various
Thickness (mm)	0.32–0.37
Weight (g/m^2)	220–240
Strength (kg/5 cm)	280–320
Elongation (%)	33–36
Tear strength (kg)	10–12

Sources: various

time. Table 12.3 illustrates mechanical performance of silicone-coated fabrics extensively used by the automotive industry for airbag application.

In the early to mid-1990s, airbag fabrics were largely coated with neoprene or silicone elastomers. These coated fabrics were found extremely stable in containing hot gas during deployment. However, the shortcomings of coated fabrics were established and they were proven to be their excessive thickness (particularly for neoprene coating), inability to be folded into small spaces and the tendency to degrade rapidly over time. Also concerns were expressed on possible unpredictable degradation behaviour over a period of time that could lead to catastrophic failure if the bag was deployed several years after it was fitted.

For uncoated airbags, the fabric needs to be particularly dense and of extremely high quality. Definitely in manufacturing that kind of fabric, significant pressure can be created on pricing of the materials. However, performance of high quality uncoated fabric for airbags can be more predictable over a period of time. The uncoated airbag releases the gas mainly through the fabric pores and hence it can be deployed at a relatively lower pressure with a cooler gas. Thus it minimises chances of burn injuries and face/neck injuries during deployment.

Considering technical performance relating to safety and environmental needs, the automotive industry prefers driver airbags to be made of lighter silicone coated fabric and the passenger airbags are made of relatively heavier uncoated fabric. The weight of coated fabrics for driver airbags ranges from 180 to 240 g per square metre and for the uncoated fabrics for passenger airbags, the weight ranges from 240 to 260 g per square metre. All in all a passenger airbag is somewhat stiffer and thicker (fabric thickness 0.3–0.4 mm) and a driver airbag is relatively more pliable and finer (fabric thickness 0.24–0.27 mm). Perhaps it is worthwhile to mention here that over the years airbag inflator technology has changed radically with the gradual introduction of augmented compressed gas and solid pyrotechniques other than sodium azide. The new inflators generate hotter gas than sodium azide which in turn has created a greater consumption of silicone coated fabrics in Europe.

The finished airbag fabric is cut into panels and sewn. The best method of cutting either coated or uncoated fabric is to use lasers. The technique is efficient and accurate. It fuses the edges of the fabric to prevent fraying. The normal design of the driver size bag is two circular pieces of fabric sewn together whilst the passenger bag is tear-drop shaped, made from two vertical sections and a horizontal panel. Airbags are normally sewn with sewing threads made of either nylon 6.6 or nylon 4.6 yarns. The sewing pattern and stitch densities are chosen carefully to maximise the performance of the bag. High density stitching has been developed to provide less leakage through seams. Also special 'seal and sew' assembly technology has been developed particularly for side curtains to meet air pressure retention requirements. In this process an adhesive is applied to the area to be stitched before the stitch is made. Adhesive in this case acts as a gasket on a coated or immersed fabric.

Like a parachute, airbags are folded with care and efficiency to ensure smooth deployment. A variety of folds are found suitable (examples: accordion fold, reversed accordion fold, pleated accordion fold, overlapped folds, etc.). Once the bag is folded, it is coupled with the necessary electronics and the deployment system within the casing for successful activation in the event of an accident. In the airbag supply chain pipeline, yarn manufacturers, weavers, bag manufacturers and system manufacturers are the major players. Table 12.4 outlines some of the major global players in the airbag supply chain.

12.3 Airbag deployment and performance criteria

Normally a car fitted with airbags is mounted with a pair of sensors in the front bumper. In the event of an accident, the sensors evaluate the severity of the impact and make an intelligent decision whether to deploy the airbag

Table 12.4 Airbag supply chain

Yarn manufacturers	Weavers	Airbag manufacturers (cut and sew airbag)	System manufacturers (assembled airbag kit)
INVISTA	BST	ASCI	TRW
PHP	UTT	TRW	AUTOLIV
SOLUTIA	MILLIKEN	AIL	PETRI (TAKATA)
TORAY	NCV (PORCHER)	AERAZUR	BREED
TOYOBO	FOV	SVENSKY	
RHODIA	AIL	WOODVILLE	
Others		BREED	
		TAKATA	

module. Mostly the cars with airbags are designed such that if they hit an object at a speed normally exceeding 22 km/h, the system will be activated and the bags will be deployed. In actual operation, an impulse is sent, within 5 ms following an accident, to the ignitor which in turn lights a propellant and releases a stream of hot gas rich in nitrogen. The gas surges through the inflator into the bag. The bag deploys with high force through its mountings. Bags are stored behind the mountings in the steering wheel (for driver), in the glove compartment (for front seat passenger) and in the back of the front seats (for passengers in the back seats) and on the window ledges for curtains (for rollover protection). The deployment occurs within 40 ms of an accident. After inflation, the bag deflates in a controlled manner through vents and the fabric pores (for uncoated fabric only) until the inertia of the occupant's body comes to a rest. The whole cycle of airbag inflation and deflation is remarkably short and takes a little more than 100 ms. Table 12.5 shows the time sequence of an airbag's various actions following an accident.

The total situation of a bag deployment is particularly violent. The gas pressure inside a driver's airbag reaches 70kPa, and the temperature reaches (for hot gas deployment) close to 600°C (for a few milliseconds). In the airbag module, very often a fabric of high thermomechanical performance is embedded. This fabric acts as a thermal barrier and a filter. When the inflator is activated, hot gas is issued from sodium azide or similar compounds following an exothermic reaction. The high performance fabric in the inflator module not only acts as a cooler for the hot gas but it also separates any ash or debris from the gas before it enters the bag. The Institute of Occupational Health at the University of Birmingham in the UK had conducted some studies on the exposure to the toxic gas and particle phase pollutants evolved during deployment of airbags in a vehicle. It was established that during airbag deployment, a range of pollutants (total inhalable dust, respirable dust, alkaline content, anions and inorganic gases) would be emitted which could cause subsequent illness and respiratory problems. Possible modifications to currently used airbag deployment systems were considered to eliminate a significant part of this problem.

Table 12.5 Approximate deployment timings for frontal airbags and side curtain

Functions	Driver/passenger airbags	Side curtain
Total time from sensing to full inflation	25–30 ms	15–18 ms
Timing for cushion to unfold, positioning with deployment	10–12 ms	10 ms
Average speed of cushion deployment	150–160 km/h	180 km/h
Total performance time	120 ms	5 s

Sources: various

Table 12.6 Airbag key technical parameters and contributing factors

Key technical parameters	Primary contributions
Tensile performance	Yarn strength and fabric structure
Tear performance	Filament count, yarn strength, fabric structure (openness/closeness) and coating
Air permeability	Yarn structure, fabric cover factor and coating
Seam performance	Sewing thread structure and tensile strength, sewing thread loop strength, fabric structure
Inertial loading	Yarn polymer density
Energy absorption	Yarn polymer density
Thermal absorption	Yarn polymer specific heat capacity
Ageing behaviour	Yarn polymer and coating polymer's sensitivity to thermo-oxidative degradation

Sources: various

For an airbag the key technical parameters and the related contributing factors are outlined in Table 12.6.

The performance of an airbag has always been related to the airbag's deployment behaviour. It is believed that relatively slower release of gas would generate a deployed bag that would feel unduly firm. The danger of this kind of performance could be that the airbag being too firm on deployment might cause occupant's upper-body to rebound from the bag causing whiplash injuries in the process. Scientists have enhanced the performance of an airbag by fine-tuning the appropriate interactions between seatbelt functioning and airbag deployment.

It is perhaps important to mention here that the effectiveness of an airbag in the event of an accident will be virtually non-existent if the driver/passengers seatbelts are not functioning at the time of an accident. Recently a mathematical-based high-speed punch technology was developed by SAE International to provide virtual testing of airbag cover design. The technology is expected to assist in material selection and predict system performance at the time of airbag deployment.

In the event of an accident, the performance of an airbag in saving an occupant's life and injuries will depend on several critical factors and these are as follows:

- on-time sensing, efficient activation of the system and effective releases of gas
- correct deployment and positioning of the bag
- structure and cover factor of the fabric used for the bag
- air/gas permeability characteristics of the fabric/bag, particularly the rate of release of gas through the fabric/bag
- balanced interactions between seatbelt functioning and deployment of the bag.

12.4 Market trends and developments

The airbag industry is one of the youngest members of the technical textile sector but since early 2000 it has been growing between 7 and 10% within the three main global regions: Western Europe, North America and Japan. Currently these three regions control approximately 80% of global airbag markets. Growth in the airbag market in Japan has declined in the last few years as fewer cars are being built there compared to the 1990s. Figure 12.3 shows an approximate breakdown of airbag market shares for Western Europe, North America and Asia for 2004. Growth for driver and passenger airbags for European and American markets were not radically significant in the last few years but side curtains in those markets have grown significantly and expect to grow steadily over the next few years. It is anticipated that a slow down in the overall airbag markets, particularly in Western Europe and North America may happen during 2008 and 2009 because of possible sluggish economic growth. Nonetheless, it is estimated that approximately 220 million airbags will be fitted in cars worldwide in 2008. Table 12.7 shows an approximate growth in the consumption of airbag yarns since 2000 in two main global car-producing regions.

Since mid 1990s the automotive airbag industry has been extremely successful, possibly due to government legislation in the USA and consumer pressure in Europe. There is a guaranteed market for airbags in the USA due to legislation that ensures that every passenger car and light truck (made after 1998) must be fitted with driver and passenger side airbag units. European automotive manufacturers have been in discussions with various governments about enforcing similar legislation in major European countries. However, markets for airbags in Western Europe and Japan have grown steadily in the last 10 years through greater awareness of safety amongst customers. It is also important to mention here that a significant upsurge in growth of the

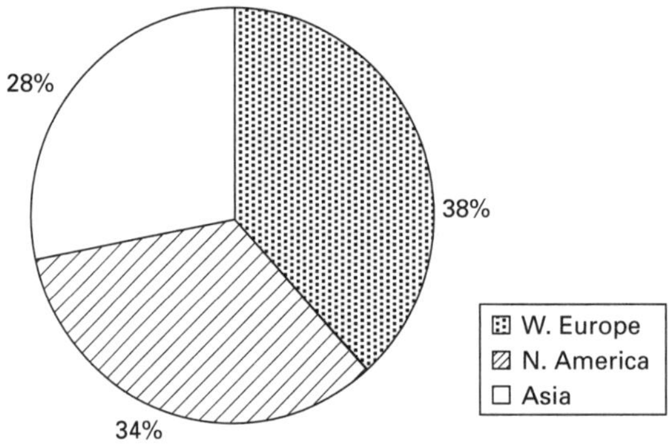

12.3 Regional share of airbag yarn demand for 2004. Courtesy: Alasdair Carmichael, PCI Fibres, UK.

Table 12.7 Approximate consumption of airbag yarns in Western Europe and North America between 2000 and 2008

	Consumption (kilotonnes)	
Year	Western Europe	North America
2000	19.75	21.65
2004	30.97	26.15
2008 (estimated)	36.85	32.30

Sources: various

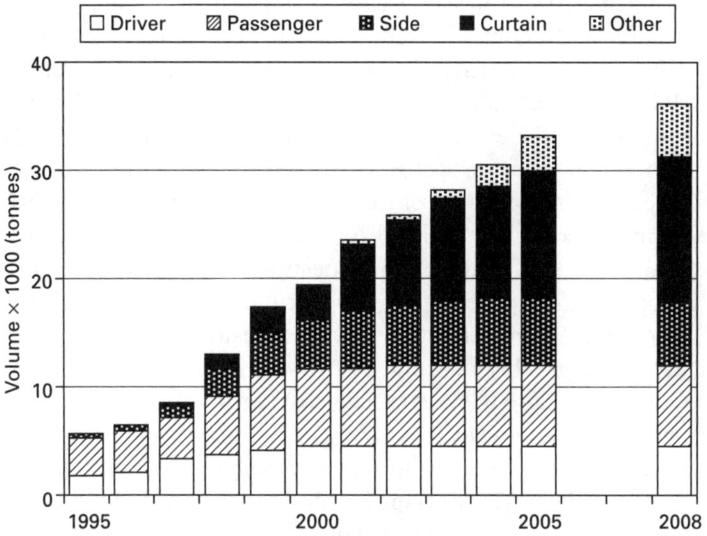

12.4 West European airbag yarn market by bag type. Courtesy: Alasdair Carmichael, PCI Fibres, UK.

airbag market since its commercial inception also came from a different angle.

Over time, the use of an airbag as a safety device spread beyond driver and passenger airbags. Now automobiles are fitted with one-piece woven side curtains and even knee bags, thorax bags (to protect chest and ribs), impact curtains, etc. Needless to say, the increased number of airbag varieties and their respective applications require a wide variety of performance functions. Figure 12.4 shows how airbag markets have grown over the years with its variety in applications. With the introduction of different types of airbag systems, different types of crash sensors have become essential. For driver and passenger frontal airbags, a central electronic module equipped with a single accelerometer is used. However, with the introduction of side curtain and head protection systems, additional side sensors coupled with an accelerometer and a microcontroller are necessary.

As discussed earlier, in Western Europe and Northern America, airbag market growth will come mainly from side curtain and thorax bags. However, large growth in the airbag business will come from China and possibly India and Korea in the coming years from driver and passenger airbags. In the next 10 years China and possibly India would be the manufacturing base for growth in the airbag industry. Both China and India have grown significantly in recent years in manufacturing automobiles. Evidence has shown that the airbag manufacturing base has followed its customers as they have relocated and expanded. Also the airbag system suppliers have outsourced either through partnerships and acquisitions or through new start-up and commission processing to take advantage of regions with lower labour and operating costs. It is not unreasonable to postulate that if production in China and India expanded beyond the volume needed to satisfy the local markets, the surplus would be exported to Europe, America and the rest of the world.

In recent years, the streets of Europe have become more congested. It has raised a major consumer demand for safer vehicles that can provide speed control and pedestrian protection. Pedestrian safety is a big issue for Japan and Europe. It has opened up new opportunities for automotive airbags in those two geographical areas. However, pedestrian safety is not an issue in the USA as the number of fatalities per year is relatively small.

In today's cars, airbag systems are activated involving crash sensors when the impact occurs. However, a new generation of pre-crash sensors have been developed with the aim to make vehicles much safer and reduce the severity of the crash. The working principle of this kind of sensor is based on the speed of light. Infrared coded laser light is sent out and reflected from objects in the proximity of the sensor. The decoder calculates the runtime between the emitted and received laser pulse, which correlates with distance of the object. The sensor, which covers the complete vehicle front, is mounted in the wiper area behind the wind screen, close to the rear view mirror.

Based on a large number of market surveys and scientific reports, it has been concluded that fatality rates are fewer in vehicles equipped with airbags than their unequipped counterparts. The concept of airbags or similar structures is currently being considered in lowering human casualties in building collapses. Simulations of collision-impact on fragile objects reveal that impact injuries can be reduced for people indoors from falling debris in earthquake and building collapses resulting from tornados, hurricanes or similar disasters.

Also experiments with motorcycle airbags to reduce injuries in frontal collisions have been conducted for many years. The Japanese company Honda has probably conducted more research, including numerous staged collisions, than any other company. The first commercial production of motorcycles incorporated with an airbag system started with Honda in 2006 with the

12.5 Airbag as a passive safety device for motorcycle. Courtesy: Honda marketing literature, 2006.

Honda Gold Wing model. Figure 12.5 shows Honda's airbag system fitted on a 2006 Gold Wing motorcycle. It is important to remember that Honda's airbag system wasn't designed to prevent the rider being ejected from the motorcycle in the event of a collision. However, a rapid deployment of an airbag after the impact can help to absorb some of the forward energy of the rider and that in essence can help to reduce the severity of an injury following an accident.

12.5 Future expectations and possible challenges

In terms of airbag yarn demand and global capacity of yarn production, currently, there is a reasonable balance between supply and demand. However, it appears that there will be an imbalance in that equation from the year 2009 onwards. Based on the current forecasts, there will be a shortage of airbag yarns of 10 000 tonnes, although new investments by Invista and PHP/Shenma in China will take care of part of that deficiency.

For the future trends in airbag developments, the industry will continue to put more emphasis in the following areas:

- More demands will be on ultra high tenacity yarns with a switch over to lower yarn linear density.

- Greater emphasis will go to higher tear strength and more controlled gas permeability of the bag fabric.
- Next generation bags are expected to have extra toughness (resilience) and better packability.
- Owing to greater use of more and more hot gas deployment, future bag fabrics are expected to have better heat strength retention and thermo-mechanical responses.
- More and more bag fabrics are expected to conform with longer heat/moisture ageing life.
- Seam performance will become more stringent for greater safety.
- Greater emphasis from the automotive manufacturers will be on zero fault quality.
- Intense pressure from the automotive manufacturers will be on lowering the prices for the bag and the deployment system without compromising quality.

Although, for the future, airbag manufacturers will continue to strive to make the bags lighter, more robust, compact and cheaper, in going forward the global automotive industry will be challenged by the consumers in the following areas:

- Unfaltering reliability of the total safety system.
- Highly competitive costs with unique safety features.
- World class service.
- Innovation for greater safety.

Finally, airbags can be a unique safety feature in an automobile but unfortunately the total system is not entirely fool-proof. The environmental fate of sodium azide which is an active ingredient for the deployment system, is unknown. Work done in the department of atmospheric science at the University of Arizona had established that since the mid 1990s demand for sodium azide has exceeded 5 million kg per year and most passenger vehicles sold in the USA contain approximately 300 g of sodium azide. This has greatly increased the potential for accidental environmental releases and for human exposure to this highly toxic, broad-spectrum biocide. Nonetheless, the problem of safely disposing of large quantities of azide will remain as the vehicle fleet ages and is left with the scrap yards.

12.6 Bibliography

Anon, *High Performance Textiles*, No. 8, February, 1994.

Anon, Nonwovens Report International, Vol. 10, 255, June, 1992.

Anon, *American J. Appl. Sci.*, Vol. 2, No. 4, 2005 (information obtained online).

J A Barnes, Private Communications, 1998.

J A Barnes and J Sun, Conference Proceedings, Technical Textiles, The Innovative Approach, Manchester, UK, April, 2004.

E A Betterton, University of Arizona, USA, 2005 (Obtained online).

A Carmichael, Conference Proceedings, The Airbag and Seatbelt Technology Conference, Michigan, USA, May, 2005.

E T Crouch, *J Coated Fabrics*, Vol. 23, 202–220, January, 1994.

T. Gornig, Conference Proceedings, The Airbag and Seatbelt Technology Conference, Michigan, USA, May, 2005.

H Gramling and R Hubbard, *J. Biometrics*, Vol. 1, S372, 2006.

J W Hetrick, US Patent 2649311, August, 1953.

Honda Corporation, Marketing Literature, 2006.

R Koehnen, Conference Proceedings, Polyamide 2004, Germany, June, 2004.

B Konrad, *Technical Textiles*, Vol. 47, August, 2004.

M C H Lee and G E Novak, SAE Technical Paper Series Presented at SAE World Congress in Michigan, USA, April, 2006.

S Manley, Conference Proceedings, The Airbag and Seatbelt Technology Conference, Michigan, USA, May, 2005.

H J Mertz, SAE International Report, No. 880400, 1988.

S K Mukhopadhyay and J F Partridge, Automotive Textiles, *Textile Progress*, Vol. 29, No. 1 & 2, 1999.

J F Partridge, Private Communications, 2000.

J F Partridge and S K Mukhopadhyay, *J. Appl. Polym. Sci.*, Vol. 77, 2104–2112, 2000.

R E Resh, *Scientific American*, 103, June, 1996.

TRW, US Patent 5375878, 1994.

A D Wheatley, S Sadhra and J R Beach, *J. Indoor & Built Environment*, Vol. 6, No. 3, 134–139, 1997.

13

Key technology developments in textiles for use in automotive tires

W K WESTGATE and J G GILLICK,
The Goodyear Tire & Rubber Company, USA

Abstract: Textile reinforcements have been used in the manufacture of automotive tires since their inception. Starting with cotton, the textile types, components and technical attributes have evolved over the years. Today's automotive tire use of textiles is key in obtaining performance parameters such as longevity and comfort. The textiles most widely used in automotive tires are polyester, polyamide and rayon, but they are highly tailored to our industry's needs and are more technologically advanced than normal textile grades. This chapter will discuss the evolution of textile fibers for automotive tire use and expound on the more unconventional textiles being utilized to enhance tire performance.

Key words: tires, industrial fibers, elastometric reinforcement.

13.1 Introduction

In today's fast paced, mobile society, just about every mode of transportation involves the use of a wheel. A simple machine, the automotive wheel (or more commonly referred to as a tire) is typically a pneumatic (air supported) composite structure whose main function is to help carry a load (e.g., people, material) from point "A" to point "B" reliably and with a high degree of comfort. To accomplish this task, the tire needs to transmit driving, braking and steering forces from the vehicle through the tire to the road surface. In addition, the tire is expected to help absorb road surface irregularities, providing a smooth, comfortable, and quiet ride for the occupants. The automotive tire is one of the most complex, highly engineered, composite structures, widely utilized and generally underappreciated by the end user.

In this chapter, we will explore, in further detail, the properties and performance requirements of automotive tires, how the various reinforcement materials and components contribute to tire performance, and review some recent tire developments using new, novel, advanced fiber/reinforcement technology. As we will see, there is more than one way to achieve a desired tire characteristic by utilizing various materials and design options. The closing section will take a look at trends in the automotive world and how this may impact future tire development and hence fiber reinforcement opportunities.

13.2 Properties and performance requirements of tires

13.2.1 Tire performance requirements

A tire is one of the most complex soft composite structures in wide commercial use today. More than 5000 years ago, the Sumerians invented the wheel. John Boyd Dunlop is the recognized inventor of the first pneumatic tire (i.e., bicycle tire) in 1888.[1] In 1895, Andre Michelin was the first person to use a pneumatic tire on an automobile. It was not until 1911 that Philip Strauss[2] invented the first commercially successful automotive tire: a combination tire and an air-filled inner tube.

Technically speaking, the automotive tire geometrically can be considered a torus[3] (Fig. 13.1); that is: a surface of revolution generated by revolving a circle in three-dimensional space about an axis coplanar with the circle, which does not touch the circle. Mechanically, the automotive tire is a flexible membrane pressure vessel. Structurally, an automotive tire is a high performance, soft composite. From a material science point of view, a tire is a compliant, viscoelastic cord/rubber reinforced structure which is subjected to continuous deformations during its life. It is in each of these views that the desired properties and resultant performance criteria are based.

Overall, the tire must meet a vast array of performance criteria to be acceptable by the end users. The automotive tires must support the weight of the entire vehicle, provide efficient mobility, dampen out road textures and irregularities, generate cornering forces for steering control and transmit all traction and braking forces. All this needs to be accomplished through direct contact with the road surface in an area roughly the size of a human hand (Fig. 13.2). In addition, the tire must meet these criteria over a wide range of

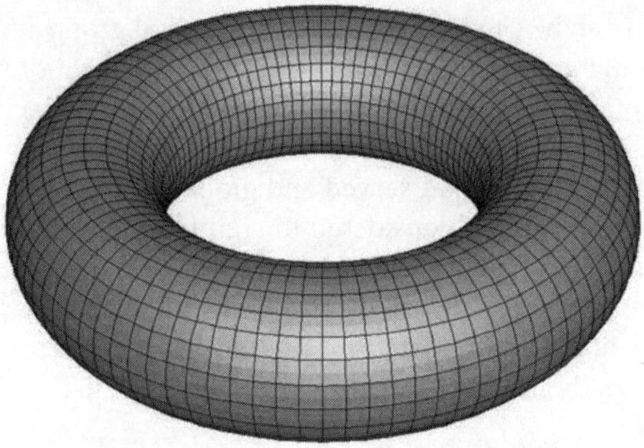

13.1 Automotive tire geometric representation as a torus. (Source: http://commons.wikimedia.org/wiki/image:Torus.png).

13.2 Quintessential automotive tire foot print.

thermal (–20°C to 50°C) and environmental conditions (e.g., dry, wet, snow, ice).

Figure 13.3 depicts a variety of vehicle applications and their respective design load and speed ranges. Comparatively speaking, the automotive tire is designed to accommodate relatively low loads moving at very high rates of speed, in contrast to earth-moving vehicles which carry very high loads at relatively low rates of speed. These basic requirements are then merged with external customer needs such as weather traction, comfort, responsive handling and durability to develop what can be referred to as tire attributes (Fig. 13.4). These tire attributes can be categorized into three general groups: (1) vehicle mobility, (2) performance and (3) comfort/aesthetics. Each set of properties and/or performance requirements are tailored to help achieve the objectives of the vehicle and/or end-customer. The level of importance for each objective will vary depending on the market served and global drivers. For example, European legislation[4,5] is being considered to significantly reduce the "air-borne" noise generated by tires by 2012; hence, new designs/materials that can help reduce tire noise would be a high priority in the European market. Increased handling and response are key drivers for the North American market. In Japan, aesthetics, performance and stopping distance[6] are top drivers.

In addition to the broad market needs, there are several niche applications where specific automotive tire designs are needed; for example: run flat, on/

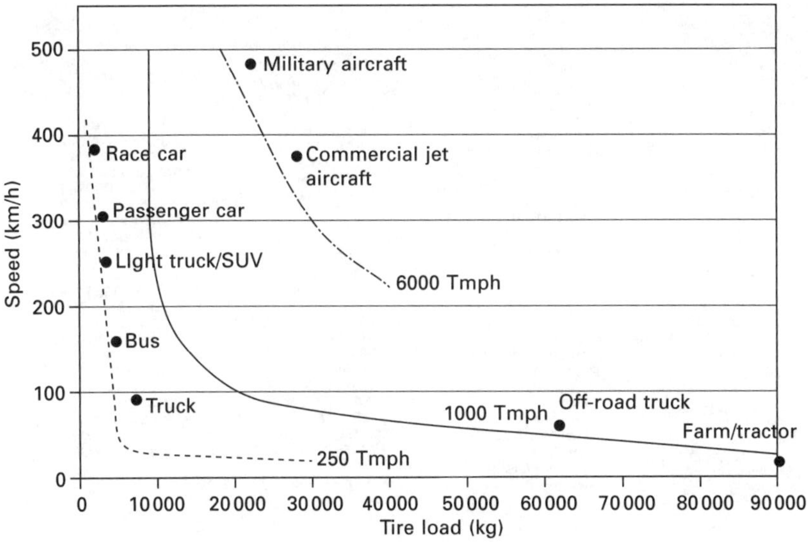

13.3 Tire load carrying capacities and speeds.

13.4 General automotive tire attributes.

off-road applications, and all-season. The combination of needs and drivers governs how tire companies focus their resources to devise, design, and develop new, improved automotive tires with measurable performance enhancements for our customers. All told, there are arrays of product performance characteristics that an automotive tire could be asked to satisfy prior to being released into the marketplace (Fig. 13.5). The exact performance

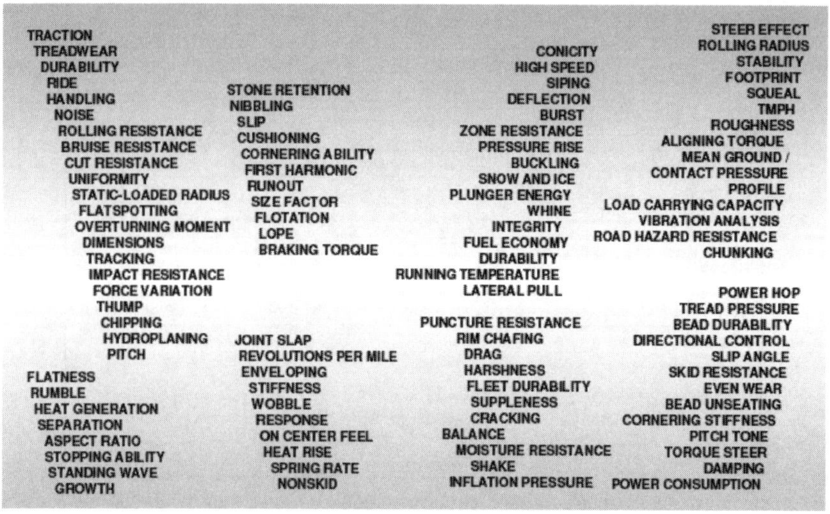

13.5 Automotive tire performance characteristics.[2]

characterization matrix is dependent on the end-use requirements for the tire. These characterizations have evolved over the past 100+ years and are designed to simulate end-use conditions. The top, generalized tire performance areas today are: reduced fuel consumption, increased load-carrying capabilities, better durability and retreadability, increased tread wear, improved ride and handling, increased traction, and lower noise.

13.2.2 Tire designs in use today

In the automotive vehicle market, there are three types of unique automotive tire designs in use today: bias belted, run flat, and radial (Fig. 13.6). Bias belted tires are named after the two or more carcass plies which cross at a bias angle to the centerline of the tread.[7] Bias belted tires usually have a rounded tread face which increases tread wear and hence shortens the in-service tire life. Bias belted tires are one of the more durable tire designs due to the ability of the bias-angled carcass to help absorb most impacts through pantographing. Nowadays, bias belted automotive tires are typically used in areas where the road infrastructure is more demanding or for some specialty applications such as antique cars.

The run flat tire (either self-supporting or ring-supported design) represents the latest commercialized automotive tire design. The self-supporting tire is designed to minimally deflect (upon air loss) through the inclusion of stiff, sidewall inserts. The ring-supported tire is also designed to minimally deflect with air loss by contacting the tire onto a structural insert component which is attached to the inside of the rim. Each run flat tire design offer features that provide continued mobility in the event of an air-loss for a specified period of kilometers at a given maximum rate of speed.

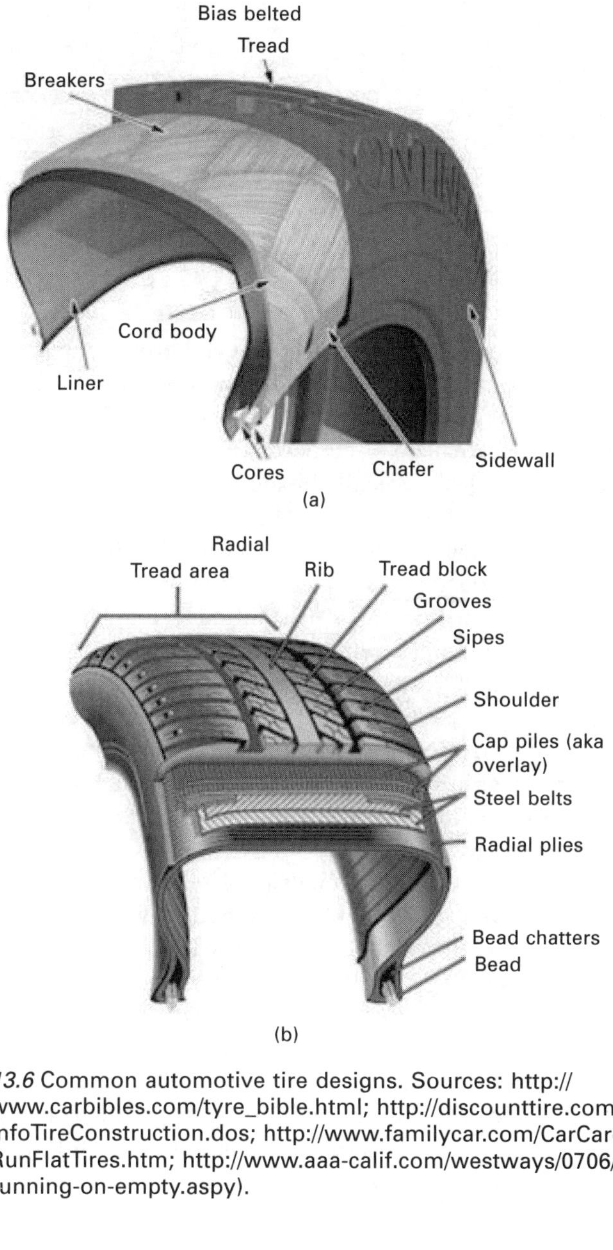

13.6 Common automotive tire designs. Sources: http://www.carbibles.com/tyre_bible.html; http://discounttire.com/dtcs/infoTireConstruction.dos; http://www.familycar.com/CarCare/RunFlatTires.htm; http://www.aaa-calif.com/westways/0706/features/running-on-empty.aspy).

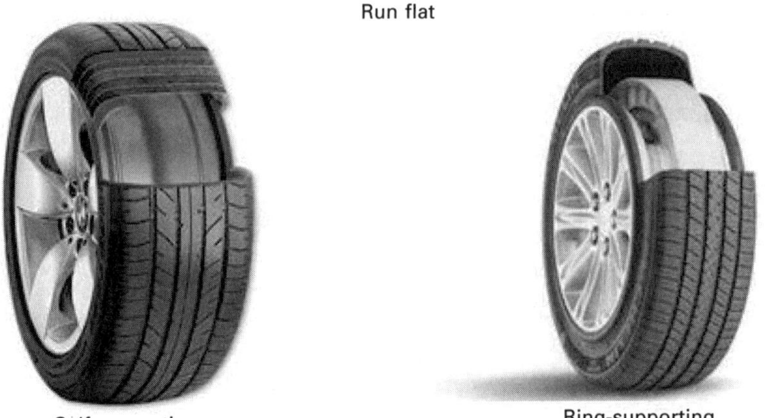

13.6 (Continued)

The radial tire is by far the most prevalent and broadly applied design in the automotive arena with over 1,000 MM units sold annually.[8] The radial tire[9] gets its name from the direction that the carcass plies run; i.e. radially across the tire, perpendicular to the beads. Radial tires have multiple belt plies at opposing angles (typically made-up of steel cords) which run circumferentially around the tire, under the tread, to stabilize the tread area. This additional restriction also increases the tread life (over the bias/belted tire design) through reducing squirm and friction with the road surface. The radial automotive pneumatic tire was first introduced by Michelin in 1949[10] but didn't get broad acceptance until decades later.

The typical automotive radial tire is composed of approximately 18 individual components, at least 12 different elastomeric-based compounds, 1 or 2 plies of fabric, two steel belts, and a bead.[11] Approximate compositions by weight can be seen in Fig. 13.7. While the typical automotive tire uses minimal amounts of textile reinforcements (i.e. ~5 wt%), their contribution towards the overall tire performance and behavior can be significant. In general, the automotive tire is a composite of relatively low strength, high extensible elastomeric matrix and a high strength, low extensible reinforcements. Primarily, the reinforcements give the tire shape, overall size, stability, and load carrying capability. To a lesser extent, the reinforcements can also impact ride, handling, noise, tread wear and vehicle fuel economy.

General cord properties are strength, toughness, dimensional stability, modulus and linear density (i.e. mass). Figure 13.8 compares/contrasts the most commonly used reinforcing fibers for some of these key properties (normalized for equivalent cord structure). The fiber/cord selected, for each tire type and tire component, is dependent on these properties coupled with

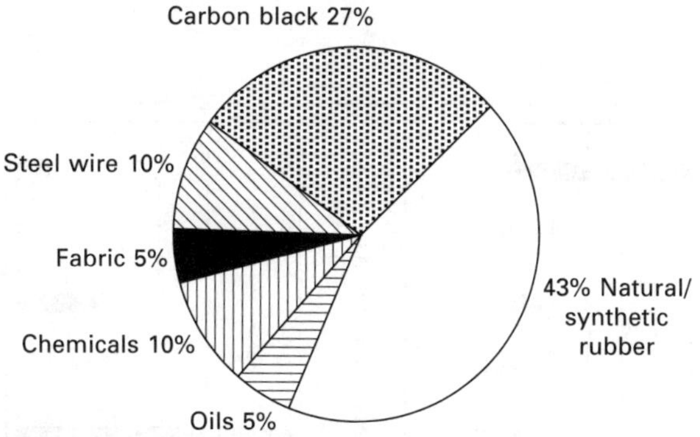

13.7 Typical automotive tire compositional breakdown (by weight).[10]

additional design considerations such as: the fibers' tensile and/or bending stiffness, thermal/chemical/mechanical stability, hysteretic behavior, cord structure, surface reactivity (e.g., for adhesion), end-use requirements and performance characterisitics. In most cases, the use of a textile-grade fiber is inadequate for tires. It is common to better describe the reinforcement fibers, used in tires, as industrial fibers. Industrial fibers generally offer significant physical property and chemical resistance enhancements over standard textile-grade fibers. In the tire industry, most fibers are used in the form of a twisted-cord. A twisted-cord structure can readily be varied to provide a wide range of properties desirable for in-tire use (Fig. 13.9).

Textile-based cords, in the automotive tire today, are most commonly used as the carcass (aka ply) reinforcement. Depending on tire design and usage, textile cords may also be used as flipper/chipper, overlay (aka cap ply) and toeguard (Fig. 13.10). The carcass (aka plies) is a highly engineered, twisted-cord structure made from PET polyester (predominately) or viscose cellulose (aka rayon). They are placed radially from bead-to-bead and are the primary reinforcing and strength member of the tire. Chippers and/or flippers are typically made from polyamide 6.6 or aramid fibers, twisted and woven into a uni- or multi-directional (e.g., square woven) structure and placed in the lower sidewall of the tire at a 30° to 65° angle (relative to the carcass angle). They provide additional composite stiffness in the lower bead-area to increase handling, help absorb deflection, reduce rim chaffing, and provide fretting protection between the steel bead and the carcass components. The overlay is made up of a series of unidirectional oriented twisted cords that run circumferentially around the tire, on top of the steel belts. They are used to enhance tire performance particularly for high speed applications. The most commonly used fiber type is polyamide 6.6, while aramid is also used especially in ultra high performance tires.

Pneumatic tires have undergone continuous evolution for the past 100

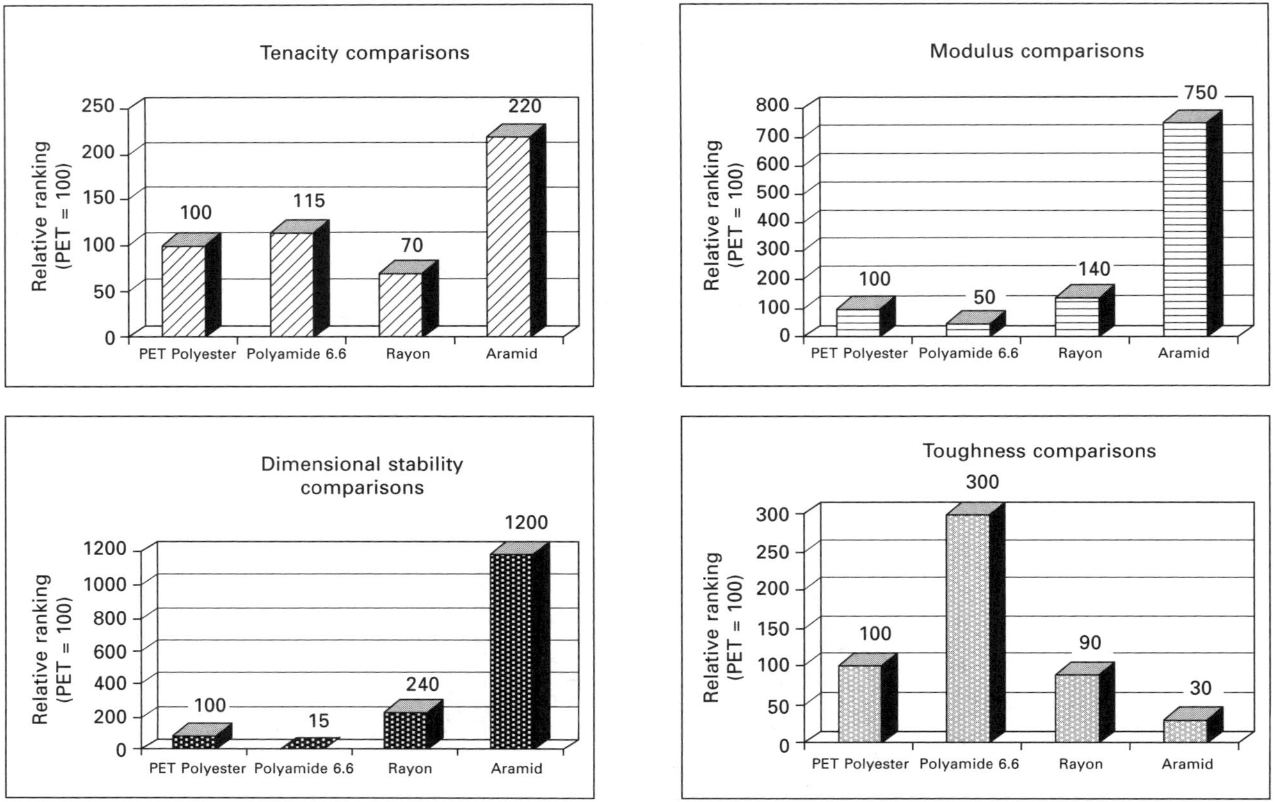

13.8 Relative property rankings (normalized for structure).

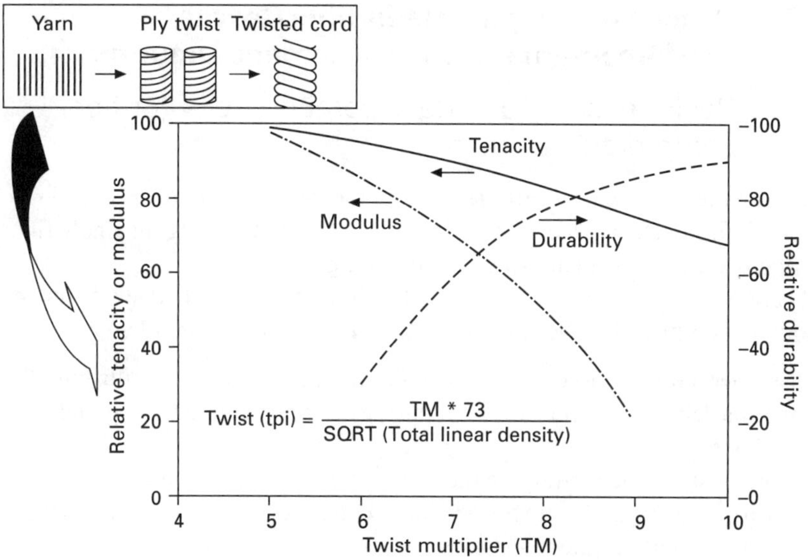

13.9 Cord properties as a function of twist level.

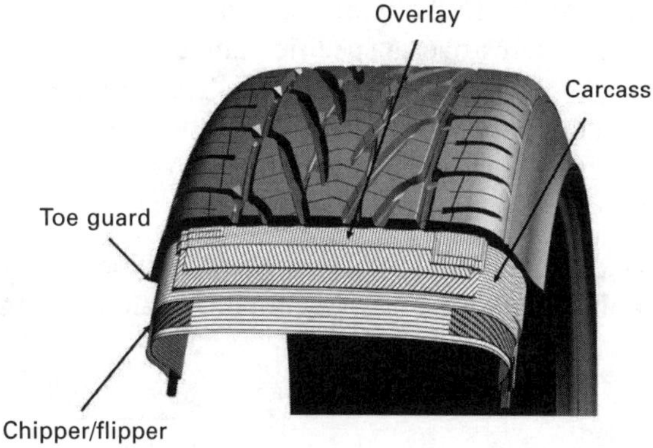

13.10 Textile-based reinforcement areas sometimes used in automotive tires.

years. Although the tire for the most part consists of rubber, the reinforcing materials do play critical roles in the overall tire performance. Today's tire reinforcement requirements are varied and demanding giving the designer/engineer an array of options and combinations to be used to enhance tire performance. Lighter weight materials, improvements in rolling resistance, uniformity and toughness continue to provide significant technological challenges.

13.3 Recent developments in fiber/textile reinforcements used in tire manufacturing

13.3.1 Basic research and development steps for fiber commercialization in tires

Fibers are the main reinforcing elements in the tire and provide a critical role in overall tire performance. Hence, it is no surprise that bringing a new fiber into the tire design is somewhat of a daunting task.

There are four basic research and development steps that must first be accomplished prior to fiber commercialization in tires (Fig. 13.11):

1. Engineered morphology – modify the fiber's structure to withstand the thermal/chemical exposure seen during the tire manufacturing and over its operational life.
2. Cord design – depending on the targeted tire component application, the textile fibers typically are formed into a cord structure to optimize strength, modulus, and fatigue resistance.
3. Adhesive science – a designed adhesive chemistry is used for each fiber type providing chemical and mechanical bonding to both the fiber/cord surface and surrounding matrix (i.e. rubber).
4. Processing science – where the reinforcement is subjected to elevated temperatures and tensions over a specified amount of time to stabilize the material for subsequent plant processing. This process, while both time consuming and costly, is necessary for the successful translation of intrinsic fiber properties and behaviors into positive "in tire" responsive characteristics.

Key fiber physical properties, for use in tires, include: linear density, flexibility, and modulus (Table 13.1). These properties, along with structural

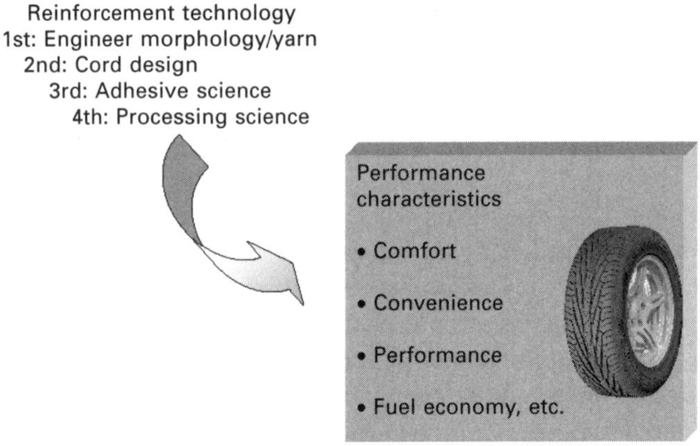

13.11 Tire reinforcement development methodology.

Table 13.1 Key intrinsic fiber properties for in-tire use

Linear density	Flexibility	Initial modulus	Specific tensile stress
$\lambda = \rho A_f$ where: λ = linear density ρ = mass density f = fiber length	Flexibility = $[EI]^{-1}$ where: E = initial modulus I = moment of inertia	$E = \left.\frac{d\sigma}{d\varepsilon}\right\|_{\varepsilon=0}$ where: σ = stress ε = strain	STS = σ/ρ where: ρ = mass density σ = stress

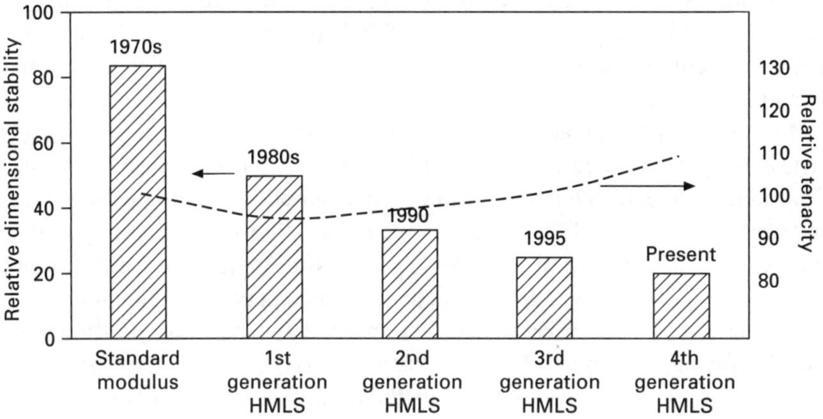

13.12 PET polyester tire yarn development history.

(e.g., Tm, Tg, MW), morphological (e.g., crystallinity, amorphous orientation), and thermal/chemical/dimensional (e.g., creep, modulus response, contraction) characteristics, are used to guide the designer in fiber selection and subsequent structural optimization to address a specific characteristic by targeting a specific tire component. These newly designed components are then encompassed into the overall tire design and optimized for specific tire performance characteristics.

13.3.2 Recent fiber and cord developments

Market demands for increased raw materials and technological advancements (mainly in processing) have been the main drivers for R&D of tire reinforcement materials. Throughout the 1970s, 1980s, and 1990s, advancements in tire reinforcement technologies could be easily described as derivative; that is to say, incremental physical property improvements in the areas of steel cord, polyester, polyamide 6 & 66, rayon and aramid.

As an example, Fig. 13.12 depicts the technology advancements of PET polyester over the past 20 years. PET polyester fiber development mainly focused on improvements in dimensional stability. Dimensional stability, measured in terms of shrinkage, creep, modulus retention, etc., is a measure of reinforcement's ability to maintain its physical properties and dimensions when subjected to an elevated temperature environment (e.g., tire cure, operating conditions). PET polyester fiber improvements advanced from standard modulus to 1st generation high modulus/low shrink (i.e. HMLS (higher modules, lower shrinkage), invented by Celanese[12] in the late 1970s) to the current 4th generation HMLS. These advancements have greatly improved tire uniformity (e.g., tire growth, appearance, sidewall indentations, and vibrations) and streamlined the tire manufacturing process by helping to reduce the need for a post-cure inflation (PCI) process. As an added benefit, fiber manufacturers' productivity level was also greatly increased thus reducing fiber manufacturing costs.

More recent polyester advancements have been in the form of new polyester derivatives; for example PEN (Fig. 13.13). PEN (poly 2,6 ethylene naphthalate) offers additional enhancements in dimensional stability (e.g., higher modulus, lower shrinkage) over PET. Intrinsically, PEN vs. PET offers not only higher dimensional stability but also higher Tg (glass transition temperature) and Tm (melt temperature) which can lead to additional benefits in manufacturing as well as in-service performance. Commercialized by several companies (e.g., Performance Fibers, Teijin, Invista and Hyosung), PEN[13] polyester fiber has seen limited success in tires. In 2004, Pirelli commercially introduced PEN as the carcass reinforcement for their new Dragon SuperCorsa PRO motorcycle tire.[14] There, PEN offers reduced in-tire weight for improved handling with excellent response and control. In 2006, Sumitomo Rubber introduced a tire with PEN[15] as the overlay component providing improvements in tire flat spotting and reduced tire noise. The significantly higher modulus

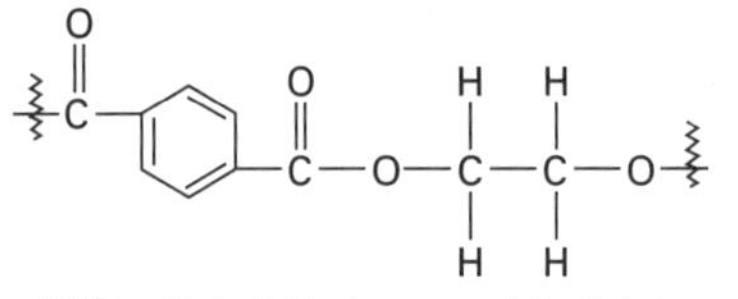
PEN = Poly (ethylene naphthalate)

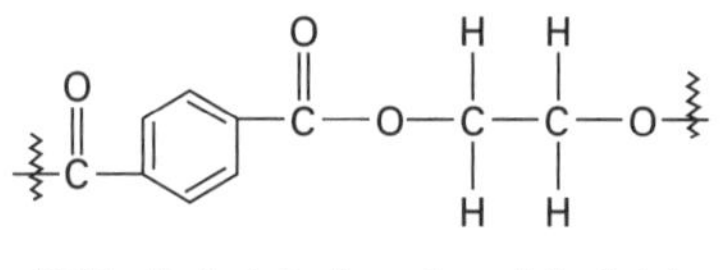
PEN = Poly (ethylene terephthalate)

Intrinsic properties:	PEN	PET
Melt point, °C	268	250
Tg, °C	125	80
Specific gravity, g/cc	1.3–1.4	1.3–1.4

13.13 PEN vs. PET structure and intrinsic property comparisons. (Sources: http://www.bp.com/liveassets/bp_interest/globalbp/STAGING/global_assets/downloads/pdfs/acetyls_pta/N_17_PEN_Homopolymer_Prep.pdf; http://www.azom.com/details.asp?ArticleID=1933, ArticleID=2047).

of PEN vs. conventional polyamide 6.6, helped to reduce tire vibrations under dynamic operating conditions.

Towards the end of the 1990s, new aramid tire yarns, offering enhanced properties, were being developed by both E.I. du Pont de Nemours and Teijin (formally known as Akzo Nobel).[16] In the mid 1990s, Dunlop[17] introduced an ultra light weight passenger tire (Dunlop SP Sport 200 ULW) that was manufactured with all aramid reinforcements: belts, carcass and beads. The significant weight reduction translated into improved fuel economy and reduced wear. However the tire's unique structural components and its low benefit/cost ratio did not resonate well with consumers during this time frame.

In 2004, Kumho[18] pioneered the use of Lyocell as a new carcass reinforcement material in their ultra high performance tire line called ECSTA. Lyocell fiber, a subclass of rayon, is commercially manufactured by Hyosung Corporation. Lyocell fibers offer similar in-tire responsive characteristics as viscose-based rayon. However, Lyocell is considered more environmentally green than viscose rayon because the manufacturing process is considered more environmentally friendly (e.g. reduced effluents, less harmful solvents).

In 2005, The Goodyear Tire & Rubber Company introduced a new line of radial light truck tires called "Silent Armor™" that featured DuPont's Kevlar® in a merge/hybrid cord structure.[19] The cord structure was engineered to provide additional modulus and stiffness for enhanced tire performance while significantly reducing the tire noise level. Later that same year, a similar merge cord construction was launched by Goodyear in Europe for the ultra high performance passenger tire market. The merge/hybrid cord structure concept enables one to optimize reinforcement properties using two or more dissimilar fiber types. The addition of Kevlar® to a polyamide 6.6 fiber (Fig. 13.14) produced a cord which has high toughness, mid-modulus level with moderate ductility. Also in 2005,[20] a new advanced polyester tire cord was introduced in on-highway radial light truck tires under the name Fortera Triple Tred®. The specialized polyester tire cord helps reduce flatspotting, while maintaining a smooth, comfortable ride. As one can see, there is no one catch-all approach that can satisfy all our customers' needs. Hence, the tire designer selects the best known combination of reinforcements, compounds and structural design to address each specific performance characteristic.

In 2006, The Goodyear Tire & Rubber Company introduced the incorporation of carbon fiber tire cord[21] as a chipper in a high performance passenger touring tire called Eagle ResponsEdge®. The carbon fiber chipper, placed along the lower bead area of the tire on the outboard side of the tire, offers enhanced sidewall stiffness when needed (e.g., during an evasive maneuver) without sacrificing ride and comfort (Fig. 13.15). Carbon fiber's high stiffness/weight ratio was the key intrinsic property used to achieve the desired tire response without adversely impacting tire balance and inducing non-uniformities (e.g., vibrations).

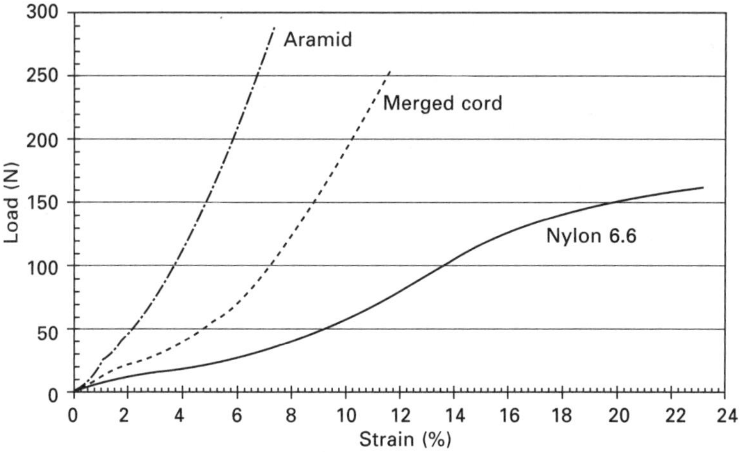

13.14 Tire cord's deformational response when subjected to an applied axial strain.

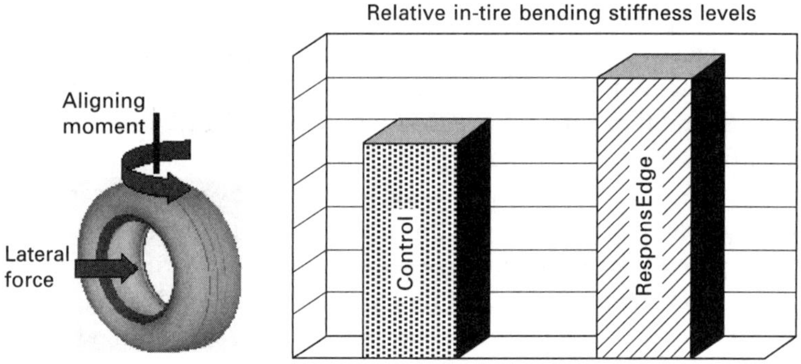

13.15 In-tire stiffness comparisons.

In 2007, The Goodyear Tire & Rubber Company introduced carbon fiber micro-technology in an ultra high performance passenger tire called Eagle® F1 All Season.[22] The addition of micro-carbon fibers, in both inner and outer lower sidewalls, further enhanced the tire's handling characteristics through increasing sidewall stiffness (tension, torsion) without compromising other tire performance characteristics.

Recent advancements in reinforcement technology have enabled tire companies to further enhance tire performance characteristics. In addition, these new reinforcements have enabled tire producers to design highly engineered, highly tailored, and highly differentiated automotive tires that offer significant value to both the end-use consumer and manufacturer. Continued new and improved fiber technologies should result in further enhancing tire performances to meet the ever changing needs in the market place.

13.4 Fiber-rubber adhesion in tires

13.4.1 Bonding of fiber reinforcement to tire matrix

Bonding of continuous fiber reinforcements to the elastomers used to form the matrix of the tire requires the meshing of different chemistries to form a link stronger than the cohesive strength of the matrix. In the original pneumatic tires, where cotton duck fabric was used as the reinforcement, the mechanical bond of the rubber to the fabric was sufficient to exceed the relatively weak internal strength of the cotton staple fibers. With the introduction of continuous rayon fiber in the 1930s, the need for a fiber adhesive coating became apparent since the higher strength of rayon fiber shifted the focus area of improvement to the fiber/rubber interface.

Early adhesives were based on naturally derived protein glues such as casein,[23] but these were quickly supplanted by synthetic phenolic resins, primarily based on the reaction of resorcinol with formaldehyde.[24] Resorcinol can be readily reacted with formaldehyde in aqueous solution at room temperature to form a three-dimensional resin network. Extensive studies were conducted to determine the optimum ratio of formaldehyde to resorcinol and the resin to rubber ratio to provide the proper balance of strength and flexibility.[25]

Natural rubber latex was used for fabric adhesives until the 1950s, when synthetic lattices became readily available. SBR latex was used at first, followed by the introduction of butadiene/styrene/vinylpyridine terpolymer (ASTM designation PSBR). Incorporation of vinylpyridine into the polymer structure results in much stronger bonding of the resin to the adhesive elastomer.[26] This became particularly important with the introduction of nylon 6 and 6.6 fibers for tire use. The higher internal strength of the adhesive and greater polymer functionality was well suited to the higher fiber strength and lower surface reactivity of nylon relative to rayon. PSBR quickly became the standard latex for tire cord adhesives and remains so today (Fig. 13.16).

Modification of the rubber matrix to promote fiber adhesion was also studied after the introduction of nylon fibers. Addition of silica, resorcinol, and a formaldehyde donor to the rubber compound was found to improve adhesion to nylon. The level of adhesion obtained and the change in compound properties caused by the high level of additives required have limited the utility of this approach for tire applications, although it has been widely used in mechanical rubber goods.[27]

13.4.2 Adhesion activation chemistry

The development of polyethylene terephthalate (PET) polyester fiber for tire cord in the 1960s required new adhesive chemistry to match the level of adhesion obtained with resorcinol formaldehyde latex (RFL) adhesives on

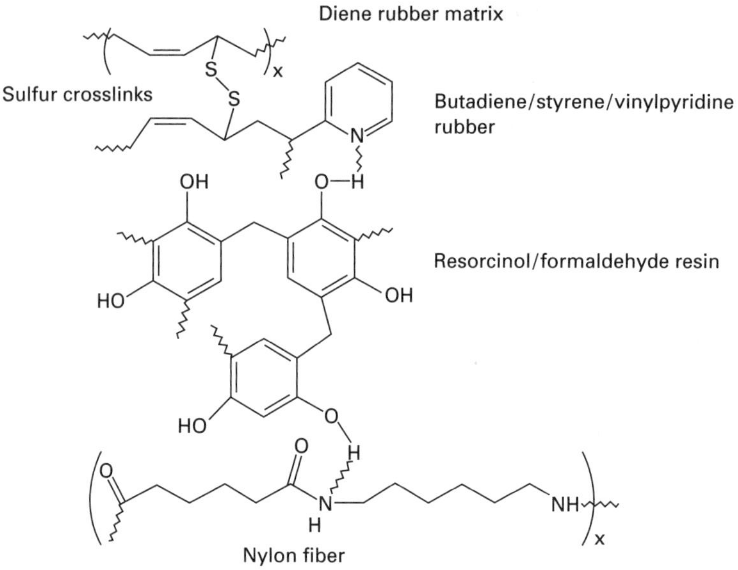

13.16 Schematic of RFL (resorcinol formaldehyde latex) fiber to rubber bonding.

rayon and nylon. Polyester lacks the high wettability and reactive surface functionality of rayon and nylon. As a result, the simple RFL adhesives which worked well for the latter two fibers showed lower adhesion to polyester.[28] Various alternative chemistries were explored in an effort to increase polyester adhesion to rubber, including modification of the resorcinol resin to increase interaction with the fiber, use of solvent-based reactive resins to promote surface reactions, and increases in processing temperature to increase reaction rates. Theoretical studies by Iyengar at Dupont demonstrated that matching the solubility parameter of the bonding resin to that of the fiber resulted in the highest level of adhesion and offered an explanation for the particular effectiveness of epoxy/isocyanate resin combinations in bonding to polyester[29] (Fig. 13.17).

The demand for improved polyester to rubber adhesion also led to the development of reactive sizings applied by the yarn manufacturers during spinning or drawing stages of the yarn.[30] The term "adhesive activation" was coined to distinguish such treatments from the standard spin finishes applied as lubricants to protect the fiber during subsequent mechanical operations such as twisting and weaving. Each manufacturer developed and in many cases patented their own chemistry for adhesive activation, with the result that performance can vary from one supplier's yarn to another depending on the adhesive subsequently applied and the process conditions used. Modifications to adhesive activation chemistry continued into the 1990s to address toxic emissions and cord fatigue from filament fusion.

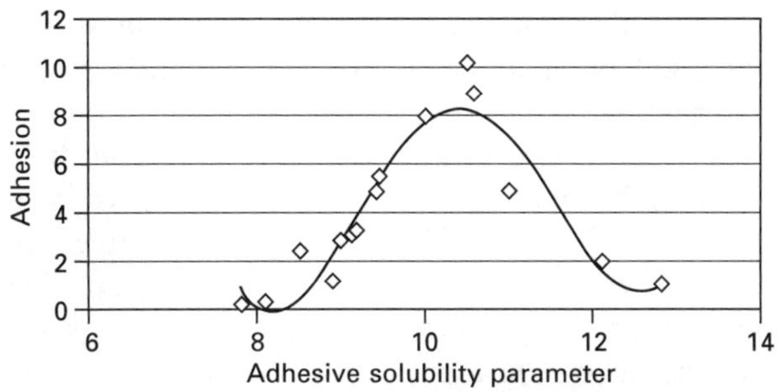

13.17 Adhesion of coatings to PET film vs. solubility parameter.

Another approach to improving polyester adhesion has been to modify the yarn composition. This can be accomplished in several ways, including bicomponent fibers, in which a core of PET is surrounded by a sheath of another polymer, polymer blends, and copolymers. Each of these alternatives presents a challenge to retain the desired yarn physical properties while enhancing surface adhesion. In practice such yarns have not found wide use in the tire industry because of unfavorable tradeoffs in either performance and/or price.

Fiberglass was introduced as a tire reinforcement in the 1970s for use in belts. A radically different process for adhesive coating was developed to deal with the surface chemistry and physical properties of glass cord. The yarn is treated with a functional silane and heat treated to bond it to the fiber surface.[31] The silane functionality is chosen to provide reactive groups suitable for interaction with RFL chemistry. The resulting treated yarn is then coated with a customized RFL adhesive in a process designed to provide complete penetration of the yarn bundle with the RFL. The dipped yarn is then twisted and woven into fabric. The RFL adhesive here serves as both a bonding agent and a protective coating to prevent damage to the brittle glass filaments.

Aromatic polyamide or aramid fibers were also introduced as tire cord reinforcements in the 1970s. Poly-p-phenylenediamine terephthalamide, produced commercially as Kevlar® by Dupont and Twaron® by Akzo (later Teijin), has seen the widest use in tires of this general class. At first examination it would seem that the RFL adhesives used successfully for nylon fibers should function well for bonding aramids, but this is not the case. Aramids differ from nylons in that they are fully crystalline and have little reactive functionality present on the outer surface of the cord. A variety of reactive chemicals have been used to bond to the fiber surface, followed by coating with an RFL adhesive to provide bonding to the rubber coat compound. High chemical reactivity to promote bonding has to be weighed against potential damage to the fiber and concomitant loss of tensile strength. Low molecular weight polyepoxides are most widely used for the initial adhesive coating.[32]

Adhesive activation coatings similar to those used in polyester have been developed and marketed, but increased yarn cost and limited availability has limited their use for tire cord.

Carbon fiber was proposed for tire cord use in the 1980s, but only recently has the "in-tire" value become favorable for its use. A reactive resin is applied by the fiber manufacturer to functionalize the relatively inert fiber surface. The sized yarn is then coated with RFL adhesive in a process similar to that used for fiberglass.[33] As with fiberglass, the adhesive serves a dual function as a bonding agent and a protective coating to prevent filament fretting.

Polyethylene naphthalate (PEN) polyester, although developed in the 1980s, did not become widely available commercially until the 1990s due to availability of the naphthalene diamine monomer. Bonding to PEN can be achieved with chemistry similar to that used for PET and relies on the same principle of mutual solubility of the adhesive resin and the fiber.

Many other fibers that have been proposed for tire use, including polyvinylalcohol (PVA), nylon 4.6, solvent spun cellulose (Lyocell), and polyolefin ketone (POK), have sufficient surface polarity and reactivity to bond well to conventional RFL adhesives. High strength crystalline fibers such as poly(bis)benzoxazole (PBO) typically require adhesive coatings similar to those used for aramids.

13.4.3 Alternatives to the classical resorcinol formaldehyde latex adhesives

Alternatives to the classical RFL adhesives for tire cords have been proposed, especially recently given the regulation of formaldehyde and resorcinol. Several commercial resin products have been marketed as resorcinol alternatives, including melamine resins[34] and naturally derived alkyl-resorcinols.[35] Formaldehyde alternatives that have been suggested include furfural and glyoxal, but both these aldehydes present issues similar to those associated with formaldehyde.

Radically different coating chemistries have also been studied. Elimination of the epoxy and isocyanates used for PET and aramid bonding has been proposed through the use of reactive surface treating, including low temperature vacuum plasma,[36] corona, ozone/UV, and fluorination. These methods are aimed at increasing the surface reactivity of the fibers through oxidation, so that the resulting hydroxyl, ketone, and carboxylic acid groups are available for bonding to RFL adhesives. The resulting fiber surface then exhibits reactivity similar to rayon. Direct bonding of fibers to rubber elastomers has also been attempted through the use of surface grafting. In this case reactive monomers are polymerized onto the fiber surface after activation by strong base, electron beam, or plasma treatment.

13.5 Recent advances in tire designs

13.5.1 Self-supporting run flat tires

Continual advancement in tire designs has enabled tire manufactures to not only incorporate new materials but to take full advantage of their intrinsic characteristics and optimize the translation into tire performance characteristics. The majority of these advancements involved new polymers, fillers (like silica), oils, etc., coupled with new tread/sidewall designs and overall tire contour shapes. New fiber reinforcement materials have historically played a smaller role in these developments due in part to the role each reinforced component plays in maintaining the tire's overall performance. However, as discussed earlier, new fibers/structures are becoming more and more important in meeting these tire performance needs.

In 1978, The Goodyear Tire & Rubber Company developed the industry's first self-supporting run flat tire at the New York auto show.[37] First commercialized for Chevrolet Corvette, the self-supporting tire design (now called RunOnFlat™) is built to maintain its shape and driveability even after losing all the air pressure through incorporation of special reinforced-rubber inserts in the sidewall. When a standard tire deflates, it simply collapses under the weight of the vehicle, allowing the beads to be unseated and the sidewalls to be in contact with the rim. Within a very short distance, the tire quickly reaches its end of serviceability. The reinforced sidewalls in the Goodyear RunOnFlat™ tire keeps the tire on the rim and carries the vehicle load for up to 80 km @88 km/hr after a complete air-loss event. The RunOnFlat™ tire operates almost undetectably when flat and hence must be fitted with a pressure monitoring system to alert the driver when air loss occurs. Currently fitted on many high performance vehicles all over the world, the RunOnFlat™ tire also offers many benefits including: piece of mind, continued mobility, space savings, and convenience. Additional tire manufacturers have designed similar run flat technologies including: Dunlop DSST, Michelin ZP, Bridgestone RFT, Pirelli EUFORIA, and Continental SSR.

13.5.2 Modular tire manufacturing

In the late 1980s, Michelin introduced a radically new kind of tire building process called C3M (Carcass, Monofil, Moulage, Mechanique).[38] The C3M process is a modular, custom designed process that enhances accuracy for the placement of multiple components during tire manufacturing. The overall footprint of a C3M machine (tire building/curing module) would fit comfortably in a standard commercial trailer vs. today's tire plants that encompass many acres. The C3M is not specifically engineered to be a cost saving process, but instead to maximize manufacturing flexibility in the form of short production runs of specialized tire orders (e.g., Scorcher T/A's) and OE-quality ultra high performance and SUV tires. In addition, C3M provides the flexibility of physically placing the equipment virtually anywhere in the world. Michelin has already placed 15 million original equipment and replacement C3M-made tires on the road since 1997 (e.g., Pilot Sport A/S Ultra-High Performance All-Season tire) while the others are still in the pre-mass market stages. Goodyear (IMPACT), Pirelli (MIRS), Bridgestone (ATMSS), and Continental (MMP) have also developed or are developing modular tire manufacturing systems.

13.5.3 Ring-supported run flat tire

Ca. 1996,[39] Michelin introduced a ring-supported run flat technology system called PAX®. Requiring a special wheel and a polyurethane inset to function, the PAX® is designed to keep the tire on the rim even if all the air is removed. With the loss of air, the tire collapses on the insert and thus preventing the sidewalls from coming in contact with the rim. Fully inflated, the PAX® tire offers similar performance features as the conventional pneumatic radial tire (e.g., ride, handling, fuel economy) due in part to the compliant sidewall design. Under a total air-loss condition, the PAX® tire design enables the operator to continuing traveling at 55 mph for up to 125 miles. The Michelin PAX® tire was first fitted on the 2001 Renault Scenic automobile. Since then, O.E. fitments have grown and now include Audi A6 Quattro and upcoming Honda and Nissan models. In addition to Michelin, Continental, Pirelli, Goodyear and Sumitomo have also been engaged in the development of a PAX® tire system.

13.5.4 "Self-sealing" run flat tires

Another form of run flat tire is the "self-sealing" type which came on the scene in the 1970s. These tires are specially designed with an extra lining (located inside the tire) which contains a pre-described amount of puncture sealant polymer. Upon puncturing, the sealant flows around the penetrating object, helping to minimize the loss of air. If the object fully penetrates the cavity, the void is filled by the same sealant polymer. This technology is very effective for small objects such as nails, bolts or screws up to ~3/16″ in diameter in the repairable area of the tread. Uniroyal (Tiger Paw NailGard) and Continental (AST Self-Sealing) both manufacture tires to this market segment.

Run flat tire summary

The use of run flat tires has been most successful in high-end "niche" applications like ultra high performance vehicles and mini-vans. All of the major tire companies, including Goodyear, Bridgestone/Firestone, Michelin, Continental, and Pirelli offer one or more run flat tire designs. To date, no one design will satisfy all customer requirements. Some automotive vehicles come fitted with run flat tires including: BMW Mini Cooper S, Chevrolet Corvette, Lexus SC430. BMW has stated plans to make run flat tires standard equipment on all its models in the very near future. Despite their success, run flat tires still need to address additional hurdles[40] before becoming more broadly accepted. These hurdles include further improvements in ride and comfort (especially for self-supporting designs), uniformity (primarily self-sealing tires), reduced weight, increased fuel economy, increased tire size selection, fitment on conventional wheels (notably ring-supported designs), and reduced costs.

13.5.5 Multiple tread-zone tires

In 2004, The Goodyear Tire & Rubber Company introduced a new, broad market automotive tire design called Assurance® Triple Tred,[41] featuring three distinct tread zones. The Assurance® tire's three tread zones help provide continual performance when subjected to dry, wet or icy conditions. The ice zone incorporates small amounts of fiberglass in the compound providing added grip and traction. Many other tire companies have followed suit with similar features that provide enhanced tire performance under various road surface and environmental conditions.

13.5.6 Pneumatic and non-pneumatic polyurethane tires

In 2003, Amerityre announced its Arcus™ pneumatic passenger car tire concept.[42] The tire uses conventional reinforcement materials and incorporated their proprietary polyurethane elastomer technology. By November 2007, Arcus™ successfully completed testing to meet USA Federal Motor Safety Vehicle Standards (FMVSS) 109 and 139. Arcus™ is said to offer (among other things) longer life and lower rolling resistance (resulting in improved fuel economy) than conventional rubber pneumatic tires. Amerityre also offers a wide range of other polyurethane-based tires including: non-pneumatic low speed foam tires called Flatfree™ (e.g., bicycles and golf carts), composite tires (e.g., polyurethane treads on conventional rubber tires), and solid tires (e.g., fork trucks).

13.5.7 Integrated tire and rim non-pneumatic tire

The latest new automotive tire design announced is the Michelin "Tweel" concept.[43] The heart of Michelin's Tweel innovation is its deceptively simple looking hub and spoke design that replaces the need for air pressure while delivering performance previously only available from pneumatic tires. The flexible spokes are fused with a flexible wheel that deforms to absorb shock and rebound with unimaginable ease. Without the air needed by conventional tires, Tweel still delivers pneumatic-like performance in weight-carrying capacity, ride comfort, and the ability to "envelope" road hazards. Michelin has also found that it can tune Tweel performances independently of each other, which is a significant change from conventional tires. This means that vertical stiffness (which primarily affects ride comfort) and lateral stiffness (which affects handling and cornering) can both be optimized, pushing the performance envelope in these applications and enabling new performances not possible for current inflated tires. The Tweel prototype, demonstrated on the Audi A4, is within five percent of the rolling resistance and mass levels of current pneumatic tires. That translates to within one percent of the fuel economy of the OE fitment. Additionally, Michelin has increased the lateral stiffness by a factor of five, making the prototype unusually responsive in its handling. Commercially unavailable at the moment, the Michelin Tweel as well as similar air-less automotive tire technologies will not be available to the general public for several years.

13.6 Future trends

One could ask "What drives the future?" The answer is many things: demographics, natural resources and the environment, the global economy and globalization, national and international governance, future conflicts, and last, but certainly not least, science and technology. No single driver will dominate. Each will have varying impacts in different regions. They are not necessarily mutually reinforcing. So how are they going to impact the transportation business? More directly, how are they going to impact automotive vehicles of the future? How will the future automotive vehicles impact future tire needs including materials, reinforcements, and designs?

For the automotive markets, one can visualize convergence of several technologies including wireless, nano and sensors. These technologies should converge and create embedded connectivity, robotic-like behaviors, 3D video projections and interactive holography all in the name of improving vehicle capabilities, passenger comfort and convenience. Future vehicles look to be smaller, use alternative fuels (e.g., hydrogen, electricity), and carry higher loads coupled with higher fuel efficiency.

As for tires, our requirements will continue in the path already in progress. In the near term, demand will continue to increase primarily driven by significant growth in China, India, and Eastern Europe. Increase in tire size and complexity will help advance the need for additional modular tire building manufacturing like Michelin's C3M or Goodyear's IMPACT.

In the longer term, significant reduced tire weight (approaching 50%+ lower) will become a necessity in order to achieve the fuel economy needed for the most advanced automotive vehicles. In helping to achieve these needs, new, advanced materials and fibers will be discovered and incorporated into new tire designs to not only achieve the advanced performance requirements imposed, but to make the tire more ecologically friendly through the incorporation of more evergreen, non-petroleum based materials.

Future reinforcement opportunities for tires (especially automotive tires) remain positive. Driven by customer needs, the general fiber industry trends can be seen in Fig. 13.18. Higher modulus and higher strength materials are where all the activities are. Key market drivers are aerospace, military and ecology which will place huge pressure to further advance fiber properties as well as to make them more bio-friendly. Smart/responsive materials, at all levels including macro, micro and nano, will be incorporated into tires making them intelligent and auto-responsive to the changing environment. Novel reinforcement materials and structures will be discovered to protect tires from environmental hazards and improve tire life.

New fibers on the horizon are many and varied. Some will be directly applicable to tires, while others may require novel approaches to best take advantages of their features. Fibers such as polyvinyl alcohol, polyamide 4.6, poly(diimidaza) pyridinylene (dihydroxy) phenylene (aka PIPD or M5), polyethylene ketone, poly p-phenylene benzobisoxazole (PBO) and the like offer enhanced features that will require advanced tire designs and knowledge to be incorporated as a structural member in the tire. Other fibers such as nanocomposite and spider silk offer potential for extreme advancements in

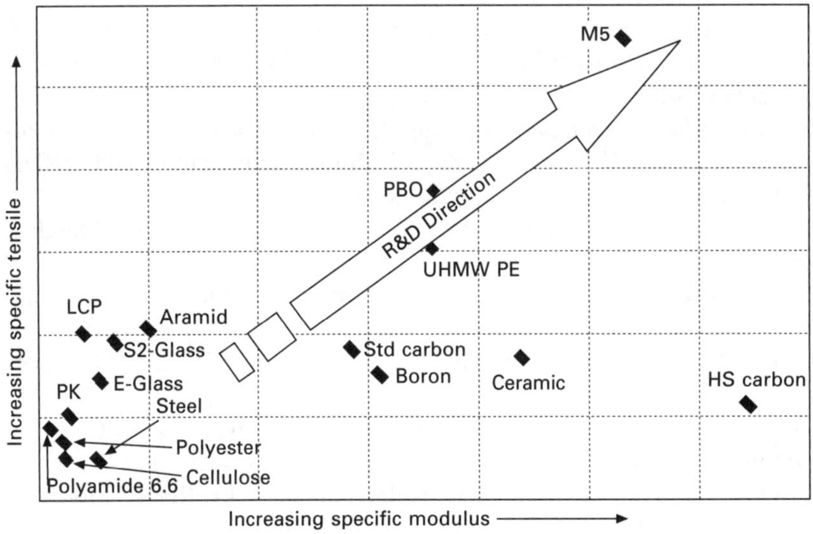

13.18 New fiber technology R&D path.

tire reinforcement science and design. Still other bio-fibers such as bamboo, polylactic acid, hemp, etc., may offer surprises in advancing the reinforcement capabilities of elastomeric matrix under both static and dynamic conditions. In addition, the expansion of composite-like cord structures (e.g., merging two or more dissimilar fibers into one cord structure, core-sheath fibers) will offer unique reinforcing behaviors for maximizing "in-tire" value.

In summary, today's tire reinforcement fiber requirements are varied and demanding. Lighter weight materials, improvements in rolling resistance and uniformity along with increased toughness continue to place significant technological challenges. New emerging fibers, while not initially designed for tires, offer significant potential to address these challenges and provide further options to tire designers. Methodologies for the development of new fiber reinforcement materials must include cure/production and in-service/dynamic composite simulations to fully assess and maximize "in tire" performance and value.

13.7 Sources of further information and advice

"*New Fibers*"; Hongu, Tatsuya and Phillips, Glyn O; Woodhead Publishing Limited; Cambridge, England; 2nd Edition; Copyright 1997.

"*A Symposium on Advanced Materials and Processes*"; The Fiber Society 2004 Fall Technical Conference; October 11–13, 2004.

"*Journal of Applied Polymer Science*"; Baer, Eric and Hiltner, Anne (eds); Volume 83; Number 3; John Wiley & Sons, Inc; January 18, 2002.

"*Global Tends 2015: A dialogue about the future with Nongovernment Experts*"; December 2000 NIC 2000-02.

"*Smart Materials: Emerging Markets for Intelligent Gels, Ceramics, Alloys, and Polymers*"; Frost & Sullivan; 2nd edition; circa 2005 (www.technical-insights.com).

"*Foresight Vehicle Technology Roadmap, Technology and Research Directions for Future Road Vehicles*"; Society of Motor Manufacturers and Traders Ltd; Version 2.0; Crown Copyright 2004.

"*Horizon 2010*"; Presentation by Michelin @ Geneva Conference, March 6, 2007.

"*2020 Science*"; Stephen Emmott and Stuart Rison; Microsoft Research; Copyright 2006.

"*Strategic Research Agenda*"; Rudolf Kunze – Chairman; ERTRAC; December 2004.

"*Twaron News*"; Pages 10–11; Teijin Corporation; November 2006.

"*Winning the Oil End Game: Innovations for Profits, Jobs, and Security*"; Lovins, Amory B and Datta, E. Kyle; Published by Rocky Mountain Institute; 1st edition, 2nd printing; Copyright 2005.

"*Adding Value To The Tire: Next Generation Inside Tire Possible Solution*";

Alfonso S Di Pasquale (Schrader Electronics Ltd.), F. Mancosu (Pirelli); Tire technology EXPO 2007 Koln Messe; March 13, 2007.

"*Trends in Automotive – Electromotive Forces*"; Duncan Leslie, Deputy Editor; European Automotive Design; AMI Semiconductor; Copyright 2005.

"*Managing Information Strategies*"; Zampetakis, Helene; Moller International; Australia, December 17, 2004.

ESA news release, June 16, 2005 and various web sites.

"*Nanocomposites in the automotive industry – November 2004*"; Lownsdale, Gary; Knoxville, Tenn, USA.

"*Automotive Telematics: The One-Stop Guide to In-Vehicle Telematics and Infotainment Technology and Applications*"; Foy, Dennis; Cheshire, UK: Red Hat, 2002. 196 pages. ISBN: 0954334000.

"*Fiber reinforcement in tires; an overview*"; Causa, AG; The Goodyear Tire & Rubber Company, Corporate Research Division; Published at Tire Technology International Symposium, 1993.

"*An Overview of Polymeric Fiber Reinforcements for Tires*"; Causa, AG, Kim, S and Westgate, WK; Tire Technology Expo 2002; Hamburg, Germany; February 21, 2002.

"*The road to quieter tires*"; Sandberg, Ulf; Swedish National Road and Transport Research Institute (VTI); Tire Technology International; 2003.

TECHTEXTIL Symposium, North America; March 31, 2004.

"*The Importance of Dimensional Stability in the Polyester Tire Carcass*"; Brown, DL; AlliedSignal, USA; Published in RubberTech China; Paper #23; 1998.

Kirk-Othmer Encyclopedia of Chemical Technology, Fourth Edition; Volume 24; ISBN 0-471-52693-2; John Wiley & Sons, Inc; Copyright 1997.

"Effect of Material Properties on Tire Performance Characteristics – Part 1. Tire Cords, Tire Science and Technology"; Futamura, Shingo; *TSTCA*; Volume 15, No. 3; pp. 198–206; July–September 1987.

"Global Trends in Fibers, Textiles Businesses"; Ward, Derek; *International Fiber Journal*; May 26, 2000.

"*Extreme Textiles*"; Maynard, Barbara; www.chemisrty.org; February 7, 2007.

"*Role of Tire Reinforcements on Composite Static and Dynamic Characteristics*"; Vorachek, JJ; The Goodyear Tire & Rubber Company; Presented at ASTM Symposium on Tire Reinforcement and Tire Performance; Akron, Ohio, USA; October 23–25, 1978.

"Technical Innovations: Torvec's icey idea for treads and tires"; March 8, 2005 (www.sae.org/ohmag/techinnovations).

"Toward autonomous cars"; Weisman, Robert; Boston Globe Staff; September 27, 2006 (www.boston.com/business/blog/onsite).

"Inside the Mystery"; Smith, Jim; October 14, 2007 (www.Tirerack.com and www.tirereview.com).

www.4crawler.com/diesel/tires.shtml; "All About: Tires (and Wheels)".

13.8 References

1. http://inventors.about.com/library/inventors/bltires.htm.
2. http://inventors.about.com/library/inventors/btires.htm.
3. "*Tire Technology*"; Kovac, F.J.; The Goodyear Tire & Rubber Company; 4th edition; Copyright 1973.
4. Report on March, 2001 WP.29 Meeting Geneva, Switzerland, Submitted by Doug Read, Managing Director, SAE Washington Office; www.sae.org/globalization/harmonization/wp-29mar01.doc.
5. UN Regulation 51, ISO361.
6. http://unece.org/trans/doc/2003/wp29grrf/PCGTR-2003-06e.doc.
7. "Tire reinforcing systems"; Dague, M.F.; The Goodyear Tire & Rubber Company; Eleventh Annual Rubber Lecture Series, "*The Science and Technology of Rubber Systems*", The Akron Rubber Group, Inc.; January 14, 1974.
8. 2006 LMC World Tyre Forecast Service and Global Products Planning Analysis.
9. "*Mechanics of Pneumatic Tires*"; Clark, S. K. (ed.) US Department of Transportation; NHTSA; U.S. Printing Office, 1982 (0-363-904).
10. Wikipedia, The Free Encyclopedia; http://en.wikipedia.org/wiki/Wikipedia.
11. "*New Reinforcement Materials*"; Westgate, W. K. Carbon Fiber 2007 Conference, Washington, D.C., USA; December 5–7, 2007.
12. USP 4,101,525; Davis, H. *et al.*; Issued July 18, 1978.
13. "*Modern Polyesters: Chemistry and Technology of Polyesters and Copolyesters*"; Scheirs, J. and Long, T. E. (eds); John Wiley & Sons, Ltd; Copyright 2003.
14. http://www.ridercn.com/pirelli/1/01.php.
15. http://www.sumitomo.gr.jp/english/discoveries/special/92_04.html.
16. http://www2.dupont.com/Kevlar/en_us/index.html, http://www.twaron.com.
17. http://findarticles.com/p/articles/mi_m0EIN/is_1995_Oct_11/ai_17594107.
18. http://www.kumhotyre.co.uk/news/news061201.html.
19. http://www.goodyearwrangler.com/flash.html.
20. http://www.tirerack.com/tires/tires.jsp?tireMake=Goodyear&tireModel=Fortera+TripleTred&.
21. http://news.thomasnet.com/fullstory/481398.
22. http://www.popularmechanics.com/blogs/automotive_news/4216978.html.
23. Gardner, E. R. and Williams, P. L. *Proc. 2nd Rubber Technol. Conf.*, 1948, 478.
24. Charch, W. H. and Maney, D. B. US Patent 2,128,229 (1938), to E. I. du Pont de Nemour & Co.
25. Moult, R. H., in *Handbook of Adhesives* (I. Skeist, ed.), Reinhold, New York, 1962, p. 495.
26. Wolfe, W. D., US Patent 2,816,616 (1957), to The Goodyear Tire & Rubber Company.
27. Albrecht, K. D., *Rubber Chem. Technol.* **46**, 981 (1973).
28. Kovac, F. J. and Kersker, T. M., *Text. Res. J.* **34**, 69 (1964).
29. Iyengar, Y. and Erickson, E., *J. Appl. Polym. Sci.* **11**, 2311 (1967).
30. Schroeder, W. A. and Prettyman, I. B., *Rubber Age* **99**(1), 72 (1967).
31. Vanderbilt, B. and Clayton, R., *Rubber Chem. Technol.* **38**, 379 (1965).
32. Iyengar, Y., Technical Symposium, The Akron Rubber Group, Rubber Division, ACS, 100–115, April 14, 1978.
33. Gillick, J. G. and Bennett, C., US Patent 6,077,066 (2000) to The Goodyear Tire & Rubber Company.

34. Winkler, A., US Patent 7,214,758 (2007) to Cytec Technology Corp.
35. http://www.carboshale.ee.
36. Shuttleworth, D., Mowdood, S. K., Waddell, W. H., Richards, J. L., Ofstead, E. A. and Brenner, J. L., US Patent 5,053,246 (1991) to The Goodyear Tire & Rubber Company.
37. http://findarticles.com/p/articles/mi_qn4155/is_20001105/ai_n13878985.
38. http://www.tirerack.com/tires/tiretech/techpage.jsp?techid=19.
39. http://www.michelin.com (Reperes Milestones; 2007).
40. http://www.edmunds.com/ownership/tires/articles/117588/article.html.
41. http://www.goodyear.com/media/pr/pr_2004/22832ti.html.
42. http://www.amerityre.com/index2.cfm?group=investor&function=pr&dt=20070116.
43. http://sf.orangecoat.com/michelin-tweel-gold-medal-for-innovation-intermat-2006.

Part IV

Use of textiles in automotive bodywork

14
Textile structures for load-bearing applications in automobiles

T GRIES, J STUEVE, T GRUNDMANN and D VEIT, RWTH Aachen University, Germany

Abstract: Various textile reinforcement structures are available for use in automobiles as composites. These structures can be realized using a wide range of textile technologies. One major focus of current developments is therefore load-bearing applications for structure elements. Their low weight in combination with very good impact behaviour in the event of a crash makes them an interesting alternative to conventional materials such as steel and aluminium. This chapter gives an overview of current production technologies for these kinds of textile structures, of joining and assembling techniques, and describes typical applications that are already in use.

Key words: composites, 3D textiles, preform, joining, cutting.

14.1 Introduction

Various textile reinforcement structures are available for use in automobiles as composites. The complexity of textile structures increases with dimension. These structures can be realized using a wide range of textile technologies. At present, textiles are used in automobiles for comfort, safety and structural components. This amounts to approx. 13 kg of textiles per vehicle and is expected to increase to 30 kg within the next decade.

Airbags, safety belts and seats as well as interior and acoustic structures are state-of-the-art but under constant development. Textiles for car tyres have been in use even since the 1930s. One major focus of current developments is therefore load-bearing applications for structural elements. Their low weight in combination with a very good impact behaviour in case of a crash makes them an interesting alternative to conventional materials such as steel and aluminium. Textile composite structures have already been put into practice in Formula 1 racing cars and also in several small series sports cars but are still far from being a mass production application. Considerable effort has to be put into that subject in order to reach higher productivity of the processes involved and also a higher degree of automation in order to meet the high quality demands of the automotive manufacturers.

Fibres used for load-bearing applications in vehicles are exposed to various stresses during processing (tensile stress, bending stress, fibre-fibre friction, fibre-machine friction). Bending and shear forces especially can cause fibre damage when brittle high-performance yarns made of glass or carbon fibres are used. Thus, only textile processes that induce a low deflection in the reinforcement fibres are suitable for this kind of application. Textile processes with warp- or weft-insertion (in weaves or knits), fibre-placement and overbraiding technologies are therefore well suited for the processing of carbon rovings.

Complex textile reinforcement structures in near-net-shape form (textile preforms) can be produced in one-step or multi-step processes, using one-step preforming technologies such as 3D braiding, 3D weaving, contour warp knitting or net-shape weft knitting. Usually, at least sub-preforms are manufactured that are processed in further preforming steps. In many cases 2D textile structures are used to create structures of higher complexity by forming and assembling several plies in multi-step preforming processes. An alternative technology is the use of 3D textile structures and preforming technologies. The relevance of textile and preforming technologies for use in automotive applications is discussed.

14.2 Latest advances in technologies for producing 2D and 3D textile load-bearing structures for automotive applications

For load-bearing applications in vehicles, 2D or 3D textiles can be used. To simplify the use of the expressions 2D textile and 3D textile, a definition was proposed by Gries and co-workers [1]:

> *A textile is defined as 2D if it does not extend in more than two directions, neither in yarn architecture nor in textile architecture. A textile is defined as 3D textile if its yarn architecture and/or its textile architecture extends in three directions, regardless of whether it is made in a one-step-process or a multiple-step process.*

High performance textiles for automotive applications usually consist of high performance fibres, yarns or rovings. Typical materials are carbon, glass and aramide. In most cases, twisted and untwisted multifilaments, the so-called rovings, are used. Twisting the roving slightly improves its processability significantly and decreases the risk of filament damage.

For the use of textile structures as reinforcement in composites, it is often important to realize a yarn position that is as straight as possible because non-crimped fibres can bear the highest loads and induce the highest stiffness. In other applications fibre crimp is needed, for example to achieve high damage tolerance or high energy absorption. Therefore each application requires its own textile structure.

14.2.1 2D textiles

Woven fabrics

Woven fabrics are defined as fabrics which are made of rectangular crossed yarns of at least two yarn systems, warp and weft, by shed formation [2]. The manner of crossing the weft and warp yarns is called pattern. The pattern type has a strong influence on various fabric properties such as drapability and shear stiffness. There are three basic patterns: plain weave, twill weave and atlas weave and special pattern types like leno weave (Fig. 14.1).

There are also special non-crimp woven fabrics available that use thin auxiliary yarns to avoid bending of the reinforcing yarn. The bending, which is caused by the weaving process, is induced only in the auxiliary yarn because its bending stiffness is much lower than that of the reinforcing yarn. Figure 14.2 shows a biaxial non-crimp weave with auxiliary yarn called 'Advanced Syncron Weave' from the company ECC GmbH & Co. KG, Heek, Germany. In comparison to the unidirectional reinforcement structures, the handling and drapability of fabrics is good but the wave-like yarn crimp influences the mechanical properties of the consolidated textile-reinforced structure. Strength as well as stiffness of fabric-reinforced composites are

14.1 Weave patterns.

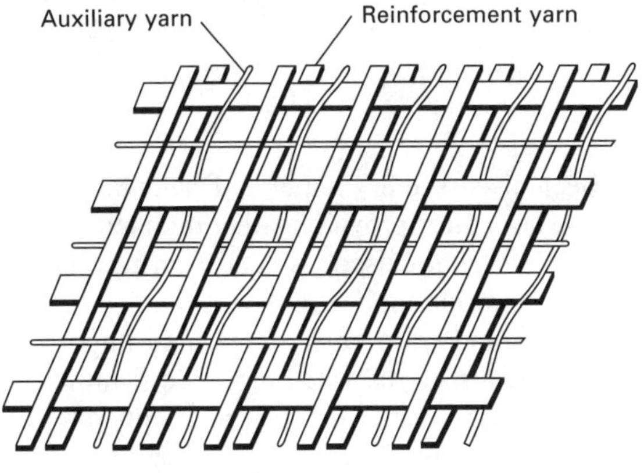

14.2 Non-crimp fabric.

therefore lower than with unidirectional reinforcements. Choosing a proper weft and warp density as well as suitable patterns, like Atlas, Advanced Syncron or leno weave, can minimize these effects by reducing the fibre crimp. Also the spreading of the yarns to a thin tape helps to reduce the yarn crimp.

The processing of brittle fibres to woven fabrics is possible but 'yarn-friendly' patterns (e.g., Atlas, Advanced Syncron or leno weave but not plain weave) and suitable machine set-ups must then be chosen.

Braids

Braids are defined as textile structures of regular appearance with closed selvedges. They consist of at least three yarns (healds) or at least two yarn systems (tubular braids), whose yarns are intertwined and cross diagonally to the edges of the textile (Fig. 14.3(a)).

The angle between the selvedges (that is the production direction) and the braiding yarns is called braid angle (Fig. 14.3(b)). This angle can be varied between 20° and 80°. Braids with a braid angle of 45° look quite similar to woven fabrics. They do have a comparable yarn architecture regarding yarn crimp due to the intertwining of the braiding yarns. The most common braid pattern is called regular braid. It is comparable to a 2/2-twill woven structure. Other braid patterns are the diamond braid (1/1) intersection repeat and the hercules braid (3/3) intersection repeat [3].

In addition to the braiding yarns, axial yarns (0°) can be added inside the braided structure. These yarns are inactive during the production process and are enclosed by the braiding yarns. While the braiding yarns are crimped, the axial yarns remain in straight position. Braids with axial yarns enclosed are called triaxial braids. Braids without axial yarns are called biaxial braids. The drapability of triaxial braids is worse than that of biaxial braids due to the friction between axial and braiding yarns.

For the reinforcement of flat, narrow automotive parts, 2D braids made of carbon fibres are used.

Knitted fabrics

Knitted fabrics have been discussed in Chapter 3. Due to the arrangement of the yarns in loops, these structures provide neither high tensile strength nor modulus. They are therefore not used as load-bearing applications in automobiles but solely in interior applications.

Nevertheless, weft-knits and warp-knits are of high importance for the reinforcement of composite structures when modified. Knitting allows the insertion of straight reinforcing yarns in warp and weft direction during production. Mono-, bi- and multiaxial weft- and warp-knitted fabrics can therefore be realized in economic production processes (Fig. 14.4).

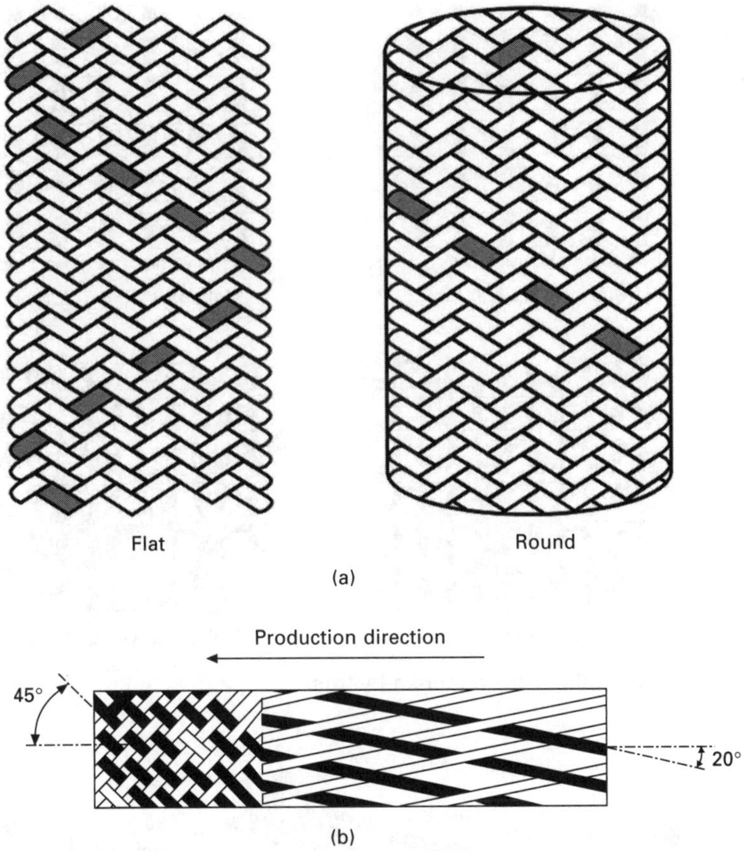

14.3 (a) Types of 2D braids and (b) braid angle.

Using knitting technology, every warp and weft yarn is bound with a single knitting loop. Compared to other technologies, the rovings therefore remain undisturbed. As knitting yarn, an auxiliary yarn made of thermoplastics like polyethylene or polyamide can be used. The insertion yarns are not subject to high load and abrasion during production, which allows the use of brittle materials such as carbon. Therefore this technology is very suitable for the production of reinforcing textiles for automotive applications.

A special weft-knitting process with warp and weft insertion which enables the production of near-net-shape knitted fabrics has been developed by Institut für Textil- und Bekleidungstechnik der TU Dresden, Dresden, Germany (ITB) [3, 4]. The innovative process is based on conventional flat-weft-knitting technology. Near-net-shape structures like disks with polarorthotrop reinforcing yarns can be realized [5].

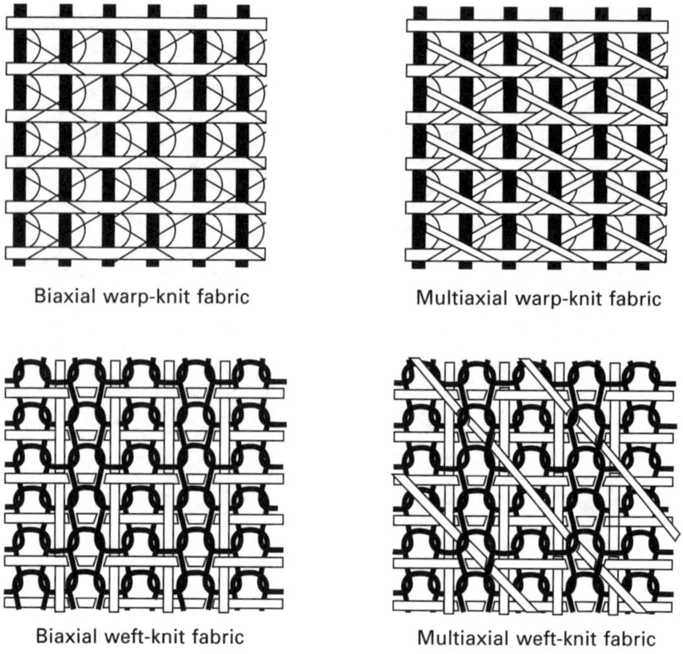

14.4 Knits with reinforcing yarns.

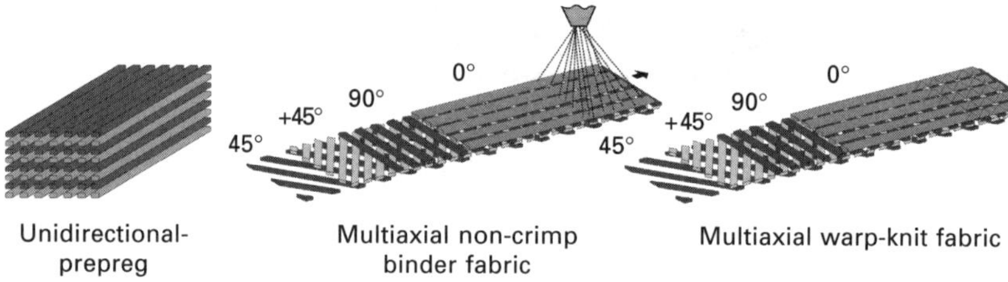

14.5 Non-crimp fabrics.

Non-crimp fabrics

Non-crimp fabrics (NCF) are defined as textile structures which consist of one or several parallel layers of straight, unidirectional yarns that are not crimped [6]. Multiaxial NCF are 2D textiles consisting of at least two layers of unidirectional reinforcing yarns (Fig. 14.5). These layers are joined by being preimpregnated with resin, by fixation with adhesives or using warp-knit seams. Adhesives for the joining of non-crimp fabrics are used in the form of fluid, powder or nonwoven hotmelt. In a multilayer non-crimp fabric the layers can have different orientations. Possible orientations are 0°, 90° and orientations between 20° and 70°. Furthermore, the yarn count and the kind of material used in each layer can be varied. Industrial production machines can today process up to eight layers plus the introduction of two surface fabrics like nonwovens. In non-crimp fabrics joined by warp-knit seams the knitting pattern can be varied. The three patterns chain stitch, tricot lap and cord lap influence the drapability of the fabric [7]. Owing to the high standard of the machine technology, the processing of brittle high-performance yarns like carbon rovings is possible without much fibre damage. Only the joining process using warp knitting can be critical for very brittle yarns because the needles punch directly through the yarns. This can create filament breakage and yarn deflection. Nevertheless, a lot of non-crimp fabrics with different yarn architecture made of glass, carbon and aramide rovings are available today. In automotive applications, mainly carbon and glass are used as roving material.

14.2.2 3D textiles

Processes for the production of 3D textile structures often allow the realization of textile preforms in one production step. Many production processes to create 3D textiles have been invented over the last two decades and development has not ceased yet. The most relevant technologies are described in the following.

3D woven structures

Different types of 3D woven structures exist. Figure 14.6 gives an overview of some types that are described in this section.

Three-dimensional woven fabrics are manufactured by inserting two picks with double shed opening, being perpendicular to each other. The yarns are fed in 0° and 90° direction and an additional yarn is positioned orthogonally towards the two others in z-direction. Three-dimensional fabrics feature quasi-isotropic properties, a high tensile and compression strength as well as a good bending stability and a very good impact behaviour [8–11]. The distribution of the reinforcement fibres is very uniform in all three dimensions. The drapability and the elongation behaviour of three-dimensional fabrics, however, is very poor. Three-dimensional fabrics are applied in fibre-reinforced composites with high thermal impacts as well as for structural parts for cars and aircraft. Another kind of three-dimensional fabric structures are the so-called *multi-layer woven fabrics*. These are manufactured by the composition of several fabric layers without any spacing between. The layers are fixed by interlocking or warp knitting. The yarns are oriented in 0° and 90° direction and the yarns in z-direction are variable. Multi-layer fabrics can be draped well, having very good elongation behaviour and good tensile, compression and bending stability. These fabrics are applied in impact charged multi-

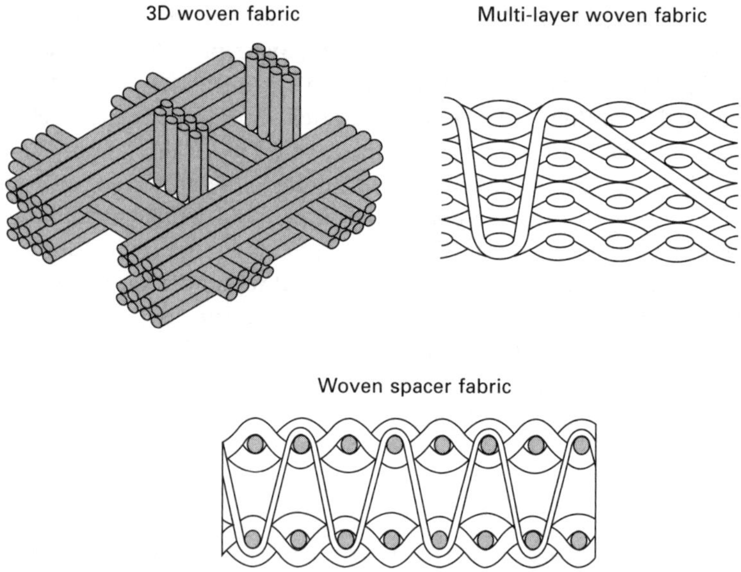

14.6 3D woven structures.

layer composite structures, like automobile floor components [12, 13]. Among others, 3Tex, Cary NC, USA produces these kinds of fabrics.

Woven spacer fabrics also have a three-dimensional structure. They are manufactured by weaving-in upright pile warps, and their properties are comparable to those of multi-layer fabrics, but their drapability is poor. The distance between the two layers can be adjusted individually. These textiles show a high resistance against perforation. They are applied in sandwich structures, for example in composite lightweight design applications [10, 14]. Another process to produce 3D-textiles is called Shape Weaving process and was developed by the company Shape 3 Innovative Textiltechnik GmbH, Wuppertal, Germany [15]. The integration of warp and weft yarns of different lengths during the weaving process is the principle of this technology. Due to these extra yarn lengths the textile expands into the third dimension. This enables the configuration of textile fabrics to a certain shape during the weaving process [16]. This technology has great potential but is currently not being used for automotive textiles.

Braided structures

Braided structures in automotive applications have mostly been used to realize cable trees. Hereby several isolated wires are joined together by surrounding them with a tubular braid. The wires are therefore protected against mechanical damage and the car assembly process is simplified. Nowadays, as the use of composites in automotive applications is increasing, braided structures are also used for structural parts like bumpers and crash absorbers. For the production of these structures, overbraiding technology is used. But the 3D rotary braiding technology might also gain importance for very specialized parts.

Overbraided structures

One possibility to create 3D textiles or textile preforms using braiding technology is overbraiding, a special kind of circular braiding. The creation of the textile is similar to circular braiding but the tubular braid is directly laid down on a net-shape mandrel. Several layers can be braided on the mandrel with individual starting points and end points of each layer. Similar to circular braiding, each layer can consist of a triaxial yarn architecture and braid angles can be varied between 10° and 80°. The orientations of the braiding yarns can be varied in each layer [17]. Also the formation of a layer that only consists of 0° yarns is possible. Because the yarn carriers can be equipped with different materials, the production of hybrid structures is possible. New developments use the initial division of the two yarn systems of braiding yarns to set up one group with reinforcing yarns and the other with auxiliary yarns. Choosing a suitable auxiliary yarn (for example, an elastic thermoplastic yarn), non-crimp structures can be realized, so-called unidirectional braids (UD braids). This technique not only leads to minimized filament damage of the reinforcing yarn, it also leads to structures which can concur with UD non-crimp fabrics.

Because the yarns support each other, orientations can be realized away from the geodesic line on the mandrel (Fig. 14.7). The braided structure remains stable as long as it stays on the mandrel. Depending on the mandrel concept, the braid can either be removed from or remain on the mandrel, which can become part of the composite. For the moving of the mandrel a multiaxis take-up can be used as well as an industrial robot. Therefore an automated production of textile preforms of complex geometry in one process step can be realized. This is of crucial importance for automotive applications where large lot sizes are common.

Recent developments of August Herzog Maschinenfabrik GmbH & Co. KG, Oldenburg, Germany (Herzog) led to a special design of the overbraiding machine. Instead of moving the yarn carriers on the front of the machine body, they are led on the inside of a circular machine body (Fig. 14.8). The orientation of the yarn carriers in the radial direction is the reason for the name of this machine type 'radial overbraiding machine'.

This design leads to reduced filament damage during the braiding process due to a minimization of yarn movement along yarn guiding elements and reduced yarn-yarn interaction. The result is a high production speed and a

14.7 Overbraiding.

14.8 Radial overbraiding machine.

reliable production process. A prominent application so far are the bumpers of the BMW M6 made of carbon fibre reinforced plastic (CFRP), that are realized using overbraided preforms [18]. Other CFRP products realized using overbraiding technology are crash tubes, rocket engine nozzles and others [19, 20].

3D braided structures

3D braided structures do not only expand in the third dimension, they also contain an integral yarn architecture with a three-dimensional yarn course. Several 3D braiding technologies do exist such as the square-braiding technology, the 4-step braiding technology and the 3D rotary-braiding technology, which has the highest potential for application in automotive engineering.

The newest generation of 3D braiding technology is 3D rotary braiding. Two different technologies exist. The ITA-Herzog 3D rotary braiding technology invented by Institut für Textiltechnik, RWTH Aachen University, Aachen, Germany (ITA) and Herzog (Fig. 14.9 (a)) and the 3TEX 3D rotary braiding technology invented by 3TEX, Cary, NC, USA (3TEX, Fig. 14.9 (b)). In both technologies the transport of the yarn carriers is realized by rotating horn gears that turn in counter direction.

With the ITA-Herzog principle, the yarn carriers are guided inside grooves that are cut into the machine table. Through switches located between the horn gears, it can be decided whether a yarn carrier is to be transferred to the next horn gear or whether it is to remain on the initial horn gear. Since every switch and every horn gear can be controlled individually, an individual path of each carrier over the machine table is possible [21]. The machine is computer controlled. Therefore complex and variable yarn architectures as well as complex and variable textile architectures can be realized in an automated production process.

The 3TEX 3D rotary braiding principle is similar to the ITA-Herzog 3D rotary braiding principle. But instead of switches, so-called gripping forks are located between the horn gears. This allows two yarn carriers to stay between two horn gears at the same time. At this position it can be decided whether they should swap places or rest on their respective horn gear [22].

The 3TEX technology can use more yarn carriers on the same area than the ITA-Herzog technology. On the other hand, the yarn carriers of the ITA-Herzog technology are bigger and have more yarn capacity than those of 3TEX. The 3TEX technology is faster than the ITA-Herzog technology but it does not offer the same freedom in design of braid design.

3D rotary braiding technologies enable the manufacture of textile preforms with integral yarn architecture and continuous changes of cross section geometry and dimension in one process step (Fig. 14.10). The processing of

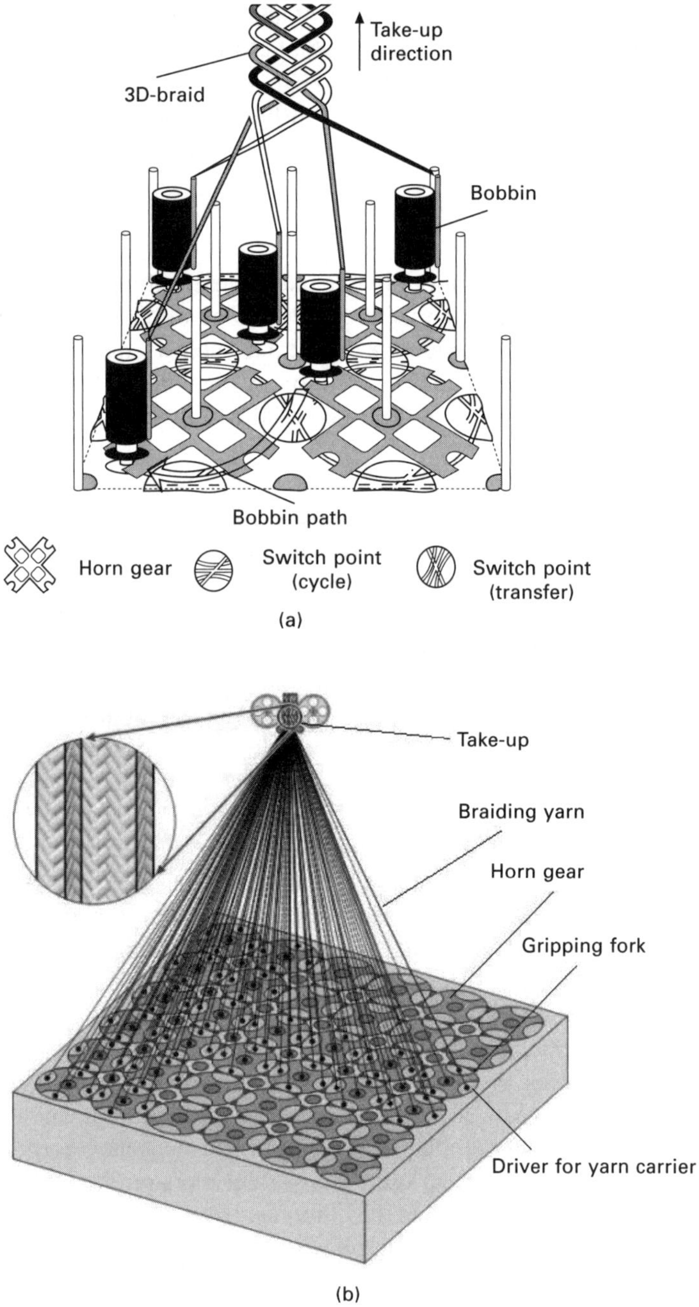

14.9 3D rotary braiding principles.

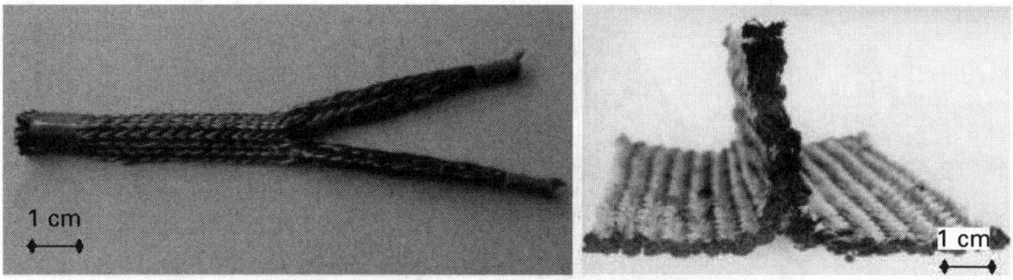

14.10 3D braids.

carbon and glass rovings is state-of-the-art. Those braided structures can be used for stiffening purposes or in energy absorption applications.

14.3 Processing textiles using preforming and prepreg technology

In today's processing of textile reinforced plastics (TRP) two ways of processing yarn architectures are mainly used: the processing of rovings or textile fabrics which are pre-impregnated with the resin (prepreg technology) and the processing of dry textile preforms which are impregnated in a second step, in the shape of the final part (preforming). In contrast to prepreg manufacturing, preforming creates a dry textile reinforcement structure which already has characteristics concerning geometry and fibre orientation required in the final part. This means that the rovings need to be formed and fixed via textile handling and assembly processes to achieve the geometry to later bear the load on the component.

For the creation of this inherently stable textile structure, several technologies are available. There are technologies which generate the three-dimensional structure from the yarns in one process step (e.g., 3D braiding, overbraiding, 3D weaving, contour knitting with weft insertion). These processes are referred to as one-step preforming. However, these technologies are limited to a certain degree of complexity concerning the reinforcing textile and thus the geometry of the final part. Examples of typical component geometries which are created with one-step preforming technologies are profiles, panels, sandwich structures and hollow parts used for, e.g., a car body.

For the realization of components of high complexity, the yarns are processed in multi-step processes. At first, 2D textiles like multiaxial non-crimp fabrics or woven fabrics and/or three-dimensional subpreforms like the above-mentioned 3D braids are manufactured. These textiles are subsequently processed to complex preforms via handling and assembly processes. A practical example for a complex composite component is the so-called rib stiffened panel (see Fig. 14.11). Rib stiffened panels consist of a flat shell (plane, unidirectional or multiaxial curved) and of stiffening profiles (e.g., top-hat-, T- or L-profile). These kinds of components are frequently used to

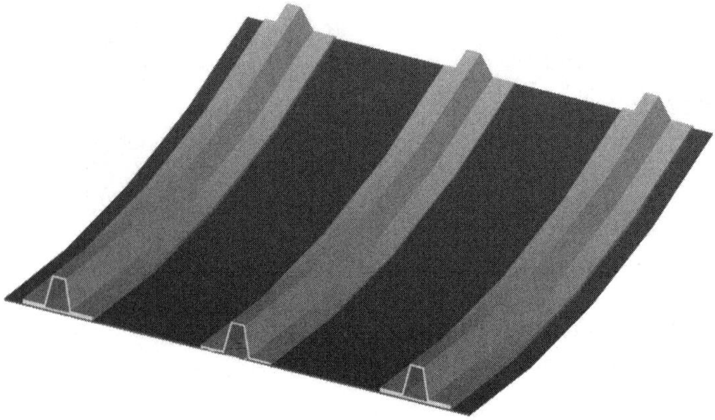

14.11 Panel with reinforcing profiles.

function as self-supporting outer skin in lightweight constructions such as automobiles and aircrafts.

Intermediate textiles for the shell are the above-mentioned flat textiles which are available as rolled fabrics. For the ribs, both 2D intermediate textiles as well as subpreforms can be used.

In order to process these textiles, the following steps are required:

- cutting
- handling and draping
- joining and conditioning.

These technologies are proven for the processing of carbon textiles for automobile parts.

14.3.1 Cutting

For an automated cutting of textile fabrics made of brittle high-performance yarns, computer controlled cutting machines which use different cutting techniques are available. Depending on the material, the fabric type and the fabric thickness that has to be cut, the suitable cutting device has to be chosen. These range from rotating cutting disks, oscillating knives to laser cutting. The cutting disks are supported by ultrasonic movement and roll along the path which has to be cut. Oscillating knives can be moved with high frequency or supported by ultrasonic movement. If necessary, the fabric can be fixed to the cutting table with the help of vacuum which is applied by covering the fabric with a sheet. Cutting using water-jet proved to be unsuitable for the cutting of dry textile fabrics. Owing to the computer-controlled movement of the cutting head, even complex contours can be cut out of the rectangular base.

14.3.2 Handling and draping

The handling and draping of textile fabrics, structures and subpreforms is at present mainly done manually. Several methods for gripping and transporting textile structures have been invented so far, e.g. needle gripping, vacuum-supported gripping, electrostatic gripping or ice gripping techniques. But no gripping or handling technology has been found so far that is suitable for the whole range of materials and applications.

14.3.3 Joining technologies

The most important joining technology for textiles besides gluing is still sewing. Sewing technologies will play an important role in the processing chain of future manufacturing of textile reinforced components. Their importance concerning cycle time reduction as well as cost savings can be regarded in line with those of qualified consolidation procedures [23]. The tasks of the sewing techniques within the preforming process range from the insertion of flat reinforcements as well as the fixation of individual components of a preform for the subsequent process steps to the manufacturing of load-compatible seam areas. Different sewing technologies are applied in differing process steps within the multi-step preform production. Sewing technologies are chosen according to the function of the seam, the preform geometry and the component loads [24].

The sewing of reinforcement textiles is both a textile and a load-compatible joining process. Thus a force-flow in form of an upright reinforcement is achieved in the seam area. Local reinforcements can be applied. Flat textiles are converted to 3D reinforcement structures. Via the application of reinforcement textiles within a component, several special mechanical properties can be realized. Additionally, the application of sewing technologies allows the production of near-net-shape textile structures. Both conventional (e.g., double-step stitch) and one-sided sewing technologies (tufting, blind stitch, ITA sewing technology) are applied [21, 25]. Figure 14.12 shows the seam architecture of relevant sewing technologies. The one-sided sewing technologies have the advantage of good automation and the possibility of sewing in a 3D mould as they are mountable to an industrial robot. The infiltration behaviour of the textile is improved because of the permeability caused by the seams.

Although seams enable high strength and drapability, there are also disadvantages, e.g. the disorientation of reinforcing yarns through the sewing thread. Therefore alternative joining processes, such as adhesive bonding, are also applied. Apart from the mentioned joining processes, the so-called tailored fibre placement (TFP) is used with reinforcement textiles. The reinforcing yarns can be put down purposefully (CAD operated) on a 2D

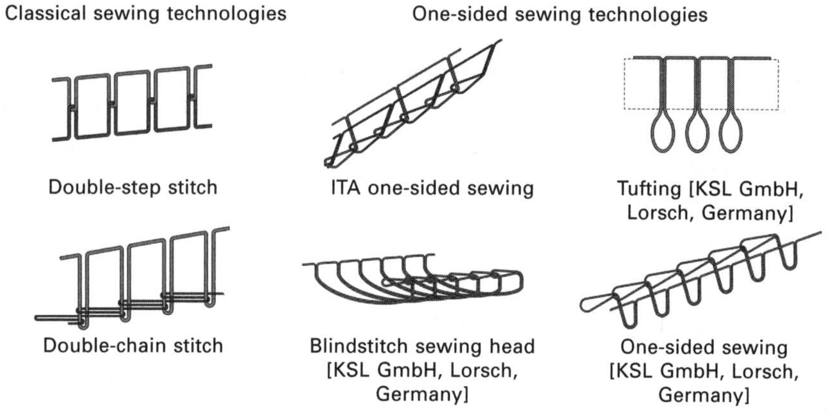

14.12 Seam types.

textile. Parallel to placing the fibres, they are stitched on the textile with a thin thread. This technology can be used for additional local reinforcements, e.g. for fittings or holes in the TRP part.

14.3.4 Mechanical properties of the final textile reinforced plastics structure

The reinforcing yarns are to be placed in a near-net-shape geometry that should be consistent with the strains that occur during use of the final structure. Different degrees of complexity in the textile structure can be realized with preforming technologies.

Disorientation or damage of the yarns in the reinforcement textile can be caused by mistakes in handling, forming or assembling processes. This decreases the mechanical properties of the final TRP structure. According to the mechanical requirements of the TRP structure, preforming technologies should be rated by their tendency to yarn disorientation and damage.

14.3.5 Productivity and production process complexity

The introduced preforming technologies are strongly differing in productivity and process complexity. According to the requirements of the TRP structure, a compromise of mechanical and geometric properties of the structure and of the productivity of the production process must be found.

There are numerous woven fabric variants of interest for textile reinforced automotive composite applications. These include amongst others multi-layer woven fabrics, three-dimensional woven fabrics as well as flat woven fabrics with three-dimensional design features. Regarding industrial usage, flat fabrics are widespread because of the good mechanical properties and the fast production process and thus their attractive price. Owing to its high productivity while creating near-net-shape textile structures, overbraiding technology is gaining importance for automotive applications.

14.4 Applications of textile structures in automobiles

Formula 1 car bodies consist of carbon textile structures. This reduces weight and improves crash behaviour considerably. Especially car noses and monocoques are made from carbon textile structures as these are crucial parts of the automobile in a crash.

Sports cars are often tuned with parts made from carbon fibres, e.g. aprons and spoilers. A serial application is the bumper which again decreases weight and improves the load take-up in case of a crash (e.g., BMW M6, Lotus Elise). These elements consist of braided carbon fibre structures [18]. The crash beams of the McLaren SLR car are located in longitudinal direction of the car, similar to crash elements in ordinary vehicles, and take up the load in a frontal and diagonal crash. They consist of foam cores overbraided with carbon rovings.

Another application is the car roof of the BMW M6 which consists of a TRP in sandwich design which reduces weight and leads to a lower centre of gravity of the car. It consists of carbon woven fabrics which are left visible from the outside on the upper surface. In Mercedes upper class cars, trunk hatches are made of SMCs (sheet moulding compound, consisting of resin and short fibres) to reduce weight and to integrate antenna for radio, etc. This is not a structural element in the true sense.

14.5 Future trends

- Development trends with reinforced structures go to more and more complex structures. This contradicts the trend to further automate the production process and to design more load-conforming TRPs. This dilemma must be solved.
- With the exception of Formula 1 racing cars, serial car frames are not made of TRPs yet. This would allow a significant reduction in the car weight and can therefore be expected to be realized in the near future.
- Another important trend is the development of processing techniques to manufacture 3D reinforcing structures from flat textile fabrics.
- New economical manufacturing processes for infiltratable and near-net-shape preforms with complex geometry become more and more essential. Great importance lies in an economic, automated preform manufacturing to be able to realize large-scale composite production. By using automated textile production technologies to manufacture complex textile preforms

made from high performance yarns, uneconomic manufacturing steps can be eliminated and production costs can be decreased dramatically.

- A further trend is the integration of different functions into composites. An example of this is the combination of energy-absorbing elements with structure composites.
- A further point could be continuous health monitoring of composites, in order to recognize structural failure at an early stage.

14.6 References

1. Roye, A., Stüve, J., Gries, T. Definition for the differentiation of 2D- and 3D-textiles. Part 1: production in one-step-processes, In: *Technical Textiles* 48 (2005), H. 4, p. E212–E2142-D and 3-D textiles.
2. Anon. *Textiles, basic terms and definitions*, DIN 60000, 1/1969, Beuth-Verlag GmbH, Berlin and Köln, 1969.
3. Cebulla, H., Diestel, O., Offermann, P. Biaxial Reinforced Multilayer Weft Knitted Fabrics for Composites. – In: Proceedings of the 6th international AVK-TV Conference, Baden-Baden, (2003). – paper B10, p. 1–9.
4. Offermann, P., Diestel, O., Godau, U. Flat knitting of biaxially reinforced multiple layer fabrics for composites. *Technical Textiles* 40 (1997), Volume 2, pp. 95, 98–99.
5. Haller, P., Birk, T., Cebulla, H., Offermann, P. Fully Fashioned Biaxial Weft Knitted and Stitch Bonded Textile Reinforcement for Wood Connections. – In: *Composites Part B: Engineering*, Elsevier B. V., Amsterdam, The Netherlands (2004).
6. Anon., Reinforcement – Specifications for multi-axial multi-ply fabrics – Part 1: Designation, German version EN 13473-1:2001, Beuth-Verlag GmbH, Berlin and Köln, 2001.
7. Hanisch, V., Henkel, F., Gries, T. Performance of textile structures determined by accurate machine settings. 14th International Techtextil-Symposium 2007, Frankfurt am Main 11–14 June 2007.
8. Bannister, M., Callus, P., Nicolaidis, A., Herszberg, I. The effect of weave architecture on the impact performance of 3-D woven composites. 'A new movement toward actual applications', Texcomp 4 – 4th Int. Symposium for Textile Composites, Kyoto/J 12–14 October 1998, Kyoto 1998, Paper Nr. P2.
9 Bilisik, A., Mohamed, M. Multiaxial 3-D weaving machine and properties of multiaxial 3-D woven carbon/epoxy composites. 'Moving forward with 50 years of leadership in advanced materials', 39th Int. SAMPE Symposium, Anaheim/USA, 11–14 April 1994, Covina 1994, 868–882.
10. Brandt, J., Drechsler, K., Arendts, F.-J. Mechanical performance of composites based on various three-dimensional woven-fibre preforms. *Composites Science and Technology* 56 (1996), 381–386.
11. Hirokawa, T. Development and application of three-dimensional woven fabric. 'From Fibre Science to Apparel Engineering', Proc. of the 26th Textile Research Symp. at Mt. Fuji, Shizuoka/J, 5–7 August 1997, Hikone 1997, 102–107.
12. Brookstein, D. Concurrent engineering of 3-D textile preforms for composites. *International Journal of Materials and Product Technology* 9 (1994), H. 1/2/3, pp. 116–124.
13. Curiskis, J. I., Durie, A., Nicolaidis, A., Herszberg, I. Developments in multiaxial

weaving for advanced composites materials. 11th Int. Conf. on Composite Materials (ICCM-11), Gold Coast/Aus, 14–18 July 1997, Melbourne 1997, Vol. 5, 86–89.
14. Swinkels, K. Low-cost production of double wall fiber reinforced plastic tanks by a winding process with a three dimensional fabric. In: 35th Internationale Chemiefasertagung, Dornbirn 25–27 September 1996, Dornbirn 1996.
15. Büsgen, W.-A. Woven 3-dimensional shapes – update and future prospects for new weaving techniques, *Melliand Textilberichte* 80 (1999), Volume 6, pp. 502–505.
16. Büsgen, A., Finsterbusch, K., Birghan, A. Simulation of composite properties reinforced by 3D shaped woven fabrics. 12th Int. Conference on Composite Materials (ECCM), Biarritz, France, 29 Aug–1 Sep 2006.
17. Ko, F. K., Pastore, C. M., Head, A. A. *Handbook of industrial braiding*, Covington, Kentucky/USA: Atkins & Pearce, 1989.
18. Kümpers, F.-J., Brockmanns, K.-J., Stüve, J. 3D-braiding and surround braiding as textile production method for textile preforms. In: *DWI Reports * Volume 130* (2006), ISSN 0942-301X, file: \Aachener Textiltagung 2006\pdf\Kuempers.pdf.
19. Canfield, A. R. Braided carbon/carbon nozzle development, AIAA-1985-1096, SAE, ASME, and ASEE, Joint Propulsion Conference, 21st, Monterey, CA, 8–10 July 1985.
20. Drechsler, K. Latest developments in stitching and braiding technologies for textile preforming. In: SAMPE, International SAMPE Symposium and Exhibition, 49 (2004) pp. 2055–2067, ISBN 0-938994-96-4.
21. Wulfhorst, B., Gries, T., Veit, D. *Textile Technology*, Munich: Hanser, Cincinnati: Hanser Gardner, 2006.
22. Mungalov, D., Bogdanovich, A. Complex Shape 3-D Braided Composite Preforms: Structural Shapes for Marine and Aerospace. In: *SAMPE Journal*, Vol. 40, No. 3, May/June 2004.
23. Herrmann, A. S., Pabsch, A., Kleineberg, M. Kostengünstige Faserverbundstrukturen – eine Frage neuer Produktionsansätze, Konferenz-Einzelbericht: 3. Internat. AVK-TV Tagung für verstärkte Kunststoffe und duroplastische Formmassen, 12–13 September 2000, Baden-Baden.
24. Weimer, C., Mitschang, P., Neitzel, M. Continous manufacturing of tailored reinforcements for liquid infusion processes based on stitching technologies, 6th International Conference on Flow Processes in Composite Materials, 15–16 June 2002, Auckland, NZ.
25. Gries, T., Klopp, K. *Füge- und Oberflächentechnologien für Textilien: Verfahren und Anwendungen*, Berlin, Heidelberg: Springer, 2007.

15
Textile composites for automotive structural components

F K KO, University of British Columbia, Canada

Abstract: In this chapter, after a review of textile structures for composites, the role of fiber architecture in the manufacturing of automotive composites components is discussed. The various textile preforms are classified according to fiber architecture and manufacturing technology. The Sunrise™ electric vehicle program is used as an example to illustrate the design procedure and the feasibility for large-scale manufacturing of automotive structural composite components at targeted weight and affordable cost. Specifically composite pillars and wheel wells are used as examples to demonstrate the procedure in the integrated design for manufacturing (IDM) of automotive structural composites based on the Fabric Geometry Model (FGM). The chapter concludes with a summary and discussion of future trends in the use of textile composites for automotive structural components

Key word: textile structural composites, electric vehicle, textile preforming, fiber architecture, integrated design for manufacturing, B-pillar, wheel well.

15.1 Introduction

Textile structural composites represent an important class of affordable advanced materials, which are reinforced by textile preforms for primary structural applications. Making use of the unique combination of lightweight, flexibility, strength and toughness, textile composites found early applications in the 1950s in space re-entry vehicles. The need for higher damage tolerance, especially through the thickness strength requirements, led to the rediscovery of the merits of textile composites in the 1980s. The popularization of liquid molding processes and the demand for affordability in the 1990s added a new dimension to the interest in textile composites. As the needs for composites are being expanded to large-scale structural components, textile reinforcements have increasingly been considered for providing adequate integrity as well as shape-ability for low cost near-net-shape manufacturing. Textile composites are uniquely suited for automotive composite components because of their capability to address the uncompromising need of automotive composites for the combination of strength and toughness; scale-up manufacturability; and low cost manufacturing.

The increasing awareness of energy shortage and environmental impact caused by the pollution generated by fossil fuel-based automobiles in recent years has re-energized the awareness of the need for clean and energy-efficient vehicles. There is a global effort to seek energy saving solutions. One of them is to use lightweight composite materials [1, 2]. Because of their wide availability at low cost, textile composites have invariably been identified as a preferred reinforcement strategy to meet automotive structural composite needs [3]. In the US, in addition to earlier technology demonstration projects such as Ford Motor's million dollar all composite car, we have witnessed at least three generations of national (government/industry) programs in the US aimed at advancing the technology of composites for automobiles starting from the Automotive Composite Consortium (ACC) [2] wherein the three major auto companies in the US worked closely together with their industry suppliers and academic research partners to target selected structural components such as the cross-members and the crash rails. While the earlier program proved the feasibility of building a composite car, the ACC program demonstrated the feasibility of cost reduction via textile composites because of significant part consolidation for selected structural components.

Following ACC was the Partnership for the New Generation Vehicles Program (PNGV) championed by then Vice President Al Gore. The goal of PNGV was to achieve three times the fuel efficiency or 80 miles/gallon (at the 1993 level) by reducing the weight of the car body and chassis by 50% [2]. This must be accomplished with reduction in manufacturing cost and emission level. The Freedom CAR program replaced PNGV in 2002 with the long-term aim to develop petroleum free cars using "leap frog" technologies that will provide America with greater freedom of mobility and energy security while lowering costs and reducing impact on the environment. A significant component of this five-year program starting in 2006 is the materials technologies subprogram aiming at the development and validation of the enabling advanced materials and manufacturing technologies to achieve the goals of Freedom CAR [4]. Composite materials are identified as key components to achieve structural weight reduction of automobile body and chassis components by at least 60% and overall vehicle weight by at least 50% compared to 1997 passenger vehicles.

It is of interest to observe that many solutions to the weight reduction problem proposed in the Freedom CAR program were already available in 1997 as demonstrated in the Sunrise™ electric vehicle (EV) program sponsored by the Department of Commerce [5]. The Sunrise™ team, through strategic use of textile composites, demonstrated that a full size electric vehicle equivalent to a Ford Taurus in interior room but weighing 40% less at 159 kg (350lb.) can be built for a market competitive selling price of $20,000 at an annual production of 20,000 units. The Sunrise™ EV has a combined city/highway range of 290 km (179 miles) and achieved 38.5 km per liter (90.5 miles per gallon) equivalent of gasoline with an average energy efficiency of 57 watt-hours per km (91watt-hours per miles) [5] proving that the Sunrise™

technology already exceeded the PNGV target of 34 km per liter (80 miles per gallon) of gasoline. Of course, with gasoline price at $1.25/gallon there was not an "energy crisis" then! Thus further scale-up production of the Sunrise™ electric vehicle was put on hold due to lack of funding from the government and from the private sector. The lesson learned from the Sunrise™ program was that technology alone is a necessary but not sufficient condition to take the clean vehicle to the consumer market. The ambitious goal and urgency of the Freedom CAR program is largely driven by the rapid rise of gasoline prices in the recent years and the increasing evidence of environmental impact of automobile pollution on health and global climate change.

In this chapter, after a review of textile structures for composites, the role of fiber architecture in the manufacturing of automotive composite components will be discussed. The various textile preforms will be classified according to fiber architecture and manufacturing technology. The Sunrise™ electric vehicle program is used as an example to illustrate the design procedure and the feasibility for large-scale manufacturing of automotive structural composite components at targeted weight and affordable cost. Specifically, composite pillars and wheel wells are used as examples to demonstrate the procedure in the integrated design for manufacturing (IDM) of automotive structural composites based on the Fabric Geometry Model (FGM). The chapter will conclude with a summary and discussion of future trends in the use of textile composites for automotive structural components.

15.2 Textile structures for automotive composites

The ultimate goal of manufacturing a composite structure for automotive components is to meet design requirements including performance and cost. How successfully the goal can be met depends on the effective use of the reinforcement material, architecture and processing technology. As illustrated in Fig. 15.1, there is a large family of textile structures available for structural reinforcement of composite vehicle components [6]. Preform fiber architecture plays a key role in composite manufacturing by facilitating the forming and resin infiltration processes. The properties of the composite will also be influenced by fiber architecture. Fiber orientation (θ) and volume fraction (V_f) are key engineering parameters influencing textile composite formability, permeability and mechanical performance. The key criteria for the selection of textile preforms for structural composites center around the capability of the technology to provide (a) in-plane multiaxial reinforcement, (b) through thickness reinforcement and (c) formed-shape and/or net-shape manufacturing. Depending on the processing and end use requirements, some or all of these features are required.

Based on the fabric formation techniques through fiber entanglement or yarn twisting, the fiber architectures shown in Fig. 15.1 are created by

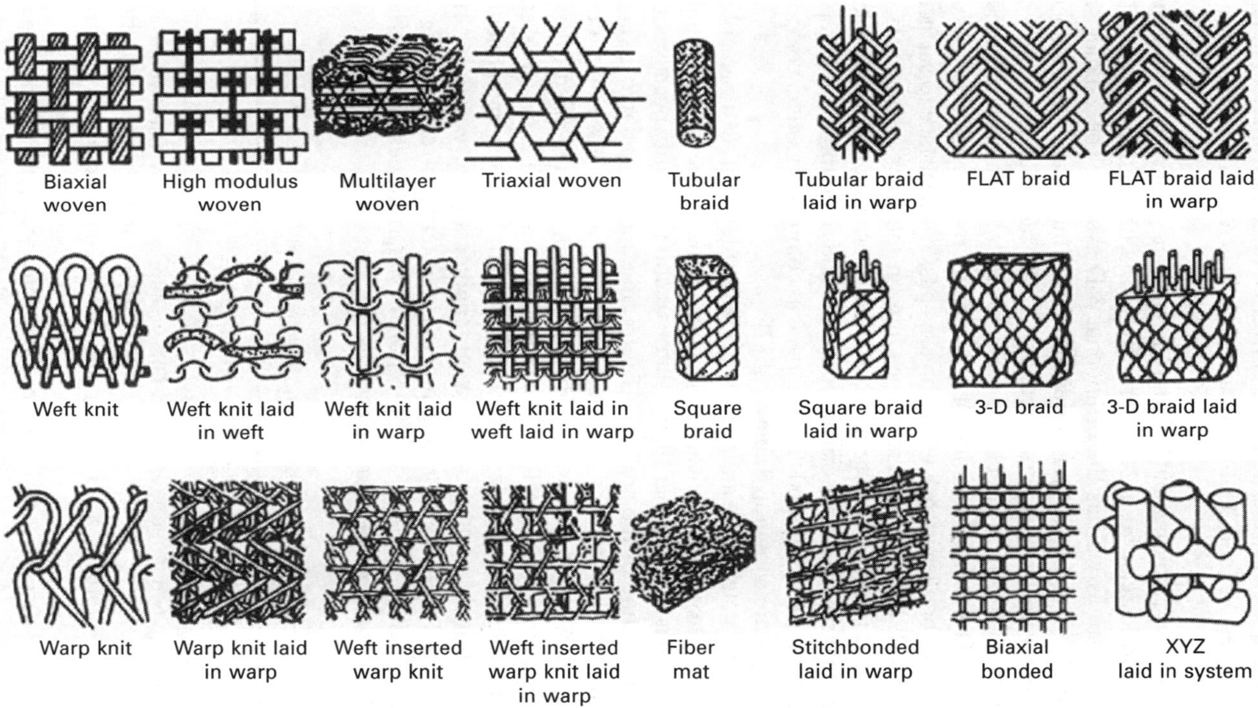

15.1 Textile structures for composite reinforcements.

interlacing, interlooping, and intertwining or multiaxial fiber placement. While most textile preforms are converted from yarn to fabric (YTF) structures, some preforms, such as fiber felts, are converted directly from fiber to fabric (FTF) structures. In Table 15.1 the four basic fabric formation techniques are compared.

While weaving, braiding and knitting can produce planar or 3-D structures; nonwoven fabrics can be a 2-D planar system with random or organized fiber orientation, as well as the orthogonal 3-D system. The 2-D and 3-D fabrics are distinguished by yarn orientation distribution and the number of yarn diameters in the thickness direction. A 2-D fabric consists of two to three yarn diameters in the thickness direction with fibers oriented in the x–y plane. A 3-D fabric, consisting of three or more yarns in the thickness direction, is a fibrous network wherein yarns pass from surface to surface of the fabric in all three directions. Depending on the fiber architecture, the orientation and packing density of fibers, the fiber volume fraction varies. The theoretical range of fiber orientation angle and fiber volume fraction for each fabric preform-system and their components commonly used for composite reinforcements are shown in Table 15.2 [7].

It should be noted that, although the achievable range of fiber volume fraction is restricted by theoretical fabric geometric limits due to yarn jamming, it is possible to achieve higher fiber volume fraction in reality because of the compressible nature of the preforms. By forcing the preform into a smaller mold, a higher fiber-volume-fraction composite can be made, but this is done at the expense of introducing fiber misalignments and dried fibers.

Through geometric modeling of textile preforms a quantitative communication link has been developed between the preform manufacturer, composite fabricators and product design engineers [7]. By defining fiber architecture and textile preforming processes in terms of engineering and processing parameters V_f, θ, and η, rational composite design procedures and process control guides can be established.

Table 15.1 A comparison of fabric formation techniques

Preforming technology	Yarn introduction direction	Formation technique
Weaving (YTF)	Two (0°/90°) (warp and fill)	Interlacing (by selective insertion of 90° yarns into 0° yarn system)
Nonwoven (YTF or FTF)	Three or more (orthogonal)	Mutual fiber placement, fiber bonding
Knitting (YTF)	One (0° or 90°) (warp or fill)	Interlooping (by drawing loops of yarns over previous loops)
Braiding (YTF)	One (machine direction)	Intertwining (position displacement)

Table 15.2 Process engineering parameters for textile preforms

Preform	Fiber orientation, θ (°)	V_f	Processing parameter
Linear assembly	θ – yarn surface helix angle	0.6 ~ 0.8 0.7 ~ 0.9	bundle tension, transverse compression, fiber diameter, number of fibers, twist level
Roving	θ = 0		
Yarn	θ = 5 ~ 10		
Woven	θ_f – yarn orientation in fabric plane θ_c – yarn crimp angle		fiber packing in yarn, fabric tightness factor, yarn linear density ratios, pitch count, weaving pattern
2-D Biaxial	θ_f = 0/90, θ_c = 30 ~ 60	~ 0.5	
2-D Triaxial	θ_f = 0/90/±30~60, θ_c = 30 ~ 60	~ 0.5	
3-D Woven	θ_f = 0/90, θ_c = 30 ~ 60	~ 0.6	
Non-woven	θ_x – fiber/yarn orientation along *x* axis θ_y – fiber/yarn orientation along *y* axis θ_z – fiber/yarn orientation along *z* axis θ_{xy} – fiber distribution on fabric plane		(2-D Non-woven) fiber packing in fabric, fiber distribution (3-D Orthogonal) fiber packing in yarn, yarn cross section, yarn linear density ratios
2-D Non-woven	θ_{xy} = uniform distribution, θ_z	0.2 ~ 0.4	
3-D Orthogonal	θ_x, θ_y, θ_z	0.4 ~ 0.6	
Knit	θ_s – stitch yarn orientation θ_i – insertion yarn orientation		fiber packing in yarn, fabric tightness factor, yarn linear density ratios, pitch count, stitch pattern
2-D Weft knit	θ_s = 30 ~ 60	0.2 ~ 0.3	
3-D MWK	θ_s = 30 ~ 60, θ_i = 0/90/±30 ~ 60	0.3 ~ 0.6	
Braid	θ – Braiding angle		fiber packing in yarn, fabric tightness factor, braid diameter, pitch length, braiding pattern, carrier number
2-D Braid	θ = 10 ~ 80	0.5 ~ 0.7	
3-D Braid	θ = 10 ~ 45	0.4 ~ 0.6	

15.3 The role of fiber architecture in composite performance

Preform fiber architecture plays a key role in composite manufacturing, influencing the productivity and quality of the composite. The properties of the composite will also vary depending on the fiber architecture. Fiber orientation (θ) and volume fraction (V_f) are key engineering parameters for textile composite formability, permeability, and performance.

15.3.1 Formability

The manufacturing of composites often requires transformation of the fiber reinforcements into various structural shapes through net shape fabrication or formed shape processing. While 3-D textile preforming is more suitable for the creation of net structural shapes, 2-D textile preforms are usually formed into shapes by molding or stitching. One of the earlier studies of fabric formability in composite manufacturing [8] showed that total available deformation could be imposed on the fabric uniformly and that the modes of deformation are important parameters for fabric formability. Potter [8] demonstrated that weft knitted fabrics are significantly more conformable than biaxially woven fabrics because deformation of the knitted fabric in the axial, transverse and bias direction are 50%, 50% and 26%, respectively, compared to 0%, 0% and 45% for the woven fabric. The same point can also be illustrated in a comparison of the shear resistance of biaxially and triaxially woven glass fabrics [9]. Comparing the strain behavior of plain woven and triaxial basic fabric of similar areal density (282 vs. 285 g/m^2), under biaxial loading, it was found that the shear deformation of the triaxial fabrics is considerably more uniform than that of the biaxial fabrics (Fig. 15.2). As a result of this study, it was found that triaxial fabric is more adaptable to 3-D draw molding than biaxial fabrics made from the same yarn. In quantifying the formability of fabrics, Dow [10] suggested that yarn slippage and low yarn jamming angles are required for fabric conformability. Accordingly, in fabric formability modeling, fiber volume fraction distribution, fiber orientation and fiber interlacing intensity as well as the limit of geometric deformation (all of which are governed by the architecture created by specific textile preforming techniques) must be considered.

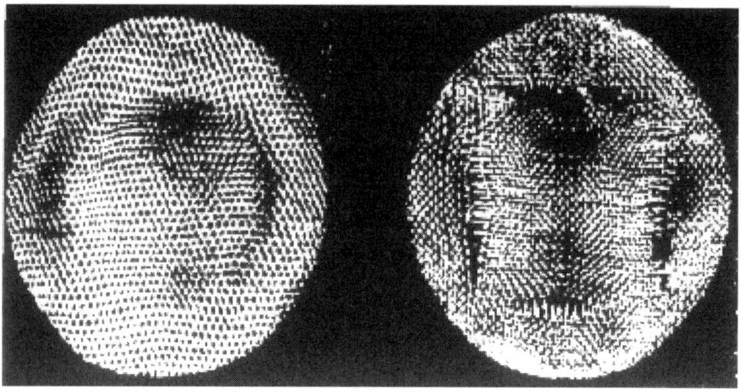

15.2 Effect of fiber architecture on formability of fabrics.

15.3.2 Permeability

Fluid flow permeability of textiles is an indication of how easily and uniformly a matrix can be infiltrated into the fibrous assembly. McCarthy and Kim [11] concluded that the permeability of textile preforms is affected by the dynamic interaction of fiber architecture and fiber volume fraction (Fig. 15.3). In the same figure, it can be noted that, for the same fiber volume fraction, ordered structure such as 0/90° woven fabrics have higher permeability than disordered structures such as discrete chopped fibers. Loos *et al.* [12] also observed the dependence of permeability on fiber volume fraction in their study of carbon multiaxial warp knit performs. It was found that the introduction of through thickness fibers significantly increased the permeability of the preforms, especially for preforms with high fiber volume fraction.

15.3.3 Mechanical properties

Fiber architecture plays an important role in the translation of fiber properties to the composites as well as controlling the level of matrix infiltration. Figure 15.4 illustrates the dynamic interaction between material (fiber), fiber architecture and processing [13]. It can be seen that fiber architecture affects fiber packing and hence the maximum obtainable fiber volume fraction. As a result the translation efficiency of fiber properties to the composite is strongly influenced by fiber architecture. It can be seen that the highest fiber

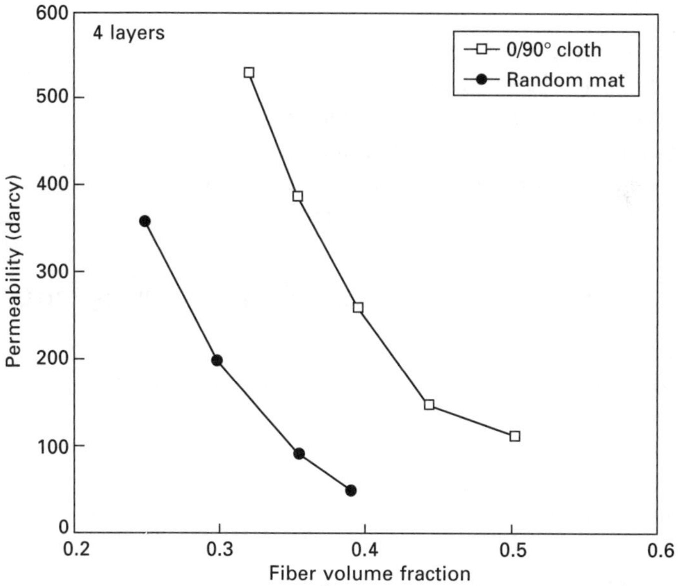

15.3 Effect of fiber architecture on fabric permeability.

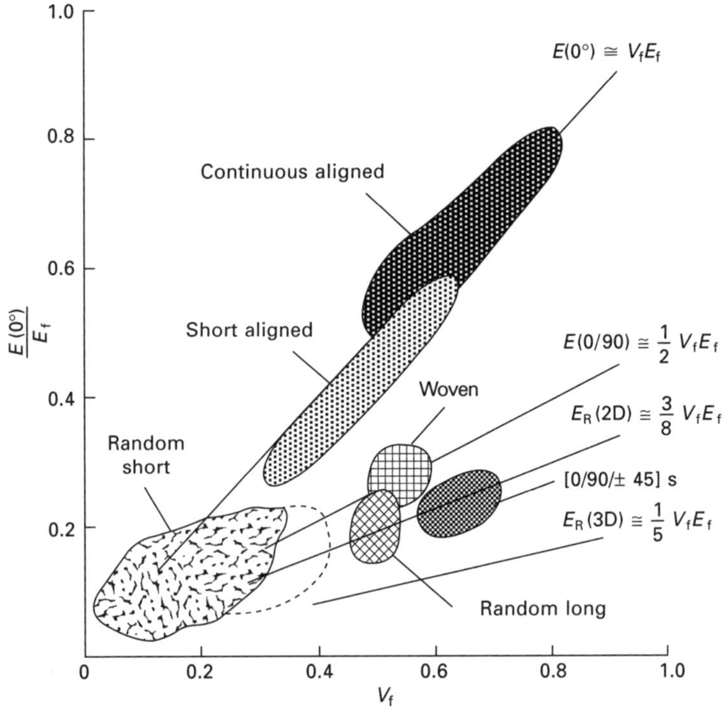

15.4 Fiber property translation efficiency as a function of fiber architecture and fiber volume fraction.

volume fraction is obtainable for aligned continuous filaments as in the case of filament wound structures. Aligned short fibers, on the other hand, have lower maximum fiber volume fraction than those of the aligned continuous filaments. For the same fiber volume fraction, the 0/90° woven and the random long fibers (as in the case of sheet molded compounds) will have lower fiber property translation efficiency.

15.4 The Sunrise™ electrical vehicle program

There are numerous examples of applications of textile composites for automotive structural components ranging from net-shape braided crash rail, to braided cross-members, to racing car seat backing, to door frames, tire belts, brake pads, to connecting rods and torsion bars as well as the monocoque body of formula racers. The most extensive integrated use of textile structures for automotive structural components is the Sunrise™ electric vehicle. Therefore the Sunrise™ is used in this article to illustrate the strategy and design methodology for producing automotive structural components. It is felt that this textile-based composite material's design for manufacturing platform is also applicable to other types of vehicles and many other non-automotive structural applications.

The Sunrise^{TM} electric vehicle program is an Advanced Technology Program (ATP) co-funded by the National Institute of Standards and Technology (NIST) in 1996 along with a consortium of eight industry partners. This program calls for the development of the technology and pilot production tools capable of producing the Sunrise^{TM} monocoque body/chassis cost-effectively at 20,000 units per year. With a target body-in-white (BIW) weight of 159 kg (350 lbs), this vehicle must also meet Federal Motor Vehicle Safety Standards (FMVSS).

One of the major tasks of this program was to organize the vehicle into components for distributed manufacturing by way of modular design. Figure 15.5 provides an overview of the extensive use of textile composites for the car body. The goal was to make the part integration as high as possible, while maintaining ease of manufacture, both in preforming and in the final composite consolidation phase.

In the design of the Sunrise^{TM} vehicle, several technical issues were addressed. These issues include the level of part integration, the integration of various textile preforms, and the integration of local fiber architecture into the global structural design. Taking advantage of the wide availability of fiber architecture as shown in Fig. 15.1, all types of textile preforms, including woven, knitted, braided and nonwoven fabrics were used along with appropriate corresponding composite processes including RTM, SCRIMP, pultrusion and thermoforming. These combinations of fiber architecture and processes are the result of the consideration of the coupling effects of vehicle structural geometry and preform fiber architecture as well as the fiber architecture-composite processing interaction.

For example, during the effort of demonstrating resin transfer molding (RTM) for the integrated lower body shell (ILBS) of the Sunrise^{TM} vehicle (Fig. 15.5), the design concept evolved from a highly integrated belly pan design to a more reasonable integration of components by modular manufacturing. Several parts of the previously existing belly pan configuration had to be removed to reduce tooling complexity. These parts include the battery box, the lower firewall, the passenger and driver kick panels and the rear cross-car beam. Some of the elements that were removed became candidates for different high rate manufacturing processes. Such an example is the battery box, which is a constant cross-section highly structural component of the vehicle body. Braiding was selected as the optimum preforming method for this part, based mainly on the off-axis fiber requirements of the lay-up and the continuous process potential of braiding. Having a constant cross-section, the battery box was also amenable to pultrusion. Combining braiding and pultrusion gives us continuous in-line preforming and high rate molding operation for the battery box. A quarter-scale battery box profile was

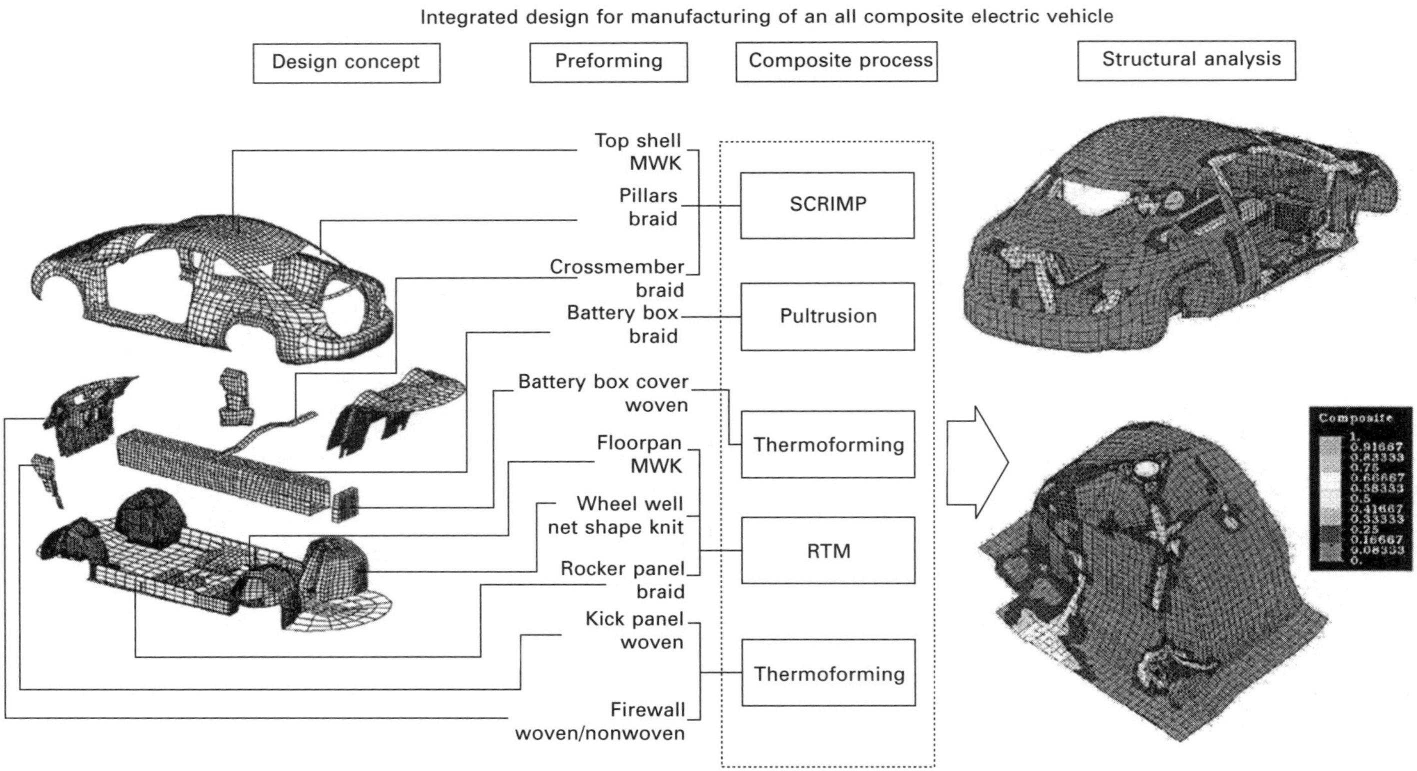

15.5 Composite parts breakdown of Sunrise™ BIW according to textile preforming processes and composite manufacturing processes.

successfully braid-truded in this program, while a full-scale battery box was braided and infused to validate the mechanical properties of the braided laminate.

15.4.1 Integrated design for manufacturing

In order to reflect the contribution of local fiber architecture effect in the structural analysis of the vehicle, an integrated design for manufacturing methodology (IDM) was used. Figure 15.6 shows the logic flow of the methodology.

The heart of IDM is a fiber-architecture-based model – the Fabric Geometry Model (FGM) that provides a quantitative link between the manufacturing process and the design strategy and structural analysis. The FGM is a volume averaging homogenization method, which takes into consideration the volumetric angular distribution associated with fiber architecture via well-established coordinate transformation methodology. The procedure of FGM and related models have been detailed elsewhere [14–16]. The system stiffness matrix generated for a given fiber architecture provides the necessary input file for finite element structural analysis. Our strategy was to establish the initial design according to the logic flow shown in Fig. 15.6 for a given laminate schedule. The fiber architecture design is then mapped to the structural performance established with the laminate schedule.

The overall structural analysis of the body-in-white (BIW) begins with the assumption of a monocoque body with stress skins [17]. The skin thickness was tailored to provide necessary stiffness and strength. The fibers in the skins were oriented to follow projected load paths. Cores were used in panels that required additional stiffening. Most importantly, the design of a centrally located battery box tube was utilized in achieving high body stiffness. Additionally, large closed cross-section rockers were incorporated in the body design for stiffness purposes. This design made the battery box and the rockers the major paths for all body loads and the laminates were developed specifically to optimize the performance of these components.

The suspension system of one of the early prototype vehicles was instrumented and tested under a wide range of driving conditions in order to obtain external driving loads to the body. These experimentally obtained suspension loads were checked against calculated load values with acceptable agreement between them. These loads were then used to develop the laminate schedules for the highly loaded parts of the BIW. The Algor [18] finite element analysis (FEA) program was utilized, along with standard composite engineering practices to optimize the laminate schedule and minimize weight in the vehicle. The highest externally loaded parts of the vehicle are the wheel wells, which directly accept the suspension loads. Several iterations of analysis and laminate design were performed (see Fig. 15.7).

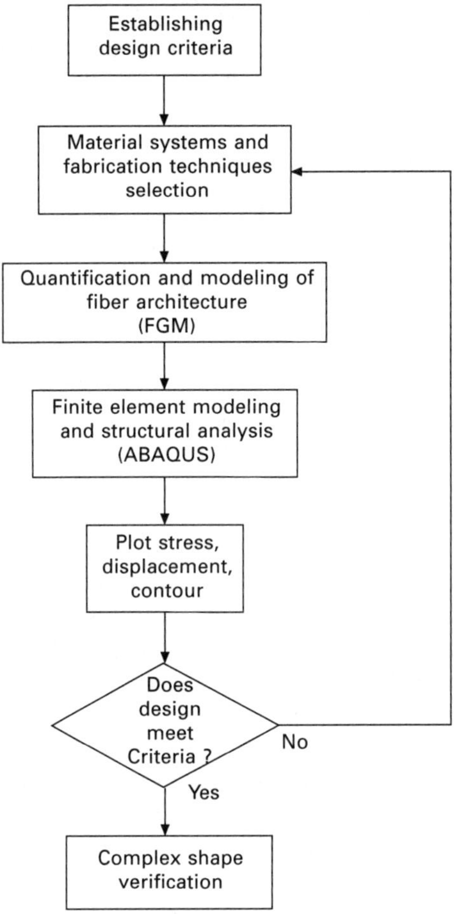

15.6 Logic flow of the integrated design for manufacturing methodology.

The BIW main load path components were designed to sustain external crash load, carry the road loads locally and to provide for the targeted torsional and bending stiffness of the BIW. FEA was used in both cases, along with test results, to validate the FEA input. In the case of the load from crash conditions through the BIW, strain gage readings from the frontal impact tests were used to compare to battery box and rocker stresses and the values were in good agreement.

In the case of the BIW torsional and bending stiffness, one of the prototypes was used to evaluate those numbers. Both torsional and bending stiffness evaluation tests [19, 20] were performed on the body shell. The tests were repeated after bonding the firewall, battery box, inner rockers, kick panels and seat rails, cross car beams and rear bulkhead to the shell.

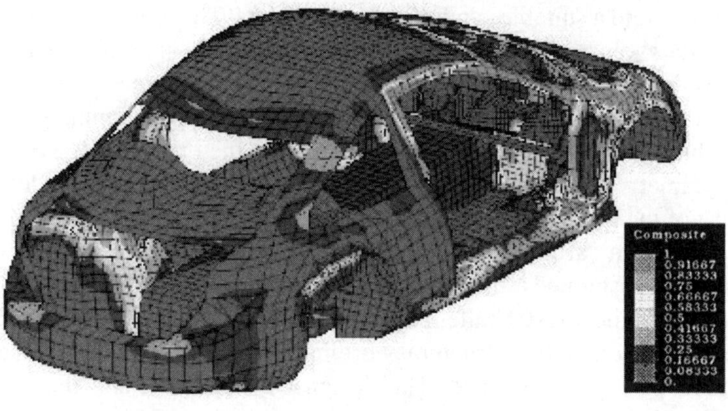

15.7 Stress contour from body torsion finite element analysis [16].

The results provide the design and manufacturing team a good understanding of the contribution of each of these components to the BIW performance. The assembled vehicle results were also compared to FEA results of body torsion and bending. The experimental and test results indicated similar contributions from the different BIW components to the body stiffness. The torsional stiffness of the assembled prototype BIW measured during these tests was 6000 ft-lbs/degree with the FEA results indicating 10 000 ft-lbs/degree. The bending stiffness was measured to be 17 000 lbs/in with the FEA results indicating 24 000 lbs/in. The torsional rigidity value exceeded the original lowest design limit of 4000 ft-lbs/degree, but the bending stiffness was lower than the lowest desired value of 35 000 lbs/in. The FEA model was reconstructed using composite material elements, and the FEA of the new BIW model resulted in stiffness of 6370 ft-lbs/degree and 18 600 lbs/in in torsion and bending respectively. These results, being within 10% of the tested values in both torsion and bending, provided the necessary confidence in the body FE model for subsequent studies. Using this model, the laminate sequence of the BIW components was reconstructed to improve both the torsional and bending stiffness of the BIW. Concurrently textile preforms enabling lower cost and simplified assembly were selected and designed to match the properties of the laminates.

15.4.2 Design and manufacturing of vehicle components

Selected components of the vehicle that are highly loaded were analyzed. The pillars and wheel wells represent two of the most highly loaded components in the top body shell and the lower body shell respectively. The mapping of material properties and fiber architecture are illustrated using the pillars and wheel wells. The requirements for high torsional and bending stiffness makes triaxial braid a suitable candidate for the pillars. The required fiber orientation and fiber volume fraction to meet the design criteria quantified in the laminate analysis were reduced to braiding machine control parameters such as braiding point location, braiding speed and take-up speed for programming the braiding machine for automated braiding process. On the other hand, the complexity in geometry and the large dimensions of the wheel well preclude the use of most of the preforming methods. Net shape fiberglass knitting with strategic placement of carbon fiber stiffeners was shown to be the most cost effective. The fully-fashioned knitting was performed on a computer-controlled knitting machine. The most challenging task for the knitted wheel well was the demonstration of their structural worthiness, for they are the highest externally loaded part of the vehicle. The design consideration and the selection of preforming strategy that is compatible with the RTM molding cycle for structural components of the ILBS have been detailed by Brachos *et al.* [21].

15.5 Examples of composite structural components

For illustration purposes the B-pillar and the rear wheel well are used as an example to show how a low cost textile preform can be designed for low cost manufacturing of automotive structural components. The pillars are the structural backbone of the upper body shell (UBS) whereas the wheel wells are the highest externally loaded parts of the vehicle [21, 22]. The wheel wells directly transfer the loads between the suspension of the wheels and the body shell. The wheel wells also play an important role in the structural performance of the entire integrated lower body shell (ILBS) and thus the overall car body. The upper body shell (UBS) was attached to the ILBS via the edges of the wheel well caps.

15.5.1 Braided B-pillar

The pillars are the structural backbones of the top body shell of the electric vehicle. For example, the B-pillar/seat-rail structure experiences a high seat-belt load in a crash situation. The laminate in that area was designed to sustain seat-belt loading resulting from frontal and rear impact based on the FEA analysis shown in Fig. 15.8. The laminate schedule for the B-pillar calls for a symmetrical lay-up of 2 mm thick carbon uni-tape and 1 mm thick fiberglass multiaxial warp knit (MWK) sandwiched with a foam core. The predicted composite properties are: E11 = 88.3 GPa (12.8 Msi); E22 = 8.3 GPa (1.2 Msi); E22 (flange) = 26.2 GPa (3.8 Msi); and G12 = 4.8 GPa (0.7 Msi).

To reduce the assembly time, triaxial braided structure was designed and built by mapping the laminate properties to the braided composite properties generated by the FGM. The first step in the FGM is to design the fiber

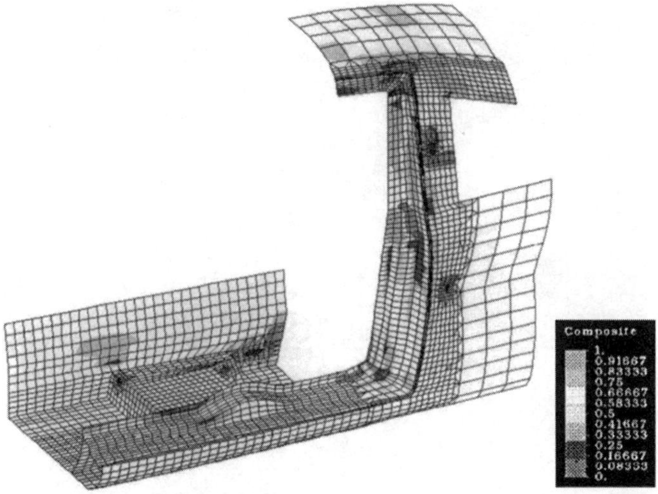

15.8 Finite element analysis of the B-pillar.

architecture according to the following relationship between fiber volume fractions and the braid angle [5]:

$$V_f = \frac{\kappa w_y N_c}{4\pi R_m \cos\theta}$$

where κ is the fiber packing fraction, w_y is the yarn width, N_c is the number of braiding carriers, R_m is the equivalent radius of the mandrel, and θ is the orientation angle of the braiding yarns.

The analysis suggested a combination of 2 mm of 15° triaxial carbon braid overbraided with 1 mm of 70° hybrid braid consisting of 80% glass braid and 20% carbon axial yarns. Figure 15.9 shows the B-pillar foam core and the braided B-pillar preform.

The following braided composite properties were generated by the FGM analysis: E11 = 85.6 GPa (12.4 Msi); E22 = 14.5 GPa (2.1 Msi); E22 (flange) = 29 GPa (4.2 Msi); and G12 = 9.7 GPa (1.4 Msi).

Figure 15.10 shows the predicted and experimental tensile moduli for the various pillars, indicating reasonable agreements between theoretical predictions and experimental results. This provided confidence in the selection of braided fiber architecture according to the processing window prescribed in the volume fraction-braiding angle relation.

15.5.2 Fully-fashioned knitting of front passenger wheel well

Both front and rear wheel wells have the form of a dome to cover the wheels. The major design loads are tensile and compressive loads in the radial direction

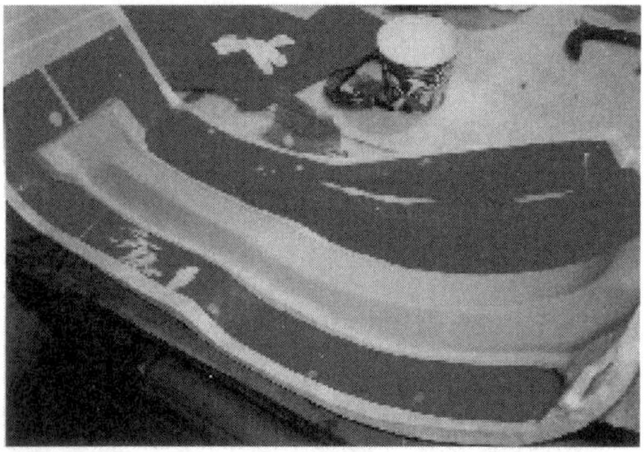

(a)

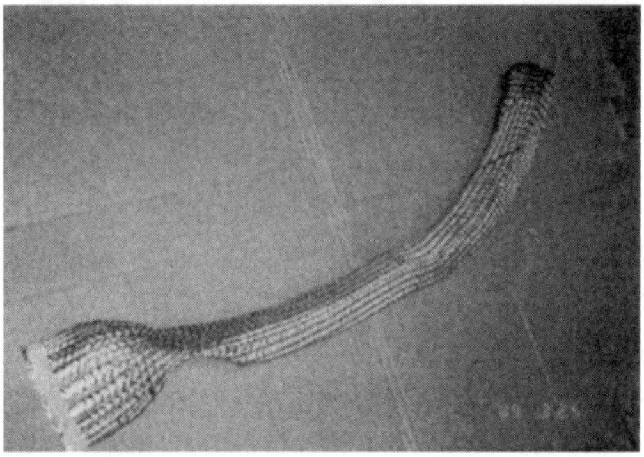

(b)

15.9 (a) Foam core fabricated by injection molding.
(b) Braided B-pillar preform.

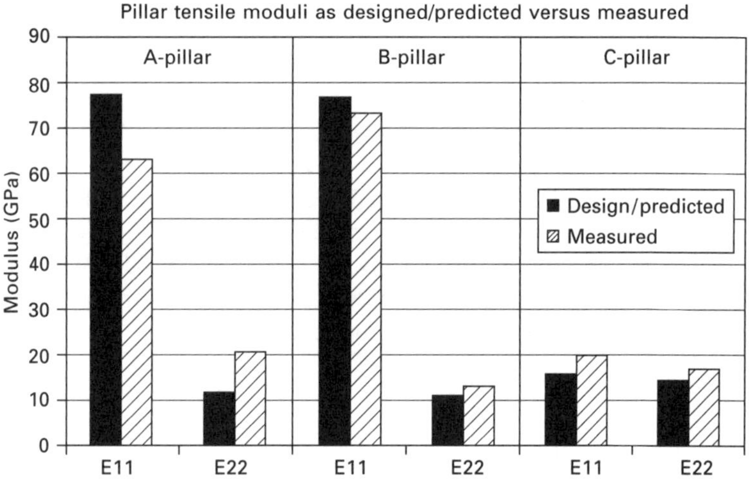

15.10 Comparison of the experimental and predicted moduli of the A, B, C pillars.

of the dome. Loads on the wheel wells are transferred to the floor pan via the mount areas, which thus have to be the highest reinforced areas. To eliminate local stress concentrations, metal inserts were used in both front and rear mounts, in locations where other connections with the suspension are present. Figure 15.11 shows an example of the front wheel well loaded in a worst case, complex, combined loading condition [17]. For the bulk of the structure a material with dominant properties in the radial direction of the dome of the wheel well is required. In the other directions minimum requirements under tension, compression and in plane shear exist. In selected locations of the wheel well, such as the corners, high stiffness and strength are required to transfer the major loads from the top of the wheel well tower to the floor mounts. The FEA results prescribed a laminate schedule based on a multi-axial-warp-knit (MWK) material for the entire structure plus unidirectional carbon strips placed in the corner area. Although the wheel well fabricated according to the laminate schedule meets the property requirements, it would be too laborious to manufacture and thus fails to meet the required production economics of the Sunrise™ program.

To improve manufacturability it was determined that a weft knitted preform be used in order to take advantage of the conformable nature of weft knitted fabrics and the net-shape manufacturing capability of fully fashioned knitting.

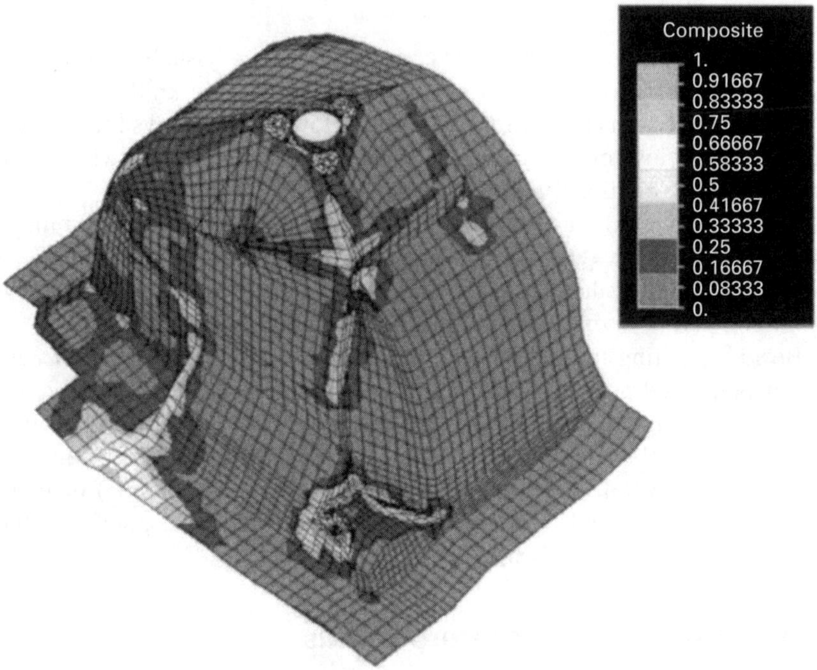

15.11 Finite element analysis of front wheel well with knitted reinforcement [17].

Using the fully-fashioned knitting technology, net-shape or near net-shape fabrics can be produced with a high productivity and reproducibility on computer-controlled knitting machines. A further advantage of the knitted material for this application is that straight (non-crimp) fibers can be laid-in in the radial direction of the dome during the knitting process to create the directional reinforcements as required. Accordingly, mapping of the laminate design to knit design was carried out on the geometric and material property level. Geometrically, as shown in Fig. 15.12, a weft knitted fabric was transformed from a planar structure to the three-dimensional wheel well geometry with the unidirectional laid-in carbon fibers placed in the selected locations. By optimization of the knitted stitch construction, sufficient mechanical performance of the knitted fabric reinforced composite was developed in the non-radial directions of the wheel well.

The added cap and mount reinforcements in the wheel well preform were produced as sub-preforms in semi-automated lay-up stations. The layers were cut automatically using carefully designed patterns. The preforms were consolidated with a binder material under vacuum pressure as in diaphragm thermoforming [21]. The complete preform can also be laid up in a semi-automated lay-up station. The knitted fabrics, sub-preforms and metal inserts were integrated to form the final wheel well assembly. A picture of a finished front and rear wheel well preform, ready for placement in the RTM mould, is shown in Fig. 15.13.

RTM production trial runs of the complete lower body shell preform with the knitted preforms integrated, have confirmed that the knitted preforms also have good permeability for liquid resin flow, which allows the removal of additional flow mats, that would have to be added to the wheel well preforms with traditional reinforcements [21, 22].

Following a similar approach shown in the design of the braided pillars, the relationship between fiber volume fraction and knit geometry were established. The predicted knitted composite properties are shown in Fig. 15.14 comparing to experimental results with good agreement.

Besides meeting the mechanical property requirements, the use of near net-shape knitted preform fabrics for the basic wheel well construction has shown promising potential for a significant cost reduction for the formation of the total wheel well preform due to the high conformability of the knitted material. The overall labor time saving (including knitting time) of more than 30% was realized in comparison to the use of broad goods for the prototype wheel well performs [21, 22].

15.6 Summary and future trends

Composite materials have long been recognized as the ideal candidate to address the need for weight reduction for automobiles. The concern with

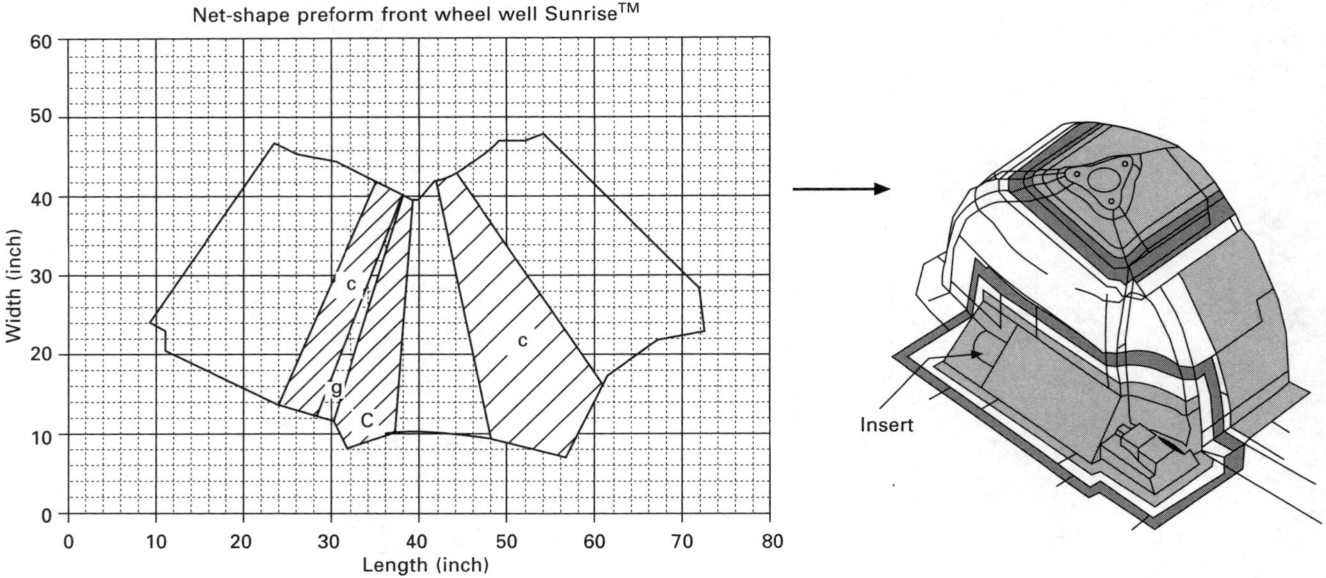

15.12 Net-shape 2D knitted preform for the front wheel wells; the fabric drapes around the mold easily produce the complex shaped part; the UD carbon reinforcements are integrated strategically into the preform during the knitting process.

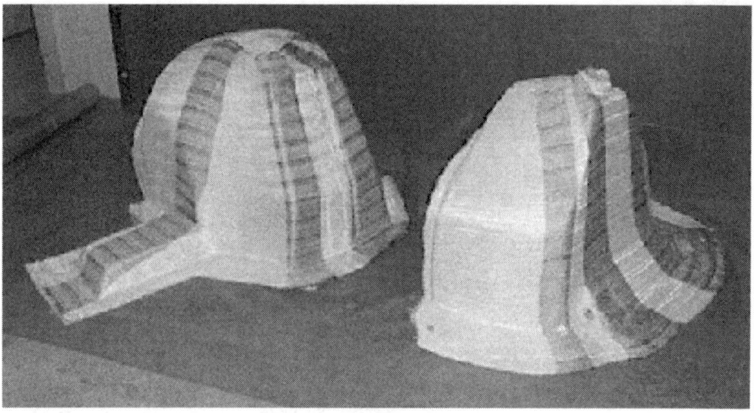

15.13 Picture of finished rear and front wheel well preforms, with near net-shape knitted fabrics encapsulating cap and mount sub-preforms. The black strips are selective carbon yarn laid-in according to structural requirements.

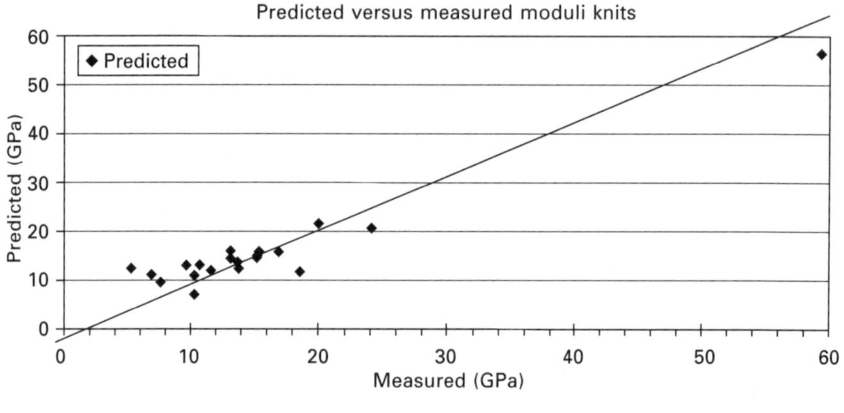

15.14 Predicted and experimental tensile moduli of the knitted composite wheel well.

production economic of composite materials and the lack of a sense of urgency in the need for clean energy source has impeded the adoption of composites for structural automotive components. With the alarming increase in gasoline price and the increasing global awareness of the environmental impact caused by fossil fuel-based vehicles in recent years, there is renewed interest in clean and energy efficient automotives. However, the uncompromising needs of the automotive industry for low cost, large-scale manufacturing capability while meeting the structural and stringent safety criteria remain unchanged.

With a rich history of textile craftsmanship and wide availability of production facilities in the industry, textile technology can be used effectively to create the fiber architecture to meet the processing, performance and cost requirements for automotive composites. The availability of structural design and process control tools coupled with a large family of advanced fiber materials has led to textile structural composites being recognized as a promising class of composites characterized by high conformability, high toughness and high affordability. Depending on the product geometry and the required fiber volume fraction and fiber orientation, there is a wealth of fiber architectures at our disposal.

To facilitate communication, the large number of fiber architectures was classified according to manufacturing technology. The role of the fiber architecture in influencing composite processing and mechanical performance was discussed. The extensive use of textile preforms in large-scale composite product manufacturing was illustrated using the Sunrise™ all composite electric vehicle as an example. The implementation of the concept of integrated design for manufacturing (IDM) was illustrated using two structurally demanding components – the B-pillar and the rear wheel well in the upper body shell and lower body shell respectively. Through the Sunrise™ program a new paradigm of agile manufacturing of structural composites was demonstrated wherein multiple suppliers, each with special capability in producing particular components for the vehicle, were organized in a way similar to how the supply chain of structural components is being managed in the commercial aircraft and electronics industries today. The question of the optimum degree of structural integration was also examined in this program. It was found that the level of structural integration depends on the coupled effect of vehicle geometry and reinforcement geometry; the process-fiber architecture interaction; and production rate compatibility of various textile preforming and composite fabrication processes.

With the increasing usage of textile composites for automotive structural components and for other structural applications, there is a need for a comprehensive database and quality control measures for textile composites. The shortening of the supply chain due to part consolidation will place more responsibility for quality assurance on the textile preform manufacturers. As the need for multifunctional products is likely to increase, future composite products will need to integrate many types of materials and fiber architectures. With the explosive growth of nanotechnology it is inevitable to include nanomaterials as a component of textile reinforcement materials. This in turn will affect composite process control and requires design methodologies capable of integrating the local effects in global structural analysis and for composite process modeling. The linkage of fiber architecture-based design with structural performance through the FGM will provide a means to bridge the communication gap between designers and manufacturers, therefore resulting in a more efficient translation of design concepts to product reality. The integrated design for manufacturing (IDM) platform of textile structural composites is applicable not only to electric vehicles but also to various vehicles powered by different energy sources ranging from methanol, ethanol, bio-fuel, natural gas, reformulated gasoline, hybrid gasoline/electric as well as other non-automotive structural components.

Finally, there are encouraging signs that there is a shift of global interest in addressing the vehicle efficiency issues as demonstrated by the US and China, two of the most energy-hungry nations in the world. This is evidenced by the signing of a five-year agreement between the US Department of Energy (DoE) and China's Ministry of Science and Technology (MoST) in September 2007 to promote large-scale deployment of next generation vehicle efficiency technologies in the US and China with a specific focus on electric, hybrid-electric, fuel cell, and alternative fuel technologies. Accordingly, it is felt that the conditions are more favorable for global market acceptance of composite intensive vehicles in general, and for electric vehicles in particular than ten years ago. With the availability of new materials such as nanomaterials, improved control processing technology, increased efficiency of alternative energy and vehicle technology, coupled with more environmentally friendly clean air legislation, there is ample opportunity for textile structural composites to contribute to the next generation of clean, smart, safe, and energy-efficient automobiles.

15.7 Acknowledgements

The author wishes to thank the National Institute of Standards and Technology who, through the Advanced Technology Program co-funded the Sunrise™ vehicle research and development effort described in this paper. Colleagues in Solectria Inc. and TPI contributed to the structural analysis and liquid molding respectively of the composite vehicle. My colleagues at APD, Rick Balonis and Aart van Vuure, played an important role in the successful design and manufacturing of the preforms for the entire Sunrise™ program.

15.8 References

1. Brylawski, M., *Uncommon knowledge: automotive platform sharing's potential impact on advanced technologies*, SAMPE-ACCR-DOE Advanced Composite Conference, Detroit, September 1999.
2. Chapman, G.B., *The need and requirements driving the development of polymer composites for automotive applications*, Proceedings of ICCM-11, Gold Coast, Australia 14–18, July 1997.
3. Rudd, C.D and Johnson, C.F., *Current trends in lightweight vehicle manufacture*, 6th International Conference on Automotive Composites, Bristol, Sept. 1999. (ISBN 1-86125-102-5).
4. http://www.uscar.org/guest/view_partnership.php?partnership_id=1

5. Fisher, J., *Solectria Sunrise^{TM} electric vehicle technology and market introduction*, Proceedings, EVS-14, Orlando, Florida, December 1974.
6. Ko, F.K., "Preform Architecture for Ceramic Matrix Composites," *Ceramic Bulletin*, Vol. 68, No. 2, February, 1989, pp. 401–414.
7. Ko, F.K. and Du, G.W., "Processing of Textile Preforms", Chapter 5, *Advanced Composites Manufacturing*, Timothy G. Gutowski (ed.), John Wiley & Sons, 1997.
8. Potter, K.D. 1979. The Influence of Accurate Stretch Data for Reinforcements on the Production of Complex Structural Moldings, *Composites,* IPC Business Press, Ltd., July.
9. Scardino, F.L. and Ko, F. K. 1981. Triaxial Woven Fabrics, *Textile Research Journal,* Vol. 51, No. 2, February.
10. Dow, R.M. 1989. *New Concept for Multiple Directional Fabric Formation*, Proceedings of the 21st International SAMPE Technical Conference, September 25–28.
11. McCarthy, S. and Kim, Y.R., *Resin flow through fiber reinforcement during composite processing*, Proceedings, 5th Textile Structural Composites Symposium, Drexel University, Philadelphia, Dec 4–6, 1991.
12. Loos, A.C., Weidermann, M.H. and Kranbuchi, D.E. 1991. "*Processing of Advanced Textile Structural Composites by RTM*," Proceedings of the 5th Textile Structural Composites Symposium, Drexel University, Dec. 4–6.
13. Gutowski, T.G., *Advanced Composite Manufacturing*, p. 18, John Wiley & Son, 1997.
14. Ko, F.K., Pastore, C.M., Lei, C. and Whyte, D.W. 1987, *A Fabric Geometry Model for 3-D Braid Reinforced Composites*, International SAMPE Metals Conference: Competitive Advancements in Metals/Metals Processing.
15. Ko, F.K., Three Dimensional Fabrics for Structural Composites, in *Textile Structural Composites*, Chou, T.W and Ko, F. K. (eds), Elsevier, New York, 1989.
16. Vendeurzen, P., Structure-performance Modeling of Two-Dimensional woven Fabric Composites, Ph.D. Thesis, Katholeike Universiteit Leuven, 1998.
17. Brachos, V., Rossi, G.A. and Kirk, W., "*Design of an all Composite Vehicle Platform – Solectria's SunriseTM$^{(TM)}$ EV*", 13th Annual Advanced Composites Technology Conference and Exposition, September 1998.
18. *ALGOR User's Manual* (1997). Algor Inc. Pittsburgh, PA.
19. Szefi, R.J., "Static Stiffness – A Cornerstone Metric", *Automotive Body International*, Spring 1998.
20. Simic, D., Jelic, L.J. and Aksic, M., "Torsion Dynamic Load Analysis of Passenger Car Body", *SAE Technical Paper Series* 851683, 1985.
21. Brachos, V., Rossi, G.A., Kirk, W., Balonis, R., van Vuure A.W., Plessinger, J. and Myers, J., *Design considerations for cost effective high rate preforming and RTM for large composite structures*, 13th Annual Advanced Composites Technology Conference and Exposition, September 1998.
22. Ko, F.K., van Vuure, A.W. and Balonis, R.J., Textile Preforming for Complex Shape Structural Composites, *SAMPE Journal*, Vol. 35, No. 3, May/June (1999).

Index